Samuel Alexander Stewart

A Flora of the North-East of Ireland

Including the Phanerogamia, the Cryptogamia Vascularia, and the Muscineæ

Samuel Alexander Stewart

A Flora of the North-East of Ireland
Including the Phanerogamia, the Cryptogamia Vascularia, and the Muscineæ

ISBN/EAN: 9783337272814

Printed in Europe, USA, Canada, Australia, Japan

Cover: Foto ©berggeist007 / pixelio.de

More available books at **www.hansebooks.com**

A FLORA

OF THE

NORTH-EAST OF IRELAND

INCLUDING

THE PHANEROGAMIA, THE CRYPTOGAMIA VASCULARIA,
AND THE MUSCINEÆ.

BY

SAMUEL ALEXANDER STEWART,

FELLOW OF THE BOTANICAL SOCIETY OF EDINBURGH; CURATOR OF THE
COLLECTIONS IN THE BELFAST MUSEUM, AND HONORARY ASSOCIATE OF
THE BELFAST NATURAL HISTORY AND PHILOSOPHICAL SOCIETY.

AND THE LATE

THOMAS HUGHES CORRY, M.A., F.L.S.,
F.Z.S., M.R.I.A., F.B.S. Edin.,

LECTURER ON BOTANY
IN THE UNIVERSITY MEDICAL AND SCIENCE SCHOOLS, CAMBRIDGE;
ASSISTANT CURATOR OF THE UNIVERSITY HERBARIUM
ETC. ETC.

Published by
THE BELFAST NATURALISTS' FIELD CLUB.

Cambridge:
MACMILLAN & BOWES.
1888.

PREFACE.

The present work is an attempt to give a full and reliable account of the native vegetation of the Counties of Down, Antrim, and Derry; an undertaking which was projected some years since by the late Thomas H. Corry, M.A., in conjunction with the surviving Editor. That it has not sooner appeared is due to the lamentably sudden and premature death of Mr. Corry, and to the fact that the writer could give but a very limited portion of time to the collection and revision of the material. As this volume will doubtless fall into the hands of many who may not know of the fate that befel him who urged on the project, a brief statement on the subject may be allowable.

Early in 1882 Mr. Corry was appointed by the Royal Irish Academy to examine and report on that rich botanical district, the Ben Bulben range of mountains in County Sligo. In pursuance of this arrangement he visited these mountains during the summer of that year, being accompanied by two enthusiastic botanical friends, Mr. Charles Dickson, solicitor, of Belfast, and Mr. R. P. Vowell, of Dublin. On this visit many specimens were collected, and a valuable series of notes was made, but the extent of the work was so great that it was found impossible to survey the district satisfactorily in one season. It was, therefore, decided to revisit the locality during the following summer, and to include more or less of the district adjacent to, though not strictly a part of, the mountains in question. In fulfilment of this plan, Mr. Corry and Mr. Dickson arrived in the town of Sligo on the evening of Wednesday, 8th August, 1883. Immediately above the town of Sligo lies Lough Gill, a romantically-situated sheet of water, whose many picturesque aspects form the chief attraction of the locality. This lake, at the base of the richest mountain district of Ireland, seemed likely to reward a botanical exploration, and it was settled that the next day should be devoted to Lough Gill. The appearance of the following morning was not at all inviting. Heavy showers of rain were frequent, and were accompanied by such sudden and fierce squalls of wind that the pedestrian was, for the time being, foiled in his attempts to proceed.

Notwithstanding the forbidding aspect of the weather, and undeterred by the warnings of the boatmen, our friends decided to carry out the programme as arranged. They were young, they were good oarsmen, as well as good swimmers, and made light of the idea of incurring danger. It is said that they were advised to take a heavier boat than the one they had arranged for, and also to engage the services of a local boatman, but these precautions, it would appear, they deemed unnecessary. Accordingly they put off in a light skiff, no doubt rejoicing in the thought that they could penetrate wherever it seemed desirable, and return at the close of a long day without having to consider the suggestions, or convenience of any but themselves. But they did not return, and the search, that night instituted, discovered the empty boat, and subsequently the bodies of the two unfortunate botanists. No eye saw the occurrence, and the exact details of the calamity will never be known, but from the few memoranda made, and the few plants collected, it would seem to have been in the early part of the day. The present writer was at that time exploring the shores of Lough Allen in the adjoining county of Leitrim. The wild gusts of wind which broke the stems of the water plants, and drove their fragments ashore were of service to him, but he little imagined their fatal significance to two fellow botanists.

Mr. Corry had not reached his twenty-fourth year, but, a diligent worker, he had already attained a position amongst rising botanists much beyond what his years seem to warrant. Had his life been prolonged there is no doubt that British Botany would have felt his influence. As regards the present work, while claiming that no pains have been spared in its compilation, the surviving Editor is not unaware that less faults would have appeared, and higher merit been attained had the co-operation of his colleague not terminated so abruptly.

In preparing the present Flora it has never been forgotten that accuracy is of immensely more value than copiousness. Nothing that is capable of being tested has been taken for granted, and the very large lists of "Excluded Plants" given at the end testify to the necessity of this method. Possibly a few still remain which should have been relegated to the Excluded lists, but where a first rate botanist has noted a plant, and its occurrence is not improbable, it must be admitted although unconfirmed. Notes by parties who, either through carelessness or incompetence, have made many mistakes are not accepted without some confirmation. With regard to localities furnished to this Flora by contemporaries the Editors hold themselves to a great extent responsible for their correctness, but the records of the older botanists, and of the published Floras have been received subject to the limitations stated above.

The statements of the altitudinal range, and of the times of flowering of plants are original, being based on notes made by ourselves within the district, and continued for many years. Where, owing to insufficient information, it became necessary to quote some outside standard the authority has been stated. The vertical range of many of our plants is imperfectly known: under this head careful observations in the future are desirable.

A local Flora has its historical aspects, and in the present case these

PREFACE. vii

have not been disregarded. It will be seen that the authority quoted for the occurrence of a species in any locality is the observer who first collected or noted the plant in that locality. The rule of priority has been followed in this respect, and the refinding of a plant in a previously known station is not mentioned here. This practice has only been departed from in the case of a few very rare species, concerning which it is desirable to have the most ample details. As respects rare plants all known stations are quoted here, but with regard to the more common such a course would have made the book too voluminous, and only the more important are given. No special localities have been assigned to such plants as are ubiquitous, or nearly so, in their distribution.

As regards the Counties of Down and Antrim it may be stated that their plants are fairly well known : in fact no equal area in Ireland has been better worked. That there are portions of these counties even yet unvisited by the botanist is, however, the fact, but the opening up of remote districts by railways, and the multiplication of local observers will doubtless bring to light ere long some plants at present overlooked, and extend the range of others. The County of Derry is, however, less known to the botanist, and were it not for the existence of the botanical notes of the late Dr. David Moore that county could not have been included in this Flora. Those notes were made during the years 1834, and 1835, and considering the shortness of the time are marvellously full. Nevertheless much remains to be done in this county before its botanical features are thoroughly known.

The sequence followed in this Flora is, as regards the higher plants, mainly that of Babington's Manual, as most convenient and natural. The *Musci* have been, as far as possible, arranged in accordance with Braithwaite's British Moss Flora, but that magnificent work being, at present, only partially completed some discordance may ultimately appear.

As to nomenclature the rule of priority has been observed, and wherever it required the substitution of a less known name this has been done. Complaints of inconvenience will, of course, be made on this head, but such rectifications are not only proper, but inevitable, and the more quickly they come the less will be the inconvenience. Persisting for convenience sake in the use of well known, but bastard, names is bequeathing a debt to our successors, who will have to meet it, and with interest too. Further, there has been an attempt here to retain, in all cases, as authority for the species, the name of its earliest describer ; want of access to some original works has prevented this point being more marked. There, doubtless, would be fewer creations of unnecessary genera, and less shifting of plants from one genus to another if the new combination did not include the name of the shifter.

This work, such as it is, could not have been completed without the assistance of many kind friends who have taken an interest in its progress. Mr. A. G. More, F.L.S., who stands at the head of Irish botanists, has often been consulted. His critical knowledge of plants, and his excellent judgment in matters pertaining to botany are well known. To several English specialists we are deeply indebted for labour bestowed on the identification of many difficult plants. Professor C. C. Babington, F.R.S., has from time to time examined the *Rubi*, and given his opinion on the specimens submitted. Mr. J. Backhouse took the trouble of identifying the *Hieracia* here catalogued, and Mr. Arthur Bennett, F.L.S., did the same for the Pondweeds. Many specimens of *Characeæ* have been diagnosed for us by the Messrs. H. and J. Groves, F.L.S. The late Dr. Boswell, of Balmuto. Mr. George Nicholson of Kew, and others have kindly assisted. In an especial manner our thanks are due to Mr. G. A. Holt, of Manchester for the labour of critically examining the specimens of *Musci* and *Hepaticæ*. These specimens were very numerous, and many of them bearing incorrect names when sent to Manchester. They have been subjected by Mr. Holt to an exhaustive scrutiny, and the names, as understood by British bryologists, may be relied upon as correct. Some species recorded here are excluded from this statement, but they stand on the credit of such very competent authorities as Templeton, Moore, and Thomas Drummond. It was considered desirable that the mosses now catalogued should be arranged in conformity with the British Moss Flora, and to Dr. Braithwaite, F.L.S., we are indebted for kindly foreshadowing the sequence to be followed in the portion of his book which is not yet published. For the genus *Hypnum*, however, as it stands in this Flora Dr. Braithwaite is not accountable. Most valuable assistance has been rendered by several local Naturalists whose names appear in the lists of species. Their kindness in supplying notes of localities, and in forwarding specimens where desired was of the utmost service. For liberty to use the MS. list of Derry plants made by the late Dr. David Moore, F.L.S., we are indebted to the courtesy of Mr. F. W. Moore, of Glasnevin. In addition, thanks are due to Messrs. W. Swanston, F.G.S., W. H. Patterson, M.R.I.A., and R. M. Young, B.A., who facilitated this work by means of information, advice, and the use of books from their libraries. The remarks on the vertical limits of cultivation in the County of Antrim were obligingly contributed by Mr. R. Glascott Symes, M.A., F.G.S., and the information on this head, as regards County Down, was kindly supplied by Rev. H. W. Lett, M.A., T.C.D.

BOOKS WHICH CONTAIN REFERENCES TO PLANTS IN THIS DISTRICT.

Ray—Synopsis Stirpium Britannicarum, ed. II., London, 1696.

In this volume are records of a few plants on the authority of Sherard, the earliest botanical explorer in the north of Ireland.

Harris—Antient and present state of the County of Down, Dublin, 1744.

Contains two small lists of county Down plants. This was the first attempt at a county Flora in Ireland. While many of the names are correct there are some which are erroneous, and others which it is impossible to say with certainty what plant was intended. For this early contribution to the botany of the north we are indebted to a former Dublin Society. The Physico-historical Society employed a collector to make a hasty survey, and the results were embodied by Harris in his history of the county.

Smith and Sowerby—English Botany, London, commenced 1790.

Occasional references to north of Ireland plants.

Smith—Flora Britannica, 4 vols., London, 1800-1804.

A few notices of plants found in northeast Ireland.

Dawson Turner—Muscologia Hibernica, Yarmouth, 1804.

North of Ireland localities scanty, and mainly supplied by Templeton.

M'Skimmin—History of Carrickfergus, Belfast, 1811.

Localities for a few plants growing about Carrickfergus.

Sampson—Chart and Survey of the County of Derry, London, 1814.

Supplies an alphabetical list of plants, which is, however, very incomplete, and includes at least 40 plants erroneously named.

Anonymous (an Irish Gentleman)—Scientific Tourist in Ireland, London, 1818.

Mentions a number of plants that occur in each county.

Hooker and Taylor—Muscologia Britannica, London, 1827.

A few localities for mosses in the north of Ireland.

Smith—English Flora, ed. II., 4 vols, London, 1828.

Refers to a few local plants.

Thomas Drummond—Musci Scotici, Vol. III., Belfast, about 1829 or 1830.

This book is of the nature of a fasciculus having no letterpress, but a number (114 species) of mosses glued down to the blank pages. There is no introduction, nor any explanatory matter, the only items written in being names and localities for the plants. Some of the specimens purport to be Scottish, but a large number have been collected in Ireland, and especially in the north of Ireland. The author was elected member of the Belfast Natural History and Philosophical Society in December, 1828, and the three volumes of his Musci Scotici are in the Society's Library. Thomas Drummond came from Forfar, and was the Curator appointed when the Belfast Botanic Gardens were formed. His stay in Belfast was short, but the materials for this volume were mainly collected here, and it is probable that his volume III. was specially prepared for the Belfast Society.

Anonymous (but known to have been prepared by Lady Kane, and the localities mainly supplied by John White)—Irish Flora, Dublin, 1833.

A considerable number of plants in Down and Antrim are first recorded in this volume.

Mackay—Flora Hibernica, and Part II. Musci and Hepaticæ by Dr. Taylor, Dublin, 1836.

Gives many localities in the north of Ireland.

Colby—Ordnance Survey of Londonderry, Dublin, 1837.

Contains small lists of plants contributed by David Moore.

Backhouse—British Hieracia, York, 1856.

Supplies localities in Antrim and Derry for several species.

Babington—Manual of British Botany, London, successive editions, 1843-81.

Notes some of the special plants of the district.

Tate—Flora Belfastiensis, Belfast, 1863.

Supplies lists of plants found within a radius of 15 miles from Belfast. A work almost entirely original, but hastily compiled, and not always correct in the nomenclature.

Dickie—Flora of Ulster, Belfast, 1864.

Contains lists of plants growing in Ulster, and a small portion of Connaught. A good Flora, with copious lists, but little attention has been given to the critical species, and not sufficiently careful to exclude errors.

Moore & More—Cybele Hibernica, Dublin, 1866.

Publishes plants from all the Irish counties. The records for Down, Antrim, and Derry are fuller, and more accurate than any previously compiled.

Babington—British Rubi, London, 1869.

Mentions the occurrence of some species in Antrim and Derry.

Belfast Naturalists' Field Club—Guide to Belfast and Adjacent Counties, Belfast, 1874.

Paragraphs 82-100 treat of the botany of this locality.

Braithwaite—British Moss Flora, Vol. I., London, 1880-87.

Localities in the North of Ireland stated for the more important mosses.

Hooker—Students' Flora, London, 1884.

A few plants are mentioned as occurring in the North of Ireland.

Rev. W. S. Smith—Gossip about Lough Neagh, Belfast, 1885.

A little book which furnishes the names of a number of plants that grow around Lough Neagh.

Belfast Naturalists' Field Club.—Systematic Lists of the Flora, Fauna, and Antiquities of the North of Ireland, Belfast, 1887.

Embodies the various classified lists which have been issued, from time to time, as Appendices to the Proceedings of the Belfast Naturalists' Field Club.

PAPERS AND SHORT NOTES CONSULTED.

J. Ball—Botanical Tour in Ireland; Annals Nat. Hist., Vol. II., 1839.
T. H. Corry—Some Rare Irish Plants; Jour. Bot. N.S., Vol. XI., 1882.
——— New Irish Rubi; Jour. Bot., Vol. XXI., 1883.
J. H. Davies—Muscology of Colin Glen; Phytologist, *n.s.*, Vol. III., 1859.
Miss Gage—Plants of Rathlin Island; Ann. and Mag. Nat. Hist., Vol. V., 1850.
H. & J. Groves—Notes on British Characeæ; Jour. Bot., Vols. XVIII.—XXV., 1880-87.
H. C. Hart—Plants of Some of the Mountain Ranges of Ireland; Proc. R.I.A., 1884.
——— Irish Hawkweeds; Jour. Bot., Vol. XXIV., 1886.
Rev. T. D. Hincks—Flora of Ireland; Ann. Nat. Hist., Vol. VI., 1841.
J. T. Mackay—Catalogue of Plants Found in Ireland; Trans. R.I.A., Vol. XIV., 1825.
——— Additions to the Plants of Ireland; Nat. Hist. Rev., Vol. VI., 1859.
Dr. J. D. Marshall—Notes on Rathlin Island; Trans. R.I.A., Vol. XVII., 1837.
W. Millen—Localities for Plants near Belfast; Phyt., *o.s.*, Vol. V., and Nat. Hist. Rev., Vol. I., 1854.
Millen and Hind—The More Interesting Plants lately found near Belfast; Phyt., *o.s.*, Vol. IV.
D. Moore—Observations on Irish Mosses; Phyt., *n.s.*, Vol. II., 1857.
——— A Synopsis of the Mosses of Ireland; Proc. R.I.A., 1872.
——— Report on Irish Hepaticæ; Proc. R.I.A., 1876.
A. G. More—Supp. to Cyb. Hib.; Proc. R.I.A., 1872.
——— Recent Additions to the Flora of Ireland; Jour. Bot., 1873.
Prof. E. Murphy—Contributions towards a Flora Hibernica; Mag. Nat. Hist., Vol. I., 1829.
Phillips & Praeger—The Ferns of Ulster; Proc. B.N.F.C., Ser. II., Vol. II., 1885-6—appendix.
John Sim—Remarks on the Flora of Ireland; Phyt., *n.s.*, Vol. III., 1869.
S. A. Stewart—Mosses of North-East Ireland; Proc. B.N.F.C., Ser. II., Vol. I., 1873. See also Systematic Lists, 1887.
——— Supplement to Mosses of North-East Ireland; Ibid, Ser. II., Vol. II., 1881-2, and Systematic Lists.
——— Botany of the Island of Rathlin; Proc. R.I.A., 1884.
Ralph Tate—Addenda to Cybele Hibernica; Jour. Bot., Vol. VIII., 1870.
John Templeton—On Rosa Hibernica; Trans. Roy. Dub. Soc., Vol. III., 1802.
Wm. Thompson—Galium cruciatum an Irish Plant; Ann. Nat. Hist., Vol. IX., 1842.
Dr. W. Wade—Plantæ Rariores; Trans. Roy. Dub. Soc., Vol. IV., 1804.

Short notes having reference to plants of this district will be found in the following publications :—

Annals of Natural History, Vol. VI., 1841 (*Babington*).
Phytologist, o.s., Vol. I., 1844 (*Thompson*).
——— n.s., Vol. II., 1857 (*D. Moore*).
Natural History Review, Vol. VII., 1860.
Proceedings Belfast Naturalists' Field Club, 1864-87; notes in all the numbers (*various*).
Proceedings Dublin Natural History Society, Vol. V., 1866-7 (*D. Moore*).
Journal of Botany, n.s., Vol. I., 1872, p. 25 (*Britten*).
——— n.s., Vol. III., 1874, p. 184 (*W. E. Hart*).
——— n.s., Vol. X., 1881, p. 312 (*A. Bennett*).
——— n.s., Vol. XI., 1882, pp. 20 (*Bennett*), and 86 (*Corry*).
——— Vol. XXII., 1884, pp. 278 (*Brenan*), and 302 (*Stewart*).
Proceedings Belfast Natural History and Philosophical Society, 1882-3 (*Stewart*).

MANUSCRIPT LISTS CONSULTED.

John Templeton—Catalogue of the Native Plants of Ireland, 1793-1814.
——— Copy of Muscologia Hibernica having marginal notes by Templeton, to whom it was presented by the author, Dawson Turner. The volume is now in the possession of Mr. R. M. Young, B.A.
F. Whitla—An interleaved copy of Flora Hibernica, with numerous notes, mainly by Mr. F. Whitla. This book was presented to the Belfast Museum Library by Mr. G. C. Hyndman.
S. A. Stewart—Notes of plants found in the North of Ireland, 1863-87.
T. H. Corry—Lists of plants noted in Ireland, 1877-83.
C. Dickson—Notes of plants found mainly in County Down, 1880-83.
Rev. H. W. Lett—List of *Musci* and *Hepaticæ* found in Down and Antrim, 1860-87.
Rev. C. H. Waddell—Notes of Mosses and Hepatics found by him, mainly in the County of Down, 1881-87.
R. Lloyd Praeger—Plants collected in Down, Antrim, and Derry, 1884-7.

INTRODUCTION.

HISTORY.—The Phytological history of the district is of very recent date, and the first botanical explorations were undertaken by strangers to the locality. The eighteenth century saw the advent of Linnæus, and in response to the stimulus of his influence many botanists of eminence appeared in England. The complicated nomenclature, and the absence of any intelligible classification had, for long ages, limited the number of botanists, and rendered their best efforts fruitless, but with the assistance of the new system it became possible to reduce the existing chaos to order. Ray's *Synopsis*, which had been the great authority on English plants, was succeeded by the works of Martyn, Wilson, and Hill, all exponents of English botany. The *Flora Anglica* of Hudson, however, which followed close on the *Species Plantarum*, superseded all these expositions of the British flora, and led the way in the introduction of the Linnean system to this country. Withering, Lightfoot, Curtis, Dickson, and Smith followed Hudson in rapid succession; local and general Floras were issued, and England took the place, which it still holds, as the best botanically investigated country in the world.

Meanwhile Ireland was, as regards its botanical characteristics, scarcely better than a *Terra Incognita*. To the Rev. Mr. Heaton, a Dublin clergyman, we are indebted for our earliest information as to Irish botany, Mr. Heaton's notes of some plants observed by him having appeared in Dr. How's *Phytologia Britannica*, 1650.

The next earliest contribution towards a knowledge of Irish plants came from Dr. William Sherard, at one time British Consul at Smyrna, a devoted botanist, and founder of the Chair of Botany at Oxford. Dr. Sherard visited several parts of Ireland in 1694, and while partaking of the hospitality of Sir Arthur Rawdon, at Moira, he explored the Mourne Mountains, and the shores of Lough Neagh. Sherard was cotemporary with the later times of Ray, and a number of his Irish plants were published by the latter in his second edition of the *Synopsis Stirpium*.

Edward Lhwyd, an enthusiastic Welsh botanist, who is said to have fallen a martyr to his zeal for botanical studies, made a tour in Ireland early in the eighteenth century. His observations on what he saw of Irish natural history and antiquities were published in the Philosophical Transactions for 1712.

Caleb Threlkeld, M.D., was an English Dissenting Minister who settled in Dublin, and combined with his clerical profession the duties of a medical doctor. His *Synopsis Stirpium Hibernicarum* was an attempt to give an account of Irish botany, but is very deficient in localities, and in fact partakes more of the character of an Herbal than a Flora. Dr. Thomas Molyneux supplied an appendix to the work, with notes of a good number of rare plants.

The Herbal of Dr. K'Eogh, of Cork, appeared in 1735, but has little scientific value. Harris's Account of County Down was published in 1744, and is interesting for many reasons, but specially so to local botanists as it supplies the earliest attempt at a list of plants for a northern district. Smith's Antient and Present State of County Kerry bears date 1766, his County Cork, 1770, and his County Waterford, 1772. Rutty's Natural History of County Dublin appeared in 1772. These books give county lists of the more important plants, but they all contain numerous errors.

It will be seen from what has been so briefly stated that when botanical science was making in England its great advance Ireland lagged behind. This lack of zeal and effort on behalf of one of the most delightful studies is to be lamented, but the causes are not far to seek. Facilities for travel were much less, and there was much more difficulty in penetrating to the remote recesses, and the mountain districts of a perturbed country such as Ireland. It is needless to say that here there has always been less wealth, and as a consequence less facilities, less leisure, and less inclination for abstract pursuits.

The wave of scientific progress did, nevertheless, reach Ireland, and the close of the eighteenth and commencement of the nineteenth century saw this country represented by a number of botanists of high repute. Dublin, of course, as the centre of the learning and the wealth of the country took the lead. Here in the north the wave was longer in running up, and Mr. Templeton was for many years the only conspicuous botanical student in Ulster. The flowing tide was running, however, and in 1821 Belfast had the honour of originating the first Provincial scientific society in Ireland. In the throng, and amid the din of commercial pursuits some citizens found time and inclination to engage in scientific research, and the Belfast Natural History Society was formed. The commencement was humble, and the eight founders of the society met in a private room for the discussion of scientific subjects. The progress of this society was not rapid. No propaganda was attempted, but there was a quiet attraction which speedily led to enlargement of numbers. Kindred societies in due time took the field, until in 1863 the Belfast Naturalists' Field Club was established with the express purpose, which to the present has been loyally maintained, of promoting local scientific investigations. The Botanical and Horticultural Society was founded in 1827, and the Botanic Gardens were formed with the avowed object of affording opportunities for the cultivation of botanical science, and providing a desirable place of recreation. That the society has succeeded in the latter object cannot be questioned, but the former purpose is at present in abeyance. The first Curator of the Gardens was Thomas Drummond, who was well known as an exploring naturalist, and a good bryologist; his successor in the Gardens, J. Campbell, was also a botanist : both did good local scientific work.

Botanical researches were carried on with increasing spirit by a number of local naturalists. William Thompson, who commenced his scientific career at twenty-one, and died at the early age of forty-seven, was the successor of Templeton, and like him a many-sided naturalist. While ardently engaged investigating many branches of the Irish fauna he maintained a love for plants, and did not neglect the native flora. He made collections in all the botanical groups, though much of this material was not used to any great extent. His knowledge of flowering plants was fair, and he held a superior position as an algologist, his collections of *Algæ*, both marine and fresh-water, being valuable. George Crawford Hyndman, who was one of the founders of the Natural History Society, was intimately associated with Thompson in his investigations of the marine fauna. He too had a lively interest in botanical pursuits, confining himself, however, to the Phanerogamia. At the head of an extensive business he could not devote the same amount of time to the work of a naturalist, and his attention was given primarily to the study of conchology. Nevertheless the flora of the district received a share of his attention, and his herbarium includes a large number of the local plants. Dr. Mateer, who was in 1836 appointed successor of Dr. J. L. Drummond as Professor of Botany in the Academical Institution, was an enthusiastic botanist, but, a worker with little system, he has left few traces behind. An intimate associate of Mateer was William Millen, who long survived his friend, and saw the advent of the new and vigorous race of naturalists trained up by Mr. Tate. Like his friend, Dr. Mateer, Millen had an enthusiastic love for the native plants of his country, and like him was most conscientious, but did not take the pains to be entirely accurate, and there are little permanent results of his work. Let us not say therefore that his labour was fruitless; he succeeded in imparting a love of nature to many pupils, and he added many bright happy hours to his own existence. David Orr, who resided for some years in Belfast, and was cotemporary with Thompson. noted many plants as occurring in the district. The suspicion of carelessness however deprives his work of much of the value it otherwise would have. Many additional names might be mentioned; Miss Hincks, of Ballycastle, Miss Gage, of Rathlin, and Miss Maffett, of Belfast, have used their talents, with good results, to promote a better knowledge of local botany. Mr. F. Whitla, Mr. Rea, Rev. Mr. Hind, Mr. W. H. Ferguson, and others also gave their services to further the same cause. Mr. Charles Dickson, a solicitor by profession who was just entering on his career, was one of the latest recruits enrolled in the ranks of the local naturalists. Legal training is often thought to be antagonistic to the love of nature, but examples to the contrary are not wanting in Belfast, and Mr. Dickson was among the most recent. How far he would have succeeded can only be conjectured, but full of energy and intelligence there is every reason to suppose that he would have taken his place amongst the foremost in promoting a knowledge of the local flora.

Space permits only this rapid review of some of the more prominent of those who in bygone times delighted in searching out our native plants. A few are included who, though still living, have finished their work amongst us, or have permanently left the locality. The major part have passed to the Brighter Land. This volume is the outcome of the labours of many workers, and without such would not have been possible in its present fulness. Its pages show that there are amongst us still those who are doing valuable work, and who are capable of still further advancing our knowledge. It will remain for some future writer to tell of their success, and to estimate their services accordingly.

To complete the foregoing historical sketch a few biographical notes of former local botanical leaders seem necessary.

*JOHN TEMPLETON.

Born in Belfast in 1766. The Templeton family had been settled at Orange Grove, Malone, near this town, since the early part of the 17th century, and there Mr. Templeton constantly resided after his father's death. To this place he gave the name of Cranmore (Crann-more, *i.e.* the Great Tree), in honour of the very fine chestnut trees in front of the house. His love of Nature was early developed, and in boyhood he took the greatest delight in a book of natural history containing pictures of birds and other animals. These he was in the habit of copying and comparing with specimens sent him by numerous friends who knew his tastes. Before he was twenty years of age he had commenced to cultivate flowers, and his fondness for horticulture would seem to have become an abiding passion. He obtained from various parts of the world rare and valuable plants and seeds, which he brought into his garden and endeavoured to naturalise in this climate. His collection of trees was amongst the most varied ever raised in so small a space.

Mr. Templeton was not, however, merely a cultivator, but he studied botany with enthusiasm as a science, and became a valued correspondent of the leading botanists of his time. By Sir Joseph Banks he was invited to New Holland, with a good salary, and a large grant of land, but his attachment to his native country and to his family caused him to decline the tempting offer. In 1795 he discovered *Rosa hibernica*, near Belfast, a plant which, though now found in several other localities, still ranks amongst the rarest British species. The Royal Dublin Society awarded him a prize of five guineas for this discovery. He it was also who first detected the red broom rape (*Orobanche rubra*) on the Cave Hill; like the preceding this was not known previously, and still remains a rare plant, though not now confined to the Cave Hill. Mr. Templeton also had the further honour of being the discoverer of a number of cryptogamic plants, and one of the mountain species of mosses bears his name — *Entosthodon Templetoni*. At the close of the last, and early in the present century,

* For many details relating to Mr. Templeton we are indebted to Anniversary Address to Belf. Nat. Hist. Soc., May 1827, by the President, Rev. T. D. Hincks, see Mag. Nat. Hist., vols. I. and II., 1828 and 1829.

great advances were made in British natural history, and the fathers of English botany issued a number of notable books. Amongst these Dillwyn's Confervæ, Sowerby's English Botany, Turner's Fuci, Turner's Muscologia Hibernica, and the Muscologia Britannica of Hooker and Taylor were indebted to Cranmore for many of the specimens figured, and species catalogued. Mr. Templeton also contributed to the Belfast Magazine a monthly Naturalists' Report, and a Meteorological Report. To Dubourdieu, and to Sampson he furnished information on natural history matters for the Surveys of Down, Antrim, and Derry. It is to be regretted that in the latter instance this was mixed with notes by less reliable informants, and the value of the work thereby diminished. He was also expected and urged to prepare a general Natural History of Ireland, but, although some progress was made in this direction, the scheme was never carried out. That such was the case is not surprising, as even at the present day, when so much more information is available, such a work would seem too much for one man to accomplish with any degree of completeness.

Mr. Templeton was a patriot of the purest type. Amongst the foremost of the pioneers of intellectual progress in his native town, he gave an ungrudging support to every movement having for its object the advancement of his fellow-countrymen. An early member of the Belfast Society for Promoting Knowledge, he continued through life to take a warm interest in its work. He was also one of the first supporters of the Belfast Academical Institution, and was one of the Visitors named in the act of incorporation. When, in 1821, the Belfast Natural History Society was founded, Mr. Templeton was in failing health, and unable to take the active part in its proceedings which otherwise he would have done. He was, however, ere the Society was one month old, elected an honorary member, and was the first on whom that distinction was conferred. When the now defunct Mechanics' Institute was established he was absent through ill health, but he heartily sympathised with the project and he advocated its cause with his pen. Mr. Templeton, in common with all genuine naturalists, was conspicuous for his humanity to the lower animals, and it is related of him that he relinquished the sport of shooting on witnessing the agonies of a wounded bird. This was in keeping with the intelligence as well as the gentleness of his nature.

His researches were confined almost entirely to Irish soil, and mainly to the province of Ulster, his furthest journey being probably that to Wicklow, if we except a trip to Scotland, of which, however, little record remains. Nearly all the great groups of the vegetable kingdom were included in Templeton's work, and his knowledge of nature did not end there, but was extended to the native fauna; not excepting sponges, of which little was then known. His death occurred near the close of the year 1825, and his burial, which was attended by the members of the Natural History Society, took place on Saturday morning, the 17th December of that year. The Natural History Society desiring to commemorate the genius and worth of their late associate, decided to institute a medal, to be offered in competition annually, as a prize for an appropriate essay. This medal was to be known as the Templetonian Medal, and to Mr. Robert Patterson, then quite a young man but subsequently well known as a naturalist, an author, and a Fellow of the Royal Society, the first award was made. As we can trace this medal no further, it is assumed that the project was shortly abandoned. Mr. Templeton had the honour of being elected an Associate of the Linnean Society. An isolated naturalist, away from all the centres of learning and of science, and with few to sympathise with his pursuits, the difficulties in the way of Templeton were great. Like all true men, however, he had the root of the matter within himself, and ultimately he was cheered with a sight of the dawn of the better day when a love of nature should be wider diffused, and his labours more fully appreciated.

DAVID MOORE, Ph.D., F.L.S., M.R.I.A.

Born at Dundee, in 1807, died at Dublin, after a short illness, in 1879.

The name of Dr. Moore will ever be remembered as amongst the ablest and most earnest of British botanists. At the age of twenty-one he came to Dublin as assistant to Mr. J. T. Mackay, the Director of Trinity College Gardens; this connection with a man zealously devoted to Irish botany perhaps originated, but, doubtless, intensified young Moore's aspirations in the same direction. About the year 1833 or 1834 he received an appointment on the Ordnance Survey of Ireland. Captain, afterwards General Portlock, F.G.S., being the head of the Geological and Natural History sections of this Survey, Mr. Moore was attached to his staff in the character of Botanist. This was a happy selection, as it afforded the botanist rare opportunities for herborizing in a region but little known, and the public secured the services of the most competent man.

The district surveyed was the County of Derry fully, and County Antrim partially, mainly the northern part of the county. Mr. Moore was now in his element, and applied himself to his work with zeal and success. At the northern end of Lough Neagh he discovered *Carex Buxbaumii*, a sedge which is not known elsewhere in Britain, and also *Calamagrostis Hookeri*, a grass that is confined to the margin of that lake. The more important of Moore's plants were recorded in Colby's Survey of Londonderry, and many notes were supplied to Dickie's Flora of Ulster, but the full results of this work were only obtained when the Cybele Hibernica, the Synopsis of Irish Mosses, and the Report on Irish Hepaticæ were published. Moore was an acute observer, and his trained eye detected many species not hitherto recognised in Ireland. He took the greatest interest in *Cryptogamic* plants. He collected *Lichens*, and *Algæ*; of the latter he had an excellent knowledge, and in addition to finding several not previously known in Britain, he discovered some species that were new to science His work amongst the *Musci* and *Hepaticæ* was not less extensive, and as the successor of Dawson Turner and Dr. Taylor he still further advanced our knowledge of these groups. Subsequently

INTRODUCTION.

Lejeunea Moorei was named in his honour by Dr. Lindberg, and, amongst *Phanerogams*, *Helosciadum Moorei* by Syme. In 1839 Moore was appointed by the Royal Dublin Society as Director of their Botanic Gardens at Glasnevin. Here again he was in his element, his success as a horticulturist being equal to his fortune as a scientific botanist. Glasnevin is not crippled, as so many provincial so-called Botanic Gardens are, by dependence on voluntary support, but has an assured income derived from Parliamentary grant. It is not necessary, as in Belfast, to compete with the Circus in tight-rope performances, or to have the grounds injured by the crowds brought together to witness balloon ascents, or displays of fireworks. It was thus in the power of an able and energetic Director to make the place an honour to the Irish Metropolis, and no one will deny to Moore the credit of having done this. His journeys to various parts of Europe were the means of introducing many new plants to British gardens, and in this and other ways the aid of the State was repaid by material and æsthetic enrichments.

In recognition of his talent and his zeal he was elected a member of the Royal Irish Academy, and was on the Council of that Society, while in England he was received as a Fellow of the Linnean Society, and in Scotland as Fellow of the Botanical Society of Edinburgh. In addition he was a Botanical Commissioner at the Moscow Exhibition in 1865, and at Paris in 1867. Distinctions were not limited to this country, the degree of Ph.D. being conferred on him by the University of Leipzig, and he was admitted to the membership of the Imperial Zoological and Botanical Society of Vienna, and appointed corresponding member of the Natural History Society of Strasbourg.

Though so much engrossed in horticultural work, and with the improvement of the Gardens, Dr. Moore did not lose his interest in scientific botany. Numerous short communications came from his pen, and in 1866 there appeared the Cybele Hibernica, which was the first Flora issued in Ireland that can be said to be at all full and satisfactory. The British Association gave a grant of £25 towards the expenses of this book, and it was prepared by Moore, in conjunction with Mr. A. G. More, F.L.S., the latter, indeed, having taken the larger part of the editorial work. It is no disparagement of Mackay to say that the Cybele was a great advance on the Flora Hibernica, for that could be accomplished in 1866 which was not possible in 1836. In 1872 there appeared in the Proceedings of the Royal Irish Academy Moore's Synopsis of the Mosses of Ireland, which not only states localities, but gives brief generic and specific characters for the plants. This was a carefully prepared list, and enumerates about 140 species beyond those known to Dr. Taylor as Irish thirty-six years previously. In 1873 Professor Lindberg, of Helsingfors, a well known authority on mosses and hepatics, visited Ireland, and was the guest of Dr. Moore during his stay. In June and July of that year they visited some of the best districts in Ireland for cryptogamic plants, and were highly successful in their work. As results of these explorations, Dr. Lindberg described three new species of Hepaticæ, besides adding to the

Irish lists several already known. Thus stimulated, Dr. Moore undertook, with the aid of a money grant from the Royal Irish Academy, to catalogue the Irish scale mosses. This work was accomplished in 1876, when the Report on Irish Hepaticæ appeared in the Proceedings of the Royal Irish Academy. This Report took the form of a descriptive list of the plants, with localities, and included 137 species, being 52 additional to those enumerated by Taylor forty years previously. Of this enumeration Dr. Moore says: "I have collected nearly every one of the plants with my own hands, having for this purpose travelled over a large portion of Ireland, from east to west, and from north to south, and from the sea level to the tops of the highest mountains."

In manner Dr. Moore was unassuming, and he gave a cordial welcome to other workers in the same field. His mental characteristics were such as are usually thought to distinguish his countrymen—shrewd common sense, and a constancy of purpose that courts success and will not be denied. There was no mistaking his nationality. After fifty years familiar intercourse there was no imitation of the mellifluous tones of the Dubliner, and the straightforward directness of his speech, as well as the accent, betrayed the North Briton.

DR. GEORGE DICKIE, F.R.S., F.L.S., etc.

Born at Aberdeen in 1813, and graduated M.A. in Marischal College in 1830. After two years study in the Medical School of Aberdeen he went to Edinburgh, and entered the Brown Square School of Medicine, in which he gained, in 1833 the Medal for Pathology and Practice of Medicine. In 1834 he became M.R.C.S. of London, and in 1842 received the honorary degree of M.D. from King's College, Aberdeen.

Dr. Dickie was born a naturalist, and though he practiced medicine for a short time in his native city, yet he lost no time in obtaining employment more congenial to his tastes. In 1839 he was appointed Lecturer on Botany in King's College, and subsequently on Materia Medica, and on Zoology. In addition he held the office of Librarian to the University. When the Government originated the Irish Queen's Colleges Dr. Dickie was selected to fill the Chair of Natural History in Belfast, and in 1849 he came to Ireland, where he was required to deliver lectures on Botany, Zoology, Geology, and Physical Geography. In 1856 he was married to Miss Low of Aberdeen, who survived him with six children.

On the union, in 1860, of Marischal College and King's College, to form the new University of Aberdeen, Dr. Dickie was appointed to the newly instituted Professorship of Botany. Soon after his return to Aberdeen he was attacked by a severe and dangerous illness which resulted in chronic bronchitis, accompanied with increasing deafness. This disease, which was caught while botanising in Braemar, with his students, in bad weather, was never shaken off, though he continued to fulfil the duties of his professorship till 1877, when he found it necessary to resign his chair. The relief, thus obtained, enabled him to perform more or less botanical work till within a short time of his death, which occurred in July 1882, from a severe attack of bronchitis.

INTRODUCTION.

Dr. Dickie was essentially a Field Naturalist, always delighted when able to add something to our knowledge of Creative Wisdom. Marine Zoology obtained a considerable share of his attention, but it is as a botanist that he will hereafter be known. To cryptogamic botany he was specially devoted, and he became a leading authority on the seaweeds. The Challenger collections of Algæ were submitted to him for revision, and he was engaged at these up to the time of his last illness.

Dr. Dickie's principal works are :—Flora of Aberdeen, 1838. Botanists' Guide to the Counties of Aberdeen, Banff, and Kincardine, 1864. Flora of Ulster, 1864. He also contributed several chapters to Dr. M'Cosh's Typical Forms and Special Ends in Creation, and he wrote the botanical appendix to Macgillivray's Natural History of Braemar and Deeside. Numerous articles from his pen appeared in the Annals and Magazine of Natural History, the Journal of Botany, etc., and many papers were contributed by him to the Transactions of the Edinburgh Botanical Society, the Linnean Society, and the British Association. Dr. Dickie was an Honorary Fellow of the Edinburgh Botanical Society, a member of the Societe des Sciences Naturelles de Cherbourg, and during his residence in Ireland an active member, and for some time Vice-President, of the Belfast Natural History and Philosophical Society. While living here he took great interest in the newly discovered art of photography, and was one of the earliest amateur photographers in Belfast. He took photos of several leading members of the Natural History Society, and was as fairly successful as the knowledge of the art at that time permitted.

PROFESSOR RALPH TATE, F.G.S., F.L.S.

Is one of an English north-country family, several of whose members have distinguished themselves as naturalists. He received his scientific education at the School of Mines, London, and having been a successful teacher of science in Bristol, he was commissioned to establish and lecture to science classes in Belfast. These classes were conducted in the Museum, and under a teacher so able and so profound were most successful. The subjects taught were Geology, Mineralogy, Zoology, Animal Physiology, Systematic Botany, Vegetable Physiology, and Chemistry. Not only was good progress made in scientific knowledge, but the teacher succeeded in embuing many of his pupils with some degree of his own enthusiasm. The Belfast Naturalists' Field Club, established by Mr. Tate in conjunction with a number of his pupils, was the direct outcome of his work, and the numerous local researches subsequently carried on by the Club are thus the indirect results of his teaching. In acknowledgment of services to the cause of Natural Science in Belfast, he was elected an Honorary Associate of the Belfast Natural History and Philosophical Society.

After leaving Belfast, Mr. Tate was for some years employed as Mining Engineer in Central America, and while there made collections of plants and shells. His bundles of dried plants were, however, so much injured by excessive heat and moisture as to become worthless. He was more fortunate in his conchological work, and several new species were established on these shells. Soon after his return to England he received the appointment which he now holds, as Professor of Natural History in the University of Adelaide, South Australia.

Not only at home, but at the Antipodes also, Professor Tate's teaching has been marked with success, and it has been his good fortune, wherever his lot has been cast, to stimulate the spirit of local scientific research. The Royal Society of South Australia has profited by his labours, and more recently he has been successful in originating the Adelaide Naturalists' Field Club.

Professor Tate's forte being the acute discrimination of closely allied fossil forms, he has made a high position as a palæontologist. This, however, has not prevented his attaining to a good knowledge of living forms, notably plants and mollusca. Since his settlement in Australia it has been his good fortune to meet with and describe many new species in both groups. The latest (1887) volume of Proceedings of the Royal Society of South Australia contains several Papers from his pen on recent and fossil shells. These communications are illustrated by figures and descriptions of many novelties. Professor Tate is the author of a large number of important papers, which have appeared from time to time in the Quarterly Journal of the Geological Society of London, the Proceedings of the Linnean Society, the Proceedings of the Royal Society of South Australia, and in other publications. He wrote an Account of the British Land and Fresh-water Mollusca, but his greatest work has been The Yorkshire Lias, of which he is joint author with Rev. J. F. Blake, F.G.S. The greater portion of this elaborate work fell to the share of Professor Tate. The Flora Belfastiensis was hastily prepared by Mr. Tate in 1863, with the trifling assistance afforded by a few helpers, who were at the time the merest tyros. This Flora does not profess to be exhaustive. It is, however, original, and, notwithstanding several errors, it was a step in advance.

PHYSICAL GEOGRAPHY.—The district to which this work refers has for its outer boundary the North Channel and the Irish Sea. It extends on the southeast to Carlingford Lough, and on the northwest is limited by Lough Foyle and the County of Donegal. Its coast line is roughly semicircular, with deep indentations at Strangford, Belfast, and Larne. There is thus a very extensive range of seashore, measuring in Down some 140, in Antrim about 100, and in Derry 28 miles. The length of the base, however, in a direct line from Greencastle in Down to Londonderry is under 90 miles. The combined area of the three counties we have grouped is close on 3,000 square miles, Antrim having the greatest, and Derry the least extension.

The surface of this region is extremely varied, and the scenery often picturesque. In Down the most striking feature is the grand series of mountains known as the Mourne Range, which extend, from Dundrum Bay, some 14 miles in a southeast direction to Rostrevor. These mountains commonly display, more or less, rounded or domed outlines, but occasionally present rugged faces of cliff, which afford habitats for sub-alpine plants. The "castles" of Donard and of Bignian are fine examples of such cyclopean structures. The highest land in Ulster is to be found here; Slieve Donard having an elevation of 2,796 feet above the sea, while several other peaks exceed 2,000 feet. The surface of the county, generally, is very uneven. The roads usually wind through a succession of low, gently rolling hills, of tolerably uniform size, where the traveller rarely sees beyond the nearest eminence. The ancient roads which passed directly over the crests of the hills, if less convenient, were, at all events, more interesting. Little lakes, dotted over the surface, and lying in the hollows of the rocks, are likewise another feature of the county.

In Antrim we get not only change of rocks, but also a different arrangement of surface. The hills commence near Belfast, in the south of the county, and continue round the coast, in an interrupted chain, to Ballycastle in the north. They usually present steep fronts to the sea, with altitudes between 1,000 and 2,000 feet, and tail off in decreasing heights inland. The continuity of this chain, which thus girdles the coast, is broken with charming effect by the streams which in their headlong rush to the sea have cut deep into the strata. This is especially the case in the north, where are located the famous Glyns, or Glens of Antrim. Commencing far up in the hills as narrow, rugged defiles, and widening out as they approach the sea, there is afforded at the same time the finest subjects for the artist's pencil and the best collecting grounds for the naturalist. The centre of the county is largely made up of moory table lands, of no great elevation, the drainage from which flows mainly inland to Lough Neagh and the River Bann.

The mountains in Derry form two extensive tracts, with elevations in most cases above 1,000 feet, but rarely exceeding 2,000. From Benevenagh in the north, one series stretch across the centre of the county until they meet the Carntogher and Sperrin Ranges in the south. These last named groups, which are of great magnitude, occupy the principal part of the southwestern boundary and extend into Tyrone. The basaltic mountain of Benevenagh, in the north, has a majestic range of cliffs which give picturesque effects, and also yield some hawkweeds and other rare plants: the same may be said of Benbradagh in a lesser degree. The Carntogher and Sperrin Mountains are of greater altitude, and more massive, but in all other respects are most uninteresting. Dart, 2,040 feet, has a somewhat rocky summit, but with this exception these mountains may be classed as heathy moorlands with rounded summits, and long, wet, boggy slopes, devoid of beauty, and the home of little but the most common plants. The lover of the picturesque, and the naturalist too, who explores the glens of the Roe and the Faughan in the north, will be much more amply recompensed for his toil.

The principal rivers in our three northeastern counties are the Bann, the Lagan, and the Foyle; the former being the most important. The Bann first appears as a rocky streamlet issuing from the heart of the Mourne Mountains near the southern limit of our district. Flowing in a northwesterly direction it enters Lough Neagh at the southern extremity of the lake, which it leaves at Toome on the north. From hence its course is nearly due north, separating the counties of Antrim and Derry, and ultimately falling into the sea near Coleraine. The Lagan has its source in the Slieve Croob Mountains, at no great distance from the source of the Bann. It takes a very circuitous course, first to the west, then curving round to the north and east empties itself into Belfast Bay, after having passed the towns of Dromore, Lisburn, and Belfast. The flow of the river is thus nearly 40 miles, though the distance accomplished is no more than 20 miles in a line direct. The Foyle is formed at Lifford, in the county of Donegal, by the union of several smaller streams. Flowing nearly due north, and dividing the counties of Derry and Donegal, it enters Lough Foyle at Culmore, after a course of over 20 miles. If traced, however, to its ultimate head this river would appear much more extensive, a great portion of its waters being collected in Tyrone and Donegal. In addition to these tidal rivers there are many others of considerable note; the district being exceptionally well watered, and its flora enriched by a number of interesting fluvial plants. Many small lakes are scattered over the country, and Lough Neagh, the largest British lake, enters into the district, the greater part of its waters being within the bounds of Antrim and Derry.

GEOLOGY. -The counties of Down, Antrim, and Derry are widely dissimilar in their geological structure, the two former especially offering strong contrasts, not only in the age of the rocks, but also in their chemical and petrological characters. Derry, while conforming to neither, is, nevertheless, made up to some extent of rocks such as occur in Down and Antrim, with the addition of others which are not at all, or only slightly developed in those counties. The county of Down has a totally different geological history, and a rock series quite distinct from that of Antrim. The masses of dark coloured rocks that prevail over the greater part of Down have suffered many vicissitudes since they were deposited in Lower Silurian times; though they seem to have remained mostly above water from the epoch of their elevation. Indurated, shattered, and contorted, the originally horizontal beds are now found inclined at high angles, sometimes vertical, and in other cases turned over and inverted. These old rocks consist of gray and black slates, too coarse to be of much economic value, and hard, massive, intractible grits. They never rise to a high altitude, and seldom present much of picturesque beauty, nor do they yield a flora so rich and varied as that found on the basaltic rocks of Antrim. This poverty is, however, compensated in some degree by the granite mountains of the Mourne range, which supply rocks of much industrial importance, and also *habitats* for many of the rarer plants. Stretching southwest, from Newcastle to Rostrevor, they form one of the most majestic groups in Ireland. A small strip of Triassic sandstone occurs close to Newtownards, and thence to near Dundonald, and unimportant patches of Carboniferous limestone appear on the shores of Belfast and Strangford Loughs, and again at Greencastle, on the extreme south of the county. These, with the addition of two little specks of Permian, and frequent deposits of Boulder Clay and superficial drift, complete the geological series as exposed in County Down.

The most striking feature in the geology of Antrim is the enormous mass of volcanic rock which covers almost the entire county. This great sheet of basalt has been erupted through wide fissures, now filled with hard rock, and known as dykes. The trap, or basalt, extends into Derry, and has a combined area, in the two counties, of some 1,200 square miles, with a thickness of more than 1,000 feet in some places. An interesting series of sedimentary rocks of *Mezozoic* age are developed in the county, and by their contrasts of colour heighten the picturesque beauty of the fine sea-cliffs for which Antrim is famous. These rocks consist of Chalk, Greensand, Liassic, and Triassic strata; but as they have been universally covered by the Trap, and are only seen in slight exposures where the igneous rock has been denuded away, they have but a subordinate influence on the native vegetation.

The county of Derry is separated by its geology into two distinct regions; the east side being part of the great basaltic plateau of Antrim, with the same underlying series of sedimentary strata, though the latter are more sparingly developed.

The hills in this part of the county are of trap, which decomposes with comparative facility, yielding a rich, loamy soil, and a native vegetation of considerable variety. To the west of the basaltic escarpments there are occasional bands of limestone, of trivial importance, and granite rocks are rare. Sandstones, however, some of mezozoic and others of palæozoic age, are developed to a considerable extent, mainly towards the centre of the county. Further west there appear great masses of Mica, Schist, or other Metamorphic and crystalline rocks, monopolising the entire surface. They cross the western and northwestern boundary, and constitute a characteristic feature in the geology of the neighbouring counties of Donegal and Tyrone. The weathering of these silicious rocks proceeds slowly, and the result of their disintegration is a less valuable soil, and a flora comparatively poor.

Over the entire region the Glacial Drift has been more or less spread, being seen especially in the valleys, and along the lower slopes of the hills. The Boulder Clay, as distinguished from moraine matter, is a marine deposit composed of fine tenacious clay, with which is intermixed an abundance of stones of various sizes, from a few grains up to many tons in weight. This drift largely influences vegetation, and is of the greatest importance to the tiller of the soil. The effect of its absence may be noted by the traveller on the railway line between Comber and Crossgar. On much of the ground passed over the drift is absent, and the treeless rocky surface seems as unprofitable to the farmer as it is uninviting to the botanist. Esker mounds, and local beds of gravel, are not uncommon, but their influence on the vegetation is not extensive.

CLIMATE-—As it is well known that the flora of a district is largely affected by the meteorological conditions which prevail, it becomes necessary to offer some general remarks on this head, although there seems to be little that is peculiar in the meteorology of the northeast of Ireland.

The means of annual temperature, hereafter stated, range from 48·5 to 48·8 degrees of Fahrenheit. This is less than the average for the remainder of Ireland by nearly one degree, but if we exclude from this comparison a limited area near the south and southwest coasts, the difference will be found to be very slight.

The average yearly temperature of the north of Ireland corresponds very nearly to that of the north of England, but in the former the thermometrical range is smaller. Our more clouded skies moderate the winter frosts as well as the summer heats. In the very limited bounds of this district the local range can of course be only slight, but the greater humidity, and more equable temperature of the sea-coast, find their botanical expression in the increased variety of plants to be met with there.

The opinion loosely formed, and freely expressed by many visitors, to the effect that Belfast and district is a very wet region is not borne out by exact observation. This popular fancy is, no doubt, due to the many wet days in which the actual volume of rain is slight. The Scottish Meteorological Society having recently collated the rainfall returns of the British Islands has displayed the results in a general way by means of coloured maps. An inspection of these maps shows that the northeast of Ireland lies in the region of least rainfall in this island. An annual rainfall of 30 to 40 inches is indicated on the maps in question by dark red shading, and this colour covers almost the entire east coast, and a great portion of the centre of the island. In the northeast the uniformity of colouring is broken by a small strip in south Down, which forms part of a greater rainfall—over 40 inches. A similar rainfall, coloured light blue on the maps, prevails in Donegal, and extending to the east, penetrates as a narrow wedge to the centre of the county of Antrim.

INTRODUCTION. xxvii

MEAN ANNUAL TEMPERATURE, 1857—1880.

Londonderry (northwest extremity of district)   48·6 Deg. Fahr.
Aghalee (centre of district, close to Lough Neagh)   48·7   ,,
Belfast (by the coast, near centre of district)   48·8   ,,
Donaghadee (on the coast, northeast end of Co. Down) 48·5   ,,

ANNUAL RAINFALL AT BELFAST, AGHALEE, WOODBURN, AND CUSHENDUN.

	Belfast.	Aghalee.	South Woodburn.	Cushendun.
			In the hills above C. fergus 630 feet above sea level.	Near northern limit of the district.
1869	35·41 inches	28·82 inches		
1870	34·37 ,,	28·86 ,,		
1871	33·17 ,,	30·18 ,,	..	..
1872	65·09 ,,	47·09 ,,	..	..
1873	31·900 ,,	31·94 ,,	..	..
1874	33·560 ,,	30·03 ,,	..	..
1875	31·195 ,,	33·63 ,,	..	..
1876	38·915 ,,	38·33 ,,	..	..
1877	43·975 ,,	41·68 ,,	..	..
1878	28·785 ,,	30· 2 ,,	..	..
1879	33·99 ,,	35·10 ,,	..	..
1880	21·024 ,,	30·11 ,,	..	..
1881	38·34 ,,	36·49 ,,	70·27 inches	..
1882	41·8S ,,	39·59 ,,	53·87 ,,	..
1883	38·36 ,,	31·92 ,,	39·41 ,,	47·71 inches
1884	36·18 ,,	32·51 ,,	38·01 ,,	46·49 ,,
1885	31 66 ,,	26·31 ,,	32·46 ,,	36·76 ,,
1886	32·23 ,,	34·61 ,,	41·73 ,,	41·17 ,,
1887	25·54 ,,	21·12 ,,	28·77 ,,	35·77 ,,

The following gentlemen kindly supplied the figures—Belfast, Mr. Maitland, Librarian, Linen Hall Library. Aghalee, Mr. Launcelot Turtle, J.P. Woodburn, Mr. George E. Reilly, Superintendent Woodburn Waterworks. Cushendun, Rev. S. A. Brenan, Glendun Lodge.

LIMITS OF CULTIVATION.—The factors that modify these limits are several, and no very definite general statement is possible as to the line where cultivation ceases. Broadly, it may be stated that cultivation is nowhere carried beyond 800 to 900 feet, and that it ceases much below that limit where the rock is not well covered with drift, and the aspect southern. The mineral characters of the rocks, and their structure, have considerable influence on vegetation, but physical features may be said to be still more potent in dictating to the husbandman. Where the long upward slopes permit soil to collect there crops will repay labour, but where the rise is rapid cultivation ceases at a much lower level. The aspect, or position of the ground towards the sun, is of course of the utmost importance. Mr. R. Glascott Symes, M.A., F.G.S., of the Geological Survey, who knows the County of Antrim so well, has kindly supplied a note on this head as respects that county. Rev. H. W. Lett. M.A., T.C.D., who has been much engaged in botanically exploring the Mourne Mountains, has also been good enough to furnish some particulars with regard to County Down. The details communicated by Mr. Lett are in substance as follow:—

About Newcastle, on the slopes of Slieve Donard and Slieve Commedagh, the cereal line does not rise above 200 feet; drift clay is absent, and the ground steep. On Tullybranagan the limit is still the same. This is exposed to the north and east, and the mountains rise somewhat abruptly behind, to the south and southwest. Proceeding south, at the Bloody Bridge, also open to the east, the limit is 350, while further south, at Tullyree, where there is an extensive moraine, the line rises to 600; but not far off, on the east slope of Spence's Mountain, it drops to 200. Along the sheltered banks of Spence's River, and on the southern slopes of Slieve Bignian, three miles inland, the cereal line is at 600 feet. There are well sheltered farms on the road that passes through the mountains, from Kilkeel to Hilltown, where the line gets up to 700 feet, its highest point on the south of the range. About Rostrevor the indigenous woods blot out the cereal line, but at Kilbroney River it stands at 700, while further north, at the favoured locality of Gruggandoo, it ascends to 900. Along the glen west of Altatagart, through which Shanky's River flows, the line is met with at 800 feet, but crops only succeed in favourable seasons; along the same northern slope, at Leitrim River, it is 600. To the southeast of Hilltown the line descends to 500, and between Hilltown and Castlewellan it is 650. Up the sides of the Shimna River, to the north of Slievemeel Beg, the limit is 700 feet; from this it descends to 400 where the Spinkwee enters the park. Slieve Croob is a good deal isolated, and the slopes on all sides are gradual; on the east, north, and west sides cultivation reaches to 700, and on the south to 800 feet. There does not seem to be much difference between the granite and the schist districts, acclivity, drift clay, and shelter from cold winds being the great modifying agents.

Respecting the limits of culture in County Antrim Mr. Symes has not paid special attention to the point. He is, nevertheless, a trained observer

INTRODUCTION.

and his conclusions are of great value. Mr. Symes says—No definite boundary line can be drawn, there being modifying influences to hinder it. There are geological facts to consider as well as aspect: as regards the former, the higher the chalk rises so will the cereals ripen at a higher altitude above sea level. This is well seen between Glenarm and Drain's Bay, on the road to Larne. The amount of drift that may be in a secluded valley has a most important influence, as for example the valley of the Braid along by Broughshane, and bordering on Slemish. The Braid valley has ripening crops at a higher altitude than Glenwherry valley which is parallel to it. Absence of drift on high ground prevents cultivation altogether. As regards slopes of hills, the northern slope, with a southern aspect is of course better. The outcrop of Basalt interferes with vegetation on account of the enormous percentage of iron; artifical grasses will soon die out, and only heather and other indigenous plants remain. This is well seen on high ground between Ballycastle and Ballintoy. Altitudes of 800 feet on the heights between Larne and Glenarm will produce good crops, whereas 200 would be the maximum in Glenwherry valley, yet the escarpment in question has to contend with the exposure to injurious east winds.

EXTENT AND CHARACTERISTICS OF THE FLORA.—Leaving out of sight the Excluded Plants, which number 271, the total here recorded amounts to 1169 species; made us thus—Flowering Plants, and Higher Cryptogams 803, Mosses 293, Hepatics 73. The reckoning of the number of plants in the flora of any district is influenced more or less by the views of authors as to species, or reputed species, and the Manual of British Botany by Prof. Babington, which enumerates 1524 plants has been taken as the standard for this Flora. It will thus appear that the ratio of the plants of the northeast of Ireland as contrasted with the entire British flora is as 803 to 1524, or rather more than half. For all Ireland there are about 1000 plants, and compared with that flora the proportion for this district will be 8-10ths. The London Catalogue of Mosses and Hepatics states the number of the former in Britain as 568, and of the latter 192; our mosses are 293, being a ratio very similar to that just shown in the case of the flowering plants. The list of Hepaticæ or scale mosses of the locality shows only 73 as against 192 British, a declining proportion which is not entirely real, but in part due to the smaller interest which these plants have excited amongst local workers. With respect to mosses Dr. Moore stated the number known to occur in Ireland at 369, an excess beyond the present list of 76, a ratio very similar to that in respect of the higher plants. The same authority gives 137 as the number of Irish hepatics, a great preponderance as compared with 73 in the present enumeration. This local poverty may be explained in part by local neglect, but the unfavourable comparison is owing, in a much greater degree, to the unparalleled richness of some districts in the south and west.

The district with which this Flora is concerned cannot be said to have any group of plants characteristic of it, or which bear the aspect of a special flora. The beautiful *Carex Buxbaumii*, now on the eve of extinction, has never been found elsewhere in Britain, and in like manner *Calamagrostis Hookeri* (if, as we believe, distinct from *C. stricta*) is special to the district. *Rosa hibernica*, and *Orobanche rubra* though discovered here, and at first thought to be limited to the north of Ireland are now found in other parts of the empire; and there was no bond of connection between them, and they in no wise constituted a local group.

On geographical considerations it might seem natural that the flora of the north of Ireland should be most nearly related to that of Scotland, and derived from thence : such a supposition is, however, scarcely borne out on closer scrutiny. The flora of this district, it is true, includes some of the Scottish alpines and subalpines, but these are not numerous, and are all, save one, plants that occur also in Wales, or in the west of England, and likely to be derived from that quarter along with the bulk of our flora. *Hieracium flocculosum* is our only hitherto exclusively Scottish plant, and this is such a very critical form that even this exception may yet disappear. On the other hand the plants of the south and east of England have failed to spread to the north of Ireland, and there is not a single representative of Watson's Germanic Type of vegetation known to occur here as a native. Here, as elsewhere, the great body of the rank and file of the flora consists of plants of Watson's British Type, which number 485 species, or three-fifths of the entire native vegetation of the district. Plants of the English Type, so named by reason of their having their headquarters and greatest development in England, number some 142 species, or rather more than one-sixth part of our flora. Some 40 of our plants, or about one twentieth of the whole may be classed as of the Scottish Type; and the Highland Type claims 25 species, being rather more than one-thirtieth. The Atlantic Type, or group of plants characteristic of the west and southwest of England numbers here only some 18 species, being less than one-thirtieth of our plants.

Of the great groups or Natural Orders of British plants we have a weak representation of *Leguminosæ*—26 out of 85 species, and *Compositæ*—68 out of 146. *Scrophulariaceæ*, and *Labiatæ* also are under the proper ratio, while in *Orchidaceæ* we have only 15 of the 46 British species, or about one-third. On the other hand *Rosaceæ*, and *Umbelliferæ* have about the same importance as in the British flora, and *Primulaceæ*, *Cyperaceæ*, *Gramineæ*, and *Filices* are in excess of their due. *Amentiferæ* are under represented, there being only 19 out of the 43 British species, which seems strange when we remember that this part of Ireland was, at no very remote period covered to a great extent with wood.

ADDENDA.

I.—Species to be added.

p. 56. Before **Crassulaceæ** insert—

ORDER PORTULACEÆ.

MONTIA *Linn.* BLINKS.

1. M. fontana *Linn.*
Wet stony and gravelly places—common. Fl. May—Aug.
Var. *minor* is very common, var. *rivularis* is frequent.

p. 81. Before **Arctium** insert—

CARLINA *Linn.*

1. C. vulgaris *Linn.*
Sandy coasts—rare. Fl. July—Sept. (*Cyb. Hib.*).
Down—Sparingly on sands at Ballykinler; R.Ll.P., 1887.
Antrim—Ballycastle (*Miss Hincks*); *Flor. Ulst.*

p. 196. To number 3 the following should be added—
A. crassinervis (*type*).
Down—On granite at top of Eagle Mountain; C.H.W.
"*This appears to me a good typical specimen*"; G.A.Holt.

p. 222. Add **G. ovata** *Web. et Mohr.*
Mountain rocks—very rare.
Down—Slieve Donard; H.W.L.

p. 289. Insert—
136. Avena strigosa *Schreber.*
Cornfields at Cherryvalley near Comber; *Flor. Belf.* Ditch banks at Drumbridge, field at Kilroot, and plentiful in fields at Ticloy in the Braid, also in a bog near Bellaghy, Co. Derry; S.A.S.
A cornfield casual.

ADDENDA

II.—Localities to be added.

p. 19. To **E. hexandra** add—
Down—Abundant in Lough Aghery, and in Loughinisland ; S.A.S.

p. 20. To **S. anglica** add—
Down—Roadside near Alexandra Dock, Belfast; R.Ll.P.
Probably only a casual in this station.

p. 30. To **G. perenne** add—
Down—Newcastle; R.Ll.P., 1887.

p. 32. To **R. linoides** add—
Antrim—Sandhills at Bushfoot ; R.Ll.P.

p. 43. To **P. reptans** add—
Down—Magheralagan ; S.A.S.

p. 62. To **C. verticillatum** add—
Antrim—Abundant on the headlands at Giants' Causeway; R.Ll.P., 1887.

p. 70. To **S. ebulus** add—
Antrim—Black Mountain ; S.A.S.

p. 83. To **C. pratensis** add—
Antrim—By the narrow gauge railway line about one mile northeast of Ballymoney; R.Ll.P.

p. 95. To **C. arvensis** add—
Down—Shore of the lake at Magheralagan ; S.A.S.

p. 102. Add **Linaria repens**—see in corrigenda for further notice of this species.

p. 104. To **M. sylvaticum** add—
Antrim—Cairncastle ; R.Ll.P., 1887.

p. 117. To **C. minimus** add—
Antrim—Sandhills at Bushfoot ; R.Ll.P.

p. 120. To **C. bonus-henricus** add—
Antrim—Bushfoot ; R.Ll.P.

ADDENDA.

p. 120. To **B. maritima** add—
Antrim—Rocks at Whitepark Bay near Ballintoy; R.Ll.P.

p. 122. To **R. hydrolapathum** add—
Down—Marshy margin of Magheralagan Lake; S.A.S.

p. 127. To **C. autumnalis** add—
Down—Plentiful in Magheralagan Lake; S.A.S., 1887.

p. 141. To **S. sagittifolia** add—
Antrim—In the Bush River near Bushmills; R.Ll.P.

p. 175. To **S. loliacea** add—
Antrim—On old wall near the quay at Carrickfergus, also on the ruins of Olderfleet Castle, pier of Ballycastle, and dry ground in Rathlin; *Templeton.* Ballintoy, and Portrush; *Cyb. Hib.* Walls of Castle Chichester, rocks at Blackhead, and on limestone rocks west of the church in Rathlin; S.A.S.
Derry—Portstewart; *Flor. Ulst.*

p. 187. To **A. marinum** add—
Down—Very sparingly on rocks by the shore at Portavo; *W. H. Patterson.*

p. 188. To **C. officinarum** add—
Down—On wall at Banbridge; *Rev. Thos. Boyd.*
Antrim—Birch Hill near Antrim; *Rev. W. S. Smith.*

p. 189. To **B. lunaria** add—
Antrim—Massereene Park; *D. Redmond.*

p. 193. To **N. translucens** add—
Down—In a lake near Slieve Croob; *Templeton.*

p. 213. To **T. papillosa** add—
Down—On trees near Loughbrickland; H.W.L.

p. 304. Add 105b **Melica nutans** *Linn.*
Among heath on the Cave Hill; *J. Sim., Phyt. N.S., vol. 3, p.* 357. Mistaken identification; probably *Aira flexuosa.*

p. 305. Add 113b **Fissidens incurvis** *Schwaeg.*
Whiterock, and Redhall Glen; *Stewart, Mosses of Northeast of Ireland.* Corrected to *F. pusillus.*

CORRIGENDA.

p. 29 **H. hirsutum.**
The suggestion that this plant is extinct near Belfast is erroneous, as Mr. Richard Hanna has specimens collected in August, 1887, from a field at a short distance from the original station.

pp. 33, 34. Creagh Bog.—A locality that is now nearly obsolete, is in County Derry, immediately bordering on Antrim, but not in the latter county.

p. 102. **L. minor** appears by inadvertence—read—**L. repens** *Aiton*, and add the following additional note :—
Down—On a stony ditch bank about one mile south of Rostrevor; *Thomas Chandlee*, 1887.

p. 176. For **F. scuiroides** read **F. sciuroides.**

p. 222. Delete **10. G. elatior.**
This species was published in Journal of Botany, for April, 1887, on the high authority of Mr. H. Boswell, and was too hastily inserted in this Flora. The specimen recorded was subsequently submitted to a most rigid scrutiny by Mr. Holt, who states that it must certainly be referred to G. decipiens *Schultz*. *Grimmia elatior* therefore remains as one of the *desiderata* of the Irish flora.

p. 288. **Acorus calamus.**
The origin of this plant was Moira, and not Hillsborough, as suggested. Harris states that Sir John Rawdon's magnificent gardens at Moira were enriched with many rare exotic plants, among which there is mention made of *Calamus aromaticus*, *vide* Harris's Down, pp. 103, 4.

ABBREVIATIONS.

Back. Brit Hier.—Backhouse, British Hieracia.
B.N.F.C.—Proceedings, Belfast Naturalists' Field Club.
Cyb. Hib.—Cybele Hibernica.
Eng. Flor.—English Flora.
Flor. Belf.—Flora Belfastiensis.
Flor. Brit.—Flora Britannica.
Flor. Hib.—Flora Hibernica.
Flor. Ulst.—Flora of Ulster.
Ir. Flor.—Irish Flora.
Jour. Bot.—Journal of Botany.
Mack. Catal. -Mackay's Catalogue.
Nat. Hist. Rev.—Natural History Review.
Mag. Nat. Hist.—Magazine of Natural History.
Musc. Brit.—Muscologia Britannica.
Musc. Hib.—Muscologia Hibernica.
Phyt.—Phytologist.
Proc. R.I.A.—Proceedings, Royal Irish Academy.
Samp's. Surv.—Sampson's Survey of County Derry.
Wade Rar.—Wade, Plantae Rariores.
T.H.C.—Thomas Hughes Corry.
H.W.L.—Rev. H. W. Lett, M.A., T.C.D., Loughbrickland.
D.M.—Dr. David Moore, Glasnevin.
R.Ll.P.—Robert Lloyd Praeger, B.A., B.E., Holywood.
S.A.S.—Samuel Alexander Stewart.
Templeton—John Templeton, A.L.S., Cranmore.
C.H.W.—Rev. C. H. Waddell, Kendal.

FLORA OF THE NORTH-EAST OF IRELAND.

ORDER I. **RANUNCULACEÆ.**

THALICTRUM *Linn.*

1. T. minus *Linn.*
On blown sands by the seacoast; usually abundant where the conditions exist. Fl. June and July.
Down—Sandhills near Newcastle; *Templeton*, 1793; and T.H.C., 1878.
   In profusion on Ballykinler sands east of Dundrum; S.A.S.
Antrim—Sparingly at Ballycastle; T.H.C. Portrush; S.A.S.
It is remarkable that this plant has not been found at Magilligan, Co. Derry, the wide expanse of sandhills being so eminently favourable.

Var. β T. MONTANUM *Wallr.*
Damp shady rocks—very rare. Fl. July and August.
Down—Mountains of Mourne, plentifully; *Harris's Down*—1744. West side of Bignian Mountain; *Templeton*, 1808. Rocks above Donard Lodge, at 1,500 to 1,600 feet; S.A.S. Kilbroney River north of Rostrevor; R.Ll.P.
Antrim—On basaltic rocks in Glenariffe; *Cyb. Hib.*
   The Portmore, and Lough Beg plants of *Flor. Ulst.* were *T. flavum.*
Derry—Artrea near Toome; D.M.

2. T. flavum *Linn.* MEADOW RUE.
Marshy ground, and wet meadows—rare; about Lough Neagh, only. Fl. July and August.
Antrim—Portmore; *Templeton*, 1800. Massereene Demesne; *Flor. Ulst.* Ram's Island; *F. Whitla.* Shane's Castle; T.H.C. Cranfield Point; S.A.S. Selshan Harbour; *G. C. Hyndman.*
Derry—Artrea near Toome; D.M.

RANUNCULACEÆ.

ANEMONE *Linn.* WIND FLOWER.

1. A. nemorosa *Linn.* WOOD ANEMONE.
Woods, and shady pastures in the hills—common. Fl. mid. March till mid. June. We have found it flowering at 1,800 feet, on Carlingford Mountain, Co. Louth.

RANUNCULUS *Linn.* CROWFOOT, BUTTERCUP.

Sect. I.—Aquatic. WATER CROWFOOT.

1. R. trichophyllus *Chaix.*
Lakes, ponds, and marshes—frequent, usually in rather clear water. Fl. June—Aug.
Down — Ballynahinch; *C. Dickson.* Islanderry Lake near Dromore; T.H.C. Ballyholme, and Castlespie; S.A.S. Craigavad; R.Ll.P.
Antrim—Massereene Park; *Cyb. Hib.* Belfast Waterworks, also Rathlin Island ("typical" C.C.B.); and abundant on shore of Lough Neagh at Selshan; S.A.S.
Derry—Marshes at Magilligan; T.H.C.
The floating leaves, which our plant frequently produces, are slightly hairy below, as are also the capillary leaves.
First published, as Irish, in Report of Belf. Nat. Field Club, 1863-4.

2. R. heterophyllus *Fries.*
Lakes, ponds, and marshes—frequent. Fl. May—July.
Down—Moneycaragh River near Newcastle, and at Castlewellan; *C. Dickson.* Copeland Island, Groomsport, and very fine in Clandeboye Lake; S.A.S.
Antrim—Shaw's Bridge, and at shore of Lough Neagh near Crumlin; T.H.C. Belfast Waterworks; S.A.S.
This, like the preceding, was first recorded as an Irish plant in the Report of the Belf. Nat. Field Club, 1863-4. Specimens of both have been identified by Prof. Babington.

3. R. penicillatus *Dum.* (R. PSEUDO-FLUITANS *Newb.*).
Frequent in quick flowing rivers; rare in lakes, and only where the entrance of a stream causes movement of the water. Fl. June—Aug.
Antrim—Doagh River, and at several points in the Bann River; S.A.S. Six-mile River at Massereene Park; *A. G. More*, and at Muckamore; T.H.C. Bush River; *Cyb. Hib.*
Derry—In the Bann at Coleraine; *G. C. Hyndman.* (*spec. in herb Hyndman*).

Floating leaves are occasionally produced, at least in some of the stations, but there seems to be no other important difference in our plants of this species.

4. R. peltatus *Fries.*
Lakes, pools, and slow streams; common in Down, less frequent in Antrim and Derry. Fl. mid. May till mid. Aug.
Down—Ballycroghan near Bangor; *Flor. Belf.* Magherascouse. Monlough, and Tullycairne; T.H.C. Downpatrick, Ballynahinch, Saintfield Lakes, Hillsborough Park, and in the Lagan near Dromore; S.A.S. Dundonald; R.Ll.P.
Antrim—Canal near Lisburn, Portmore, several places in Lough Neagh, Lough Mourne, and abundant in the lower Bann; S A.S.
Derry—Lough Beg near Toome; T.H.C. River Bann; S.A.S.
First recorded as Irish in Flor. Belf., 1863. The Dromore plant is rather curious, and may be R. ELONGATUS *Hiern.*

5. R. fluitans *Lamk.*
Deep, quick-flowing stream—very rare. Fl. June till mid. Aug.
Antrim—Plentiful in the Sixmilewater close to Dunadry station, and lower down at Muckamore, also a little distance above where the river empties into Lough Neagh; S.A.S. The only Irish station.
First found in 1865, in the river above Templepatrick, where it was plentiful for some years until destroyed by bleachworks. Our plant differs in appearance, both from English *R. fluitans*, and its *var. Bachii.* From *R. penicillatus*, which is abundant in the same stream, it is at once distinguished by its smaller, more abundant flowers, with numerous petals, and the hard rigid stems and leaves, with short stout segments. The *R. fluitans* of Flor. Ulst. was *R. penicillatus*.

6. R. hederaceus *Linn.*
Shallow pools, and wet, plashy ground, especially near springs and moving water—frequent in Down, Antrim, and Derry. Fl. May—Aug.

Sect. II.—Terrestrial, and sub-aquatic.

7. R. sceleratus *Linn.*
Stagnant ditches and muddy pools, mostly near the sea—common on the Down coast, scarce in Antrim, and rare in Derry. Fl. June—Aug.

8. R. flammula *Linn.* LESSER SPEARWORT.
Marshes and wet pastures, generally distributed throughout the district. Fl. June – Aug.

Var. R. PSEUDO-REPTANS.

Antrim—Shores of Lough Neagh; *Templeton* (as *R. reptans*). Sandy shore of Lough Neagh near Crumlin; S.A.S.

There is no specimen of Mr. Templeton's plant, but it may safely be assumed to have been the above-named form.

9. R. lingua *Linn.* GREATER SPEARWORT.

Marshy margins of rivers and lakes—rare. Fl. July and Aug.

Down—In small lakes near Echlinville in the Ards; *Flor. Hib.* Moira Demesne; *F. Whitla.* By the canal above Moira abundant; S.A.S.

Antrim—About Portmore Lough, and by the Bann above Portglenone; *Templeton,* 1796. Ram's Island; *Flor. Ulst.* Southwest margin of the lough at Portmore; S.A.S.

Derry—Wet pastures and ditches; *Sampson.* Artrea near Toome, and by the Bann at Kilrea; D.M. Magilligan; *H. C. Hart,* in *Proc.* of R.I A.

The localities about Belfast, specified in Flora of Ulster, are erroneous.

10. R. ficaria *Linn.* PILEWORT.

Moist places in woods, and damp shady spots—Common in Down, Antrim, and Derry. Fl. March—May.

11. R. auricomus *Linn.* GOLDILOCKS.

Wooded and shady places—not common; more frequent in sheltered spots along the lower slopes of the basaltic hills. Fl. April and May.

Down—Crawfordsburn; *Flor. Ulst.* Lagan-side; T.H.C. Drumbo Glen; S.A.S. Downpatrick; *C. Dickson.* Near Dundonald; R.Ll.P.

Antrim—Colin Glen, Cave Hill, &c.; *Flor. Ulst.* Near Lisburn, and along the bases of the Belfast hills, also Woodburn and Magheramorne; S.A.S. Glenarm, and Carnlough; R.Ll.P.

Derry—Frequent in the Barony of Loughinshollen; D.M.

12. R. acris *Linn.*

Roadsides, and pastures—common. Fl. mid. May till mid. Aug.

13. R. repens *Linn.* SITFAST.

Roadsides, waste ground, and pastures—very common. Fl. May—Aug.

14. R. bulbosus *Linn.*

Generally distributed in short, dry pastures, especially where hilly and gravelly, but much less common than the two preceding species. Fl. mid. May till mid. August.

NYMPHÆCEÆ.

The following localities may briefly be mentioned:—Rostrevor, Newcastle, Copeland Island, Donaghadee, Newtownards, Greyabbey, Holywood, Cave Hill, Antrim, Knockagh, Glenarm, Islandmagee, Rathlin, Benbradagh, Moneymore, Ballyronan. "Frequent in Derry"; D.M.

CALTHA *Linn.* MARSH MARIGOLD.

1. C. palustris *Linn.* MAYFLOWER.
Marshes, and margins of streams—common. Fl. mid. April till mid. June.

ORDER II. **NYMPHÆCEÆ**.

NYMPHÆA *Linn.* WATER LILY.

1. N. alba *Linn.* WHITE WATER LILY.
Lakes, and slow streams—frequent in Down, and Derry—rare in Antrim. Fl. July and Aug.
Down—Lough Leagh, 1797, and almost all lakes in County Down; *Templeton*. Ballydrain; *Flor. Belf.* Shane's Lough near Killyleagh; *Proc. Belf. Nat. Field Club*, 1876-7. Ballyalloley, Creevy Lough, and abundant in lakes near Crossgar, and Killyleagh; S.A.S. Dam near Dromore; *C. Dickson*.
Antrim—Lagan Canal, and Lough Neagh; *Flor. Ulst.* Portmore; S.A.S. Rathlin Island; *Proc. R.I.A.*, 1837, and 1884.
Derry—Ballyarnott Lake; *Ord. Surv. Londonderry.* "Abundant in most of the lakes and rivers in Derry"; D.M.

NUPHAR *Smith.*

1. N. luteum *Smith.* YELLOW WATER LILY.
Lakes, streams, and deep ditches—frequent. Fl. mid. June till mid. Aug.
The yellow water-lily is commonly found associated with the preceding species, but more frequent, and usually in greater abundance. "From 50, to 700 feet in Derry"; D.M.

ORDER III. PAPAVERACEÆ.

PAPAVER *Linn.* POPPY.

1. P. argemone *Linn.* PRICKLY LONG-HEADED POPPY.
Roadsides, and sandy or gravelly fields and waste ground—rare. Fl. June—Aug.

Down—Near Bangor; *Flor. Ulst.* Roadsides Sydenham, Belmont, and Knock, also gravel pit at Dundonald; S.A.S. Sand pit near Giant's Ring; *C. Dickson.*
Antrim—Curran of Larne, also among corn at Malone; *Templeton,* 1828; and T.H.C., 1878. Railway bank at Glynn; S.A.S.
Derry—Dunboe, and Tamlaghtard; D.M. Magilligan; S.A.S.

2. P. hybridum *Linn.* PRICKLY ROUND-HEADED POPPY.
Sandy cultivated fields—very rare. Fl. mid. June till mid. Aug.

Down—Sandy field a little below Holywood; *Templeton,* 1797. On the coast beyond Groomsport, plentiful; *Flor. Belf.* The plant continues to grow in this latter locality, having been seen by us repeatedly since Mr. Tate's record.

Mr. Templeton's note on this species is of much interest—"In a sandy field a little below Holywood, along with *P. rhœas,* and *P. dubium,* but at this time, July 15th, 1797, from its being among *Reseda luteola,* the seed of which came from Sicily, I reckon it a doubtful native." The fact that *P. hybridum* has held its ground in County Down, having neither become extinct, nor yet spreading to any extent during a period of 90 years, would seem, however, to negative the suggestion that it was an introduced plant. All our species of poppies are by some, on perhaps insufficient grounds, considered doubtful natives; as far as they occur in our district there is but little apparent ground to dispute their claim.

3. P. rhœas *Linn.* SMOOTH ROUND-HEADED POPPY.
Sandy cultivated fields—rare, and only in the southern end of our district. Fl. June and July.

Down—Common on south side of Lisburn towards Moira; *Templeton,* 1796. Near Groomsport Church; *Flor. Belf.* Fields around Killough Bay, and very sparingly at Groomsport, and Newcastle; S.A.S.
Antrim—Ballast heaps at Magheramorne (introduced); S.A.S.

4. P. dubium *Linn.* SMOOTH LONG-HEADED POPPY.
Sandy cultivated fields, and sandy or gravelly waste ground; abundant in Down, less common in Antrim and Derry. Our plant seems to be the

var. *Lecoqii* of Lamotte, but scarcely distinguishable. The colour of the sap does not appear to be a reliable guide.

MECONOPSIS *Viguier.*

1. M. cambrica (*Linn.*) *Vig.* WELSH POPPY.
Damp, shady, rocky places on the hills—very rare. Ranging from an elevation of 50 feet at Fairhead to 1000 at Rostrevor. Fl. June and July.
Down—Found in abundance in the deep valleys along the mountains above Rostrevor; *Wade Rar.* 1804. Found there also by *Templeton, Stokes,* and *Mackey,* 1808, and by S.A.S., but only very sparingly, in 1877.
Antrim—Rocks at Fairhead; *Templeton,* before 1804. Noted by *Belf. Nat. Field Club,* in same place in 1884. Between Garron Head and Glenariffe; *Cyb. Hib.* Plentiful in Cushenill glen, at Garron Point; R.Ll.P.

GLAUCIUM *Tourn.* HORNED POPPY

1. G. flavum *Crantz* (G. LUTEUM *Scop.*). YELLOW HORNED POPPY.
On loose sands by the seashore—rare, but locally abundant. Fl. June and July.
Down—About 5 miles south of Newcastle on the shore; *Templeton.* Near Greencastle; *Flor. Ulst.* Abundant on the shore from Greencastle coastguard station, for about a mile northwards; S.A.S. A single plant at Killough Bay; T.H.C. Plentiful from Kearney's Point to near Ballyhalbert in the Ards; S.A.S. The most northerly station in Ireland.

CHELIDONIUM *Linn.* CELANDINE.

1. C. majus *Linn.* GREATER CELANDINE.
Roadsides, and waste ground, widely distributed, but seldom plentiful. Fl. mid. May till mid. July. Naturalised in many places, but more frequently an escape. The following are some of the localities where this plant has been noted :—Rostrevor, Tullymurry, Tullycairne, Moira, Greyabbey, Ballyholme, Crawfordsburn, Conlig, Ballygowan, Comber, Ballylesson, Holywood, Knock, abundant at "The Court" Hillhall, Lisburn, Dunmurry, Ballinderry, Donegore Hill, Islandmagee, Ballycarry, Glynn, Randalstown, Ballyronan, Moneymore, Toome, Coleraine, Drumachose, Magilligan.

Order IV. FUMARIACEÆ.

FUMARIA *Linn.* Fumitory.

1. F. capreolata *Linn.* Ramping Fumitory.

Var. α F. PALLIDIFLORA *Jordan.*

Borders of cultivated fields, and on sandy wastes—frequent. Fl. June—Sept.

Down—Rostrevor; *W. Thompson,* 1836. Ardmillan and Castle Espie; *Belf. Nat. Field Club Rep.,* 1867-8. Mountstewart; *Ib.,* 1871-2. Fields at Newcastle, and Sandhills at Ballyholme; T.H.C. Warrenpoint, Dundrum, and Groomsport; S.A.S. Ballygowan; R.Ll.P.

Antrim—Cushendall, Kilroot, and gravel banks at Curran of Larne; S.A.S. Crumlin waterfoot, Islandmagee, and Carnlough; T.H.C.

Derry—Roadsides at Toome; S.A.S.

Var. β F. BORÆI *Jordan.*

Glenmore near Lisburn; *J. H. Davies* (*fide Eng. Bot.*). Prof. Babington thinks the Glenmore plant is *F. confusa.* Giants' Ring; S.A.S. (*fide Dr. Boswell*).

Var. γ F. CONFUSA *Jordan.*

Sandy ground—rare?

Glenmore; *J. H. Davies* (*fide Prof. Babington, as above stated*). Holywood, 1863; *Dr. Churchill Babington* (C.C.B.). Ballylesson; S.A.S. Stranmillis; T.H.C. Gravel pit near Giants' Ring; *C. Dickson*; perhaps the same plant as that from this locality, (S.A.S.) placed by Dr. Boswell under the preceding var.

Var. δ F. MURALIS *Sonder.*

Borders of fields—rare? Antrim only, so far as yet known.

Rathlin Island; S.A.S. (C.C.B.). Fields Malone; T.H.C. (C.C.B.).

2. F. officinalis *Linn.*

Borders of fields on a light soil, and on sandy or gravelly wastes—frequent. Fl. June—Sept.

Down—Hillsborough, Ballylesson, Dundonald, and Knock; S.A.S.

Antrim—Lisburn, Malone, Whitehead, Larne, Rathlin, and Bushmills; S.A.S.

Derry—Coleraine; S.A.S. Abundant throughout the county; D.M.

CRUCIFERÆ. 9

ORDER V. **CRUCIFERÆ.**

NASTURTIUM *R. Brown.*

1. N. officinale *R. Brown.* WATER CRESS.
Shallow streams of clear water, and about springs—common. Fl. mid. May till mid. Aug.

2. N. palustre (*Willd.*) *De Candolle.* YELLOW MARSH CRESS.
Sandy and gravelly marshy ground, and by streams—frequent. Fl. June and July.
Down—Moira Demesne ; *G. C. Hyndman.* Banbridge, Downpatrick, Ballynahinch, and Newtownards ; S.A.S.
Antrim—Portmore, Bog meadows, and by the Lagan canal ; *Templeton.* Lough Neagh ; *F. Whitla.* Lisburn, Whitehouse, Templepatrick, and course of the Bann to Coleraine ; S.A.S.
Derry—Frequent by Lough Neagh, and the Bann ; D.M. Toome ; *Belf. Nat. Field Club,* 1878.

3. N. amphibium *R. Brown.* (ARMORACIA *Koch*).
Lakes, rivers, and deep ditches—frequent. Fl. June and July.
Down—Witches Hole near Lambeg ; *Templeton.* By the Lagan, above and below Moira ; S.A.S.
Antrim—Portmore Lake ; *Templeton.* Massereene Park ; *Flor. Ulst.* Crumlin waterfoot ; T.H.C. Abundant by the course of the Bann, to Coleraine ; S.A.S.
Derry—Plentiful by Lough Neagh, and the Bann ; D.M.

BARBAREA *R. Brown.*

1. B. vulgaris *R. Brown.* YELLOW ROCKET.
Roadsides, waste ground, and borders of fields—frequent. Fl. May—July. Known to Mr. Templeton in 1804, but no localities stated.
Down—Holywood ; *Flor. Ulst.* Newtownbreda ; T.H.C. Warrenpoint, Moira, Ballynahinch, and Comber ; S.A.S. Newcastle ; R.Ll.P.
Antrim—Cave Hill, and Whiteabbey ; *Flor. Belf.* Malone, and Carrickfergus ; *Flor. Ulst.* Derriaghy ; *Rev. W. M. Hind.* Templepatrick ; *Herb. Nat. Hist. and Phil. Soc.* Lisburn, and Crumlin ; S.A.S.
Derry—Toome, and Coleraine ; S.A.S. Not general, but abundant about Moneymore ; D.M.

2. B. intermedia *Boreau.*
Roadsides, waste ground, and borders of fields, especially on light soil—

frequent. Fl. May—July. Supposed to have been introduced, but has as much appearance of being native as the preceding.

Down—Knock; T.H.C. Banbridge, Tollymore, Downpatrick, Ballynahinch, Saintfield, Comber, and Stormount; S.A.S. Kinnegar at Holywood; R.Ll.P.

Antrim—First gathered near Ballymena about 1836; *Cyb. Hib.* Cave Hill, Wolf Hill, Whitehouse, and near Lisburn; *Flor. Belf.* Shane's Castle; *Belf. Nat. Field Club*, 1863. Templepatrick, and abundant about Glenavy; S.A.S.

Derry—Found about Toome by S.A.S. but otherwise the distribution in this county is not known.

ARABIS *Linn.* ROCK CRESS.

1. A. hirsuta (*Linn.*) *R. Brown.*

Rocks and walls—very rare. Fl. June and July.

Down—On old walls of Dundrum Castle; *Templeton*, 1797, and found there sparingly by the *Belf. Nat. Field Club* in 1878. Murloch; *Flor. Ulst.*

Derry—From the sea to top of Benevenagh, and on the basalt to Craignashoke; D.M., 1834. Extremely scarce on Benevenagh in 1884; S.A.S.

This seems to be a species dying out in our district.

CARDAMINE *Linn.*

1. C. flexuosa *With.* (C. SYLVATICA *Link.*).

Damp shady places, especially woods and glens—common. Fl. May and June.

2. C. hirsuta *Linn.*

Roadsides, walls, and dry banks—common. Fl. March—May. Occurs on Slievegallion at 1200 feet; D.M.

3. C. pratensis *Linn.* LADY'S SMOCK. CUCKOO FLOWER.

Marshy ground, and wet pastures—very common. Fl. May and June. Occurs at 1800 feet in Derry; *H. C. Hart.*

4. C. amara *Linn.* BITTER CRESS.

Marshy ground, and beside streams—rare. Fl. May and June.

Down—Belvoir Park—sparingly; T.H.C. Marsh at Ballyalloley near Comber; S.A.S.

Antrim—Banks of Lagan; *Flor. Ulst.* Marshy thicket by Lagan canal between Shaw's Bridge and Drum Bridge, and abundant in

CRUCIFERÆ. 11

a marshy wood by the lough shore one mile south of Antrim; S.A.S.
Derry—By the river in Castledawson Demesne, and at Artrea near Toome; D.M. First recognised in Ireland by D.M.

SISYMBRIUM *Linn.*

1. S. officinale (*Linn.*) *Scop.* HEDGE-MUSTARD.
Waste ground, and roadsides—very common. Fl. mid. May till mid. Aug.

2. S. sophia *Linn.* FLIXWEED.
Waste ground—very rare. Fl. July and Aug.
Antrim—Near Carrickfergus; *Templeton*, 1812. Quarry spoil-bank at Magheramorne; S.A.S. Possibly not native, but thus known to be established in this part of Antrim for 75 years.

3. S. thalianum (*Linn.*) *Gaud.* WALL CRESS.
Walls, and dry basaltic rocks—not common. Fl. April and May. Ascends to nearly 1000 feet on the hills.
Down— Walls of ancient abbey and graveyard at Movilla; S.A.S.
Antrim—On the Deer Park wall, and rocks below the Cave Hill; *Templeton*, 1797. South face of Black Mountain, and old walls at Carrickfergus; *Flor. Ulst.* East face of the Knockagh; B.N.F.C., 1865. Cave Hill above Whitewell, rocks at Carrick Castle, and Blackhead, also sparingly on rocks at Ushet in the Island of Rathlin; S.A.S.
Derry—Frequent in the neighbourhood of Coleraine; D.M.

4. S. alliaria *Linn.* JACK BY THE HEDGE.
Roadsides and hedge banks—frequent. Fl. mid. May till mid. July.
Down—Lambeg; *G. C. Hyndman*, 1832. Rostrevor; *Miss Reid*, also, S.A.S.
Antrim—Southwest corner of the deer-park, Belfast; *Templeton*, 1797. Woodburn, and Colin Glen; *Flor. Belf.* Ballycastle, and Shane's Castle; *Flor. Ulst.* Cave Hill tramway; T.H.C. Glenavy, Antrim, Randalstown, Ballycarry Glen, Waterloo near Larne, Glenarm, and Whitepark Bay near Ballintoy; S.A.S. Carnlough; R.Ll.P.
Derry—Base of Umbra rocks; D.M.

BRASSICA *Linn.*

1. B. campestris *Linn.*
Waste ground, and borders of fields, especially where damp—common.

Fl. mid. May till end of July. Templeton says, "Common on the mud thrown up from the bed of the Blackstaff River." Still abundant in the Bog meadows, on the mud raised when cleaning, and deepening the river.

SINAPIS *Linn.* MUSTARD.

1. S. nigra *Linn.* BLACK MUSTARD.
"Growing on ditch banks, at the side of the Falls Road, about half a mile from Belfast, 1808." "Common about Carrickfergus, in flower 11th July, 1808"; *Templeton.* "Not unfrequent near Kilrea, County Derry"; D.M. Not now to be found near Belfast or Carrickfergus, and probably this, and the succeeding species, are mere casuals in this district.

2. S. alba *Linn.* WHITE MUSTARD.
Abundant in the field surrounding the old church at St. John's Point, County Down; S.A.S., 1878; probably imported with seed.
"Very common on the lands of Myroe near Magilligan, Co. Derry. It appears most on new made ditches, never in planted fields." *William Tennant*; *Templeton MS.* Perhaps naturalised; not native.

3. S. arvensis *Linn.* CHARLOCK. PRUSHUS (loc.).
Waste ground, and cultivated fields—very common. Fl. May—Aug.

DRABA *Linn.*

1. D. incana *Linn.*
Rocks, and coast sandhills—extremely rare. Fl. July and Aug.
Derry—Sandy warren of Magilligan, at 20 feet, and on Benevenagh rocks at 1100 feet; D.M., 1835. In the latter place; S.A.S., 1884. Not found recently on the sandhills, and a careful search over the cliffs revealed only a very few plants; perhaps becoming extinct in Derry.

2. D. verna *Linn.* WHITLOW GRASS.
Walls, dry rocks, and sandy or gravelly wastes—not common. Fl. April to June.
Down—Spalga Mountain Rostrevor at 1200 feet, sandhills at Newcastle, and Ballykinler, also on walls of old abbey at Newtownards; S.A.S.
Antrim—Massereene park-wall, and stone dykes on the east side of Lough Neagh, about two miles from Antrim; *Templeton.* Ballycastle, and Portrush; *Flor. Ulst.* Abundant on gravel at mouth of Six-mile River, at Antrim; *D. Redmond.*
Derry—Bannfoot, and warren of Magilligan; D.M.

CRUCIFERÆ.

COCHLEARIA *Linn.* SCURVY GRASS.

1. C. officinalis *Linn.*
Seashores—common on all our coasts. Fl. April—June.
　　Var. β C. ALPINA *Watson.*
Antrim—Rathlin Island; D.M.
Derry—Benevenagh; *Cyb. Hib.*, and subsequently; T.H.C.

2. C. danica *Linn.*
Walls, and banks by the sea—rare. Fl. May and June.
Down—Shores of Strangford Lough at Portaferry, and Castle Ward; *Flor. Ulst.* Ardglass; *Rev. W. M. Hind.* Plentiful on walls and banks by the sea at Warrenpoint, and also on the quay wall at Killough; S.A.S.
Antrim—Giants' Causeway; C.C.B., 1852. Portrush; *Cyb. Hib.* Larne, and Glenarm; *Flor. Ulst.* Not at Belfast Bay, as stated in *Flor. Ulst.* This plant does not grow on such wet muddy shores as the preceding species.
Derry—Downhill; D.M.

THLASPI *Linn.*

1. T. arvense *Linn.* PENNY CRESS.
Sandy and gravelly fields, and banks—rare. Fl. June and July.
Down—Cornfields at Dundrum, and 5 miles beyond Newcastle, also among stones at Donaghadee pier; *Templeton*, 1797.
Antrim—Field at Redbay; *Templeton.* Ballycastle, and near Belfast; *Flor. Ulst.* Borders of Lough Neagh; *F. Whitla.* Rathmore; B.N.F.C., 1884. Railway bank south of Antrim; *D. Redmond.* Glenavy, and Crumlin; S.A.S. Cushendun; *Rev. S. A. Brenan.*
Derry—Magilligan; *E. Murphy*, Mag. Nat. Hist., vol. I. Base of Umbra rocks; D.M.

LEPIDIUM *Linn.* PEPPERWORT.

1. L. campestre (*Linn.*) *Brown.*
Sandy fields—rare. Fl. June and July.
Down—(Several records—none reliable).
Antrim—Abundant in fields at Largy near the shore of Lough Neagh; *F. Whitla.* Near Lough Neagh from Antrim to Glenavy;

Cyb. Hib. Roadside Glenavy, and sandy waste at foot of Glenavy River; S.A.S. Ardmore Point north of Crumlin; *D. Redmond.*

Many other localities have been given which, however, belong to the succeeding species.

2. L. Smithii (*Linn.*) *Hooker.*

Dry banks, and waste ground on sandy soil—locally abundant. Fl. June—Aug.

Down—Common throughout the greater part of the county.

Antrim—Woodburn; *Flor. Belf.* Near Ballycarry; *Flor. Ulst.* Near Carrickfergus, and Ballymena; *Cyb. Hib.* Broughshane road near Ballymena, also at Antrim; S.A.S. Magheramorne, R.Ll.P.

Derry—Liberties of Coleraine, and by the bridge over the Roe; D.M. Portstewart; *Flor. Ulst.*

CAPSELLA *Moench.*

1. C. bursa-pastoris (*Linn.*) *Moench.* SHEPHERD'S PURSE.

Roadsides, waste ground, and borders of fields—one of the commonest weeds. Fl. April—Aug.

SUBULARIA *Linn.* AWLWORT.

1. S. aquatica *Linn.*

In shallow water on sandy lake-bottoms—very rare. Fl. July and Aug. First found in Britain, by Sherard, in the undernoted locality.

Down—Growing under water, amongst *Lobelia Dortmanna*, in an Irish lake—Lough Neagh—where it washes the gravelly shore of the townland of Kilmore near Moira; *Raii Syn. ed.* II., 1696. Under the water in Lough Neagh near Moira; *Harris's Down*, 1744. In the canal at Newry; *Flor. Ulst.*

Antrim—In Lough Neagh at Portmore Park; *Templeton*, 16 Sept 1800. Lough Neagh at Ballinderry; *Flor. Ulst.* Plentiful along the shores of Lough Neagh, between Lagan canal and Portmore, also at Selshan, and near the Creagh bog; *Cyb. Hib.*

Derry—" Side of the Bann, below the salmon leap, even where the tide raises the water"; *Templeton.* West side of Lough Beg, between Coney Island, and Church Island; S.A.S. 1870.

Subularia seems to have been one of the abundant plants of Lough Neagh, in former times. The drainage works have converted a broad stripe of the original bed of the lake, into dry land, and the plant—so far as we know—does not exist there now. Mr. Templeton's station at Coleraine, has not been verified in recent times, and the last of the localities above stated, is the only one in which it is now known to grow.

SENEBIERA *Persoon.* WART CRESS.

1. S. coronopus (*Gaert.*) *Poiret.* SWINE'S CRESS.
Waste ground, and banks by the sea—frequent. Fl June and July.
Down—Bangor; *Templeton.* Groomsport; *Flor. Belf.* Newcastle, and Portaferry; *Flor. Ulst.* St. John's Point, and Donaghadee; T.H.C. Warrenpoint, Millisle, and roadside from Mountstewart to Newtownards; S.A.S.
Antrim—Abundant near Giants' Causeway, also on Rathlin, and on road to Carrickfergus at two miles from Belfast; *Templeton,* 1796. Curran of Larne; B.N.F.C., 1869. Port Ballintrae; *Miss J. Reid.* Whitehead, and north end of Islandmagee, also several places, by the shore, from Dunluce to Ballintoy; S.A.S. Ballycastle, and Cushendun; R.Ll.P.
Derry—By the shore at Portstewart; *D. Redmond.*

CAKILE *Gaert.*

1. C. maritima *Scop.* SEA ROCKET.
Sandy seashores—common on the coasts of Down, Antrim, and Derry. Fl. July and Aug.

CRAMBE *Linn.*

1. C. maritima *Linn.* SEA KALE.
Gravelly seashores—extremely rare.
Down—Gravelly shore between Quentin Castle, and the village of Knockinelder, in the Ards. " Being informed by Rev. Jas. O'Laverty, P.P., M.R.I.A., that the sea-kale grew in this place, I visited the spot, on 25th June, 1876, and found 13 plants, occupying an area of eight square yards. There was no appearance of flower stems at that time, but I have seen flowers gathered by Rev. Mr. O'Laverty "; S.A.S.
Antrim—Gravelly shore of Church Bay, Island of Rathlin; *Templeton,* 1794. " Inserted on seeing it in Mr. Gage's garden, who transplanted it from the shore, where it is now destroyed"; *Templeton, M.S.*

RAPHANUS *Linn.*

1. R. raphanistrum *Linn.* WILD RADISH.
Sandy fields, and waste ground—not common. Fl. June and July.

Down—Banbridge, and Knock; S.A.S. Strangford, Ballyholme; T.H.C.
    Frequent about Holywood; R.Ll.P.
Antrim—Larne; *Rev. S. A. Brenan.* Field half-way between Larne and
    Glenarm; T.H.C. Railway bank at Antrim; *D. Redmond.*
Derry—Not frequent; D.M.

2. R. maritimus *Smith.* SEA RADISH.

Sandy seashores—rare. Fl. mid. June till mid. Aug.
Down—Sparingly in Mill-quarter Bay near Ardglass; T.H.C.
Antrim—About Mr. M'Neill's salt works at the Curran of Larne; *Templeton*, 1808. Rathlin Island; *Miss Gage.* In great abundance in Rathlin; S.A.S., 1882. Sparingly at Carnlough Bay; *Cyb. Hib.* By the harbour of Magheramorne; T.H.C. North end of Islandmagee; S.A.S. Cushendun; *Rev. S. A. Brenan.*

ORDER VI. **RESEDACEÆ.**

RESEDA *Linn.*

1. R. luteola *Linn.* DYER'S WEED, WELD.

Quarries, and waste ground—common. Fl. June and July. Formerly much cultivated.

ORDER VII. **VIOLACEÆ.**

VIOLA *Linn.* VIOLET.

1. V. palustris *Linn.* MARSH VIOLET.

Marshes, and wet boggy pastures—frequent; more common to the northwest. Fl. May and June. Ascends to 1800 feet in Derry; *H. C. Hart.*
Down—By the lake at Ballynahinch; *Templeton*, 1793. Kinnegar, Holywood; *Flor. Ulst.* Aughnadarragh Lake near Saintfield; S.A.S. Craigauntlet; R.Ll.P.
Antrim—Slieve Nanee; T.H.C. Divis, at 1200 feet; S.A.S. Glendun, Cushendall, and Fairhead; R.Ll.P.
Derry—Frequent through the county; D.M. Sperrin Mountains, Slieve Gallion, Benbradagh, also near Dungiven, and abundant by Lough Beg; S.A.S.

VIOLACEÆ.

2. V. odorata *Linn.* SWEET VIOLET.
Sheltered hedge banks—not uncommon. Fl. March and April. Doubtfully native, but as Mr. Templeton attached no suspicion it is included here as, at least, certainly naturalised for some 90 years. Roadside near gate of Cranmore; *Templeton*, 1797; and at same place; T.H.C., 1878. Reported from many other localities in the three counties—as Castlereagh, Dunmurry, Newforge, Lisburn, about Belfast in several places, Carrickfergus, Shane's Castle, Lignapeiste in Derry, etc.

3. V. sylvatica *Fries.* WOOD VIOLET.
Woods, hedge banks, and other shady places—very common. Fl. April —June.

Var. α V. REICHENBACHIANA *Bor.*
Not common, but occurs in Co. Antrim at Selshan shore, and in Colin Glen, close to the entrance; S.A.S.; also by Lagan canal at the 5-mile post, and between Crumlin and the waterfoot bridge; T.H.C.

Var. β V. FLAVICORNIS *Forster.*
A dwarf form with large flowers on the sandhills at Ballykinler near Dundrum, Co. Down; S.A.S. This we believe to be the "V. FLAVICORNIS *Smith*," noted in *Flor. Ulst.* as occurring on sandhills at Newcastle.

4. V. canina *Linn.* (V. FLAVICORNIS *Smith*). DOG VIOLET.
Wet boggy gravelly ground—rare. Fl. May and June.
Down—Sandhills at Newcastle, and Ballykinler; S.A.S.
Antrim—Islands and lake shore south of Toome, also margin of the lake at Shane's Castle, Langford Lodge, and Selshan; S.A.S. Portrush sandhills, and sandy shore at Cushendall; R.Ll.P. Shore of Lough Neagh; T.H.C.

5. V. lutea *Hudson.*
Represented by the var. *Curtisii*, we have not seen satisfactory specimens of the typical plant.

Var. β V. CURTISII *Forst.* SEA PANSY.
Sandy shores—common. Fl. May—Oct.
Down—Links at Newcastle; *Flor. Ulst.* Sandy shores all round the Down coast; S.A.S.
Antrim—Portrush; C.C.B. Sandy and gravelly shore of Lough Neagh at Antrim, and in another inland locality near Irish Hill; S.A.S.
Derry—Sandhills at the Bann foot; S.A.S.

6. V. tricolor *Linn.* PANSY.
Sandy banks, and borders of fields—common. Fl. mid. May till mid. Aug.
Var. β V. ARVENSIS *Murr.*
Common in sandy, cultivated fields.

ORDER VIII. **DROSERACEÆ**

DROSERA *Linn.* SUNDEW.

1. D. rotundifolia *Linn.*
Common on peat bogs throughout the district. Fl. July and Aug.

[D. INTERMEDIA. *Hayne.* There is much uncertainty as to this species, and it is better to leave it, for the present as doubtful. Templeton noted it as "common Birky moss, etc." Possibly it was found 80 years ago, but now lost by drainage. Birky moss seems to have disappeared long since. The record in Irish Flora is doubtless an error; the remark "rather larger than the last" (*D. rotundifolia*) would point to a form of *D. anglica*. The locality stated is very vague and extended—"marshy places at the foot of the Mourne Mountains." It has not been found in that region by subsequent botanists.]

2. D. anglica *Hudson.*
Wet peat bogs—widely distributed, but much less common than the round-leaved sundew. Fl. mid. June till mid. Aug.

Down—Bog between Greyabbey and Donaghadee; *Templeton*, 1794 (still found there). Cotton moss; *Flor. Belf.*
Antrim—Plentiful on Aghalee bog; *Templeton.* Bog above Dunloy in the parish of Rasharkin; S.A.S. Abundant on the Garron plateau; R.Ll.P.
Derry—Common on the "flow-bogs" of Culmore, Maghera, and Magherafelt; D.M. Bog at Bellaghy, and also by the Bann opposite Portglenone; S.A.S.

Var. β D. OBOVATA *M. & K.*

Probably not uncommon, but has scarcely been looked for. It is probable that this was the plant noted by Templeton as *D. intermedia.*

Antrim—Slogan bog near Randalstown; *Cyb. Hib.* Sparingly in bog above Dunloy; S.A.S. (! C.C.B.). In Tyrone just outside our district; *W. MacMillan.*

ORDER IX. **POLYGALACEÆ**.

POLYGALA *Linn.*

1. P. vulgaris *Linn.* MILKWORT.
Grassy heaths, short pastures, and banks—very common throughout the district. Fl. May—Sept.

ELATINACEÆ.

Var. β P. SERPYLLACEA *Weihe.*
On stony or gravelly heaths—the common form on the hills.
Down—Castlereagh Hill, also peaty and gravelly shores of Derry Lake, and Long Lake near Ballynahinch; S.A.S. Slieve Martin in Mourne Mountains; *Hart, Proc. R.I.A.,* 1884.
Antrim—Clough; *Cyb. Hib.* Sallagh Braes; T.H.C. Carn Hill, Black Mountain, and Divis, also Rathlin; S.A.S. Tannaghmore near Antrim; *D. Redmond.*
Derry—Heaths at Toome; S.A.S.

Var. γ P. GRANDIFLORA *Bab.*
Basaltic cliffs—very rare.
Derry—On the steep cliffs of Benevenagh, in small quantity; S.A.S., 1884 (! C.C.B.).

Var. δ P. OXYPTERA *Reichb.*
Down—Mealough Hill; *Dr. Mateer,* 1848 (! C.C.B.). Sandhills near Newcastle; T.H.C., 1879 (typical).
Antrim—Dry pastures on Cave Hill; *Dr. Mateer.* Dry, heathy pastures on Divis, and heath above Sallagh Braes; T.H.C. Gravelly shore of Lough Neagh at Selshan, very fine; S.A.S.

ORDER X. **ELATINACEÆ.**

ELATINE *Linn.* WATERWORT.

1. E. hexandra *De Candolle.*
Lakes, and rivers—rare. Fl. July and Aug. (*Cyb. Hib.*).
Down—Western end of Castlewellan Lake; *Templeton,* 1808. In Macaulay's Lake south of Ballynahinch; S.A.S.
Antrim—Rathlin Island; *Miss Gage.* Not found by S.A.S.
Derry—Enagh Lough, Lough Beg, and in the Bann near Coleraine; D.M.

2. E. hydropiper *Linn.*
Under water in lakes and canals—extremely rare. Fl. Aug. (*Cyb. Hib.*). Not found recently.
Down—In the canal at Newry; *Wm. Thompson.* First found in Ireland by Mr. Thompson in 1836. Lagan canal, a little above the first bridge from Lough Neagh, and more abundantly from the bridge to the lough, Sept. 1837; D.M. *in Ordnance Survey of Londonderry.* In Lough Neagh a little north of the canal; *Cyb. Hib.* The last locality is near the county boundary, and may have been in Antrim.

ORDER XI. **CARYOPHYLLACEÆ.**

SILENE *Linn.* CATCHFLY.

1. S. anglica *Linn.* ENGLISH CATCHFLY.

Sandy and gravelly fields, and waste ground—rare. Fl. mid. June till mid. Sept.

Down—Fields by the shore at Kilkeel; *Wade Rar.* Abundant about Greencastle; *Ir. Flora.* Shore north of Newcastle; *Flor. Ulst.* Abundant in gravelly places about half a mile north of Greencastle coastguard station, also by the Causeway Water, and sparingly in fields between Newcastle and the mountains; S.A.S. Near Bryansford; *C. Dickson.*
Antrim—By the new road to the ferry near Bellaghy; *R. Tate.*
Derry—Magilligan; D.M.

2. S. inflata *Smith.* BLADDER CAMPION.

Hedge banks, and dry wastes—frequent. Fl. mid. June till mid. Aug. Native, no doubt!

Down—About Clough; *Templeton,* 1795. Comber, and Holywood; *Flor. Belf.* Newtownards; *Flor. Ulst.* Tullymurry; T.H.C. In profusion in cultivated fields, from Strangford to Newcastle, also at Warrenpoint, and Greencastle; S.A.S.
Antrim—Whitehouse; *Flor. Ulst.* Templepatrick, and Donegore; *B.N.F.C.,* 1865. Crumlin; T.H.C. Ballycarry; S.A.S.
Derry—Salterstown; D.M. Near Toome, Bellaghy, Magherafelt, and Moneymore; S.A.S.

3. S. maritima *With.* SEA CAMPION.

Seashores, maritime rocks, and mountain cliffs—common. Fl. mid. April till mid. Aug. As a mountain plant it occurs on the Mourne Mountains, Cave Hill, and the entire basaltic range of Antrim, and Derry. At 1200 feet in Derry; D.M.

4. S. noctiflora *Linn.* EVENING CATCHFLY.

Dry sandy fields—very rare. Fl. Aug.

Down—Railway cutting near the base of one of the drumlins in the parish of Drumbeg, and sandpit on the Lurgan road, half a mile from Lisburn; *Flor. Ulst.* Cultivated field in Rough Island in Strangford Lough two miles east of Comber; S.A.S.

5. S. acaulis *Linn.* CUSHION PINK.

On trap rocks—very rare. Fl. June and July.

Derry—Benevenagh, facing Derry Lough ; *Templeton*, 1805. In great abundance on Benevenagh at 1000 to 1200 feet, and only there ; D.M., 1835, and still there in abundance ; T.H.C., 1878.

LYCHNIS *Linn.* CAMPION.

1. L. flos-cuculi *Linn.* RAGGED ROBIN.
Meadows, and wet pastures—common. Fl. June and July.

2. L. vespertina *Sibth.* WHITE CAMPION.
Sandy banks, and borders of fields—not common. Fl. June—Aug.
Down—Knock, and Belmont ; *Flor. Belf.* Holywood ; *Flor. Ulst.* Fields about Killough Bay ; S.A.S.
Antrim—Roadsides, and fields at Templepatrick, and Railway banks near Larne ; T.H.C. Shore of Lough Neagh opposite Glenavy, also in a wood at Glenshesk, with pale-red flowers ; S.A.S. Glendaragh ; *D. Redmond.* Drumnasole, and quarry heaps at Magheramorne ; R.Ll.P.
Derry—Castledawson ; S.A.S.

3. L. diurna *Sibth.* RED CAMPION.
Sandy banks, and shady places on rocks—frequent. Fl. mid. May till mid. Aug. Ascends to 1000 feet.
Down—Cultra ; *Flor. Belf.* Dundonald, and Knock ; S.A.S. Belvoir Park ; T.H.C. Newcastle, and Rockport ; R.Ll.P.
Antrim—Woodburn Glen, and road near Derriaghy ; *Flor. Belf.* Cave Hill ; *Flor Ulst.* Frequent all round the Antrim coast, rare inland ; S.A.S.
Derry—Frequent, especially by the coast ; D.M. Downhill ; R.Ll.P.

4. L. githago *Linn.* CORN COCKLE.
Corn fields in sandy soil—frequent. Fl. July and early part of August.
Probably not an original native, but has been known in the country since early times. Templeton considered it as perhaps introduced with wheat. The following localities have been noted, Orlock, Groomsport, Bangor, Holywood, Castlereagh, Knock, Ballygomartin, Magilligan.

SAGINA *Linn.* PEARLWORT.

1. S. procumbens *Linn.*
Waste ground, walls, and pastures—very common. Fl. May—August. Ascends to 1400 feet in Derry ; D.M.

2. S. apetala *Linn.*
Walls—not rare. Fl. June—Aug.

Down—Holywood, and Ballymacarrett; *Flor. Ulst.* Rostrevor, Newcastle, Dundrum (sand dunes), Downpatrick, and Hillsborough; S.A.S. Ballynahinch; R.Ll.P.

Antrim—Whitehouse, Carrickfergus, and Glenarm; *Flor. Ulst.* Cave Hill, Templepatrick, Muckamore, Antrim, Randalstown, Redbay, and Ballycastle; S.A.S. Ballycarry; R.Ll.P.

3. S. ciliata *Fries.*

Walls—very rare. Fl. mid. July till mid. Sept.

Antrim—On the wall by the quay at Ballycastle; *Rev. W. W. Newbould, and Prof. Babington,* Sept. 1852. In the same place, but very sparingly; S.A.S., Aug. 1882.

4. S. maritima *Don.* SEA PEARLWORT.

Rocky and gravelly seashores—frequent, but not at all general. Fl. mid. June till mid. Aug.

Down—Shores of Belfast Bay; *Templeton.* Narrow-water; *Flor. Hib.* Portaferry, and Castleward; *Flor. Ulst.* Ballycormick Point near Bangor; T.H.C. Warrenpoint, Dundrum, Strangford, and Groomsport; S.A.S.

Antrim—Shore at Ballycastle; *Templeton.* Glenarm; *Flor. Ulst.* Curran of Larne, Fairhead, and Church Bay Rathlin; S.A.S. Redbay; R.Ll.P. Wall at the quay of Ballycastle (Var. STRICTA *Fries*); S.A.S.

Derry—Not unfrequent from the mouth of the Faughan to the Bann; D.M. Portstewart; S.A.S.

5. S. subulata *Wimm.*

Rocks, and gravelly places by the shore—rare. Fl. mid. June till mid. Aug.

Antrim—On the side of Muck Island next the sea; *Templeton,* 1804. Coast of Islandmagee south of the lighthouse; *Flor. Ulst.* Fairhead, Island of Rathlin, and Portrush; *Cyb. Hib.* In many places round the north and west of Rathlin, growing on bare, dry horizontal rocks at the summit of the cliffs; S.A.S. *Proc. R.I.A.,* 1884.

First found in Ireland by Mr. Templeton.

Derry—Ballywillan, and Ballyaghran between Portstewart and Portrush; D.M.

6. S. nodosa (*Linn.*) *E. Meyer.* KNOTTED SPURREY.

Marshy ground, and borders of lakes—not common. Fl. July and Aug.

Down—Bangor, and Kinnegar at Holywood; *Flor. Belf.* By the shore below Groomsport; S.A.S. Graypoint; R.Ll.P.

Antrim—By the Lagan at the first lock, also on a gravelly bank in Colin Glen; *Templeton,* 1796. Lough Mourne; *Flor. Belf.* Bog

meadows; *Flor. Ulst.* Giants' Causeway; S.A.S. Fairhead;
T.H.C. Mouth of Sixmile River; *D. Redmond.*
Derry—Frequent on sandy warrens from Derry to Portrush; D.M.

ARENARIA *Linn.* SANDWORT.

1. A. peploides *Linn.* (HONCKENEJA *Ehr.*). SEA-PURSLANE.
In sand on seashores—common all round the coast. Fl. mid. May till mid. Aug.

2. A. verna *Linn.* (ALSINE *Wahl.*).
Shady basaltic rocks—rare, but in some places abundant. Fl. mid. May till mid. Aug. At from 700 to 1300 feet in Derry; D.M.
Antrim—Agnew's Hill, and Sallagh Braes; *Templeton,* 1808. Abundantly on the sides of steep mountains by the low glens at Ballinlig (north of Glenarm); *Ir. Flor.* Lurigethan Mountain; T.H.C. Tievebulliagh; *Rev. S. A. Brenan.* Garron Head; *Prof. Babington.* Cushendall; S.A.S. By Carnlough river, and on Fairhead; R.Ll.P.
Derry—Benevenagh, and Benbradagh; *Templeton,* 1804. Abundant on basalt from Umbra to Benbradagh; D.M.
First found in Ireland by Mr. Templeton.

3. A. trinervis *Linn.*
Shady banks—not common, though occasionally abundant. Fl. mid. May till mid. July.
Down—Rockport; *Templeton,* 1797. Kilkeel; *Ir. Flor.* Ballylesson; *Flor. Belf.* Belvoir Park; *Belf. Nat. Field Club,* 1870. Ormeau, Dundela, and Dundonald; S.A.S. Newtownbreda, and Gilhall; T.H.C.
Antrim—Belfast Deer Park; *Templeton.* Dunmurry; *Flor. Belf.* Shane's Castle; *Belf. Nat. Field Club.* Lisburn, Portmore, Glenavy, and Larne; S.A.S.
Derry—Moneymore, and elsewhere; not uncommon; D.M. Castledawson; *Flor. Ulst.*

4. A. serpyllifolia *Linn.*
Sandy seashores, and rarely on walls—Frequent. Fl. June and July.
Down—Newcastle links, Comber, and Holywood; *Flor. Ulst.* Ardglass, and Kilclief Bay; T.H.C. Sydenham, Groomsport, Cloughey, Killough, and between Greencastle and Rostrevor; S.A.S.
Antrim—Knockagh, and Glenarm; *Flor. Ulst.* Rocks at Rathlin, walls at Ballycastle, sands at Ballintoy and Portrush, and gravel at Curran of Larne; S.A.S.

Derry—Sands at Magilligan; *Sampson's Londonderry*. Abundant on the coast from the Foyle to the Bann; D.M. Downhill; *Belf. Nat. Field Club.*, 1863.

5. A. leptoclados *Guss.*
Borders of sandy, cultivated fields, and on sandy waste ground—not common. Fl. mid. May till mid. Aug.
Down—Strandtown, Knock, and Belmont; *Flor. Belf.* (*under A. serpyllifolia*). Sandy fields Groomsport, railway bank Holywood, and sandpits at Newtownbreda; S.A.S.
Antrim—Railway bank near Windsor; T.H.C.

STELLARIA Linn. STITCHWORT.

1. S. media *Linn.* CHICKWEED.
Fields, and waste ground—very common. Fl. March—Oct. Occurs at 2200 feet in Derry; D.M.

2. S. holostea *Linn.* GREATER STITCHWORT.
On bushy banks and under hedges—common. Fl. mid. April till mid. June.

3. S. graminea *Linn.* LESSER STITCHWORT.
Hedge banks, and shady places—common. Fl. June and July.

4. S. uliginosa *Murr.* MARSH STITCHWORT.
Marshy ground, and by small streams—common. Fl. May—July. At 1400 feet on Divis.

CERASTIUM Linn. MOUSE-EAR.

1. C. glomeratum *Thuillier.*
Roadsides, and dry sandy wastes—common. Fl. May—Aug.

2. C. triviale *Link.* COMMON MOUSE-EAR.
Roadsides, fields, and waste ground—very common. Fl. May—Sept.

3. C. semidecandrum *Linn.*
Sandy shores—not common. Fl. mid. March till mid. June.
Down—Greencastle; *Ir. Flor.* Sandhills at Dundrum; T.H.C. Sandy shores at Ballyholme, and Millisle; S.A.S.
Antrim—Sands at Ballycastle; S.A.S.
Derry—Sandy warrens from the Foyle to the Bann, and sandy shore of Lough Neagh; D.M. Magilligan; T.H.C.

CARYOPHYLLACEÆ.

4. C. tetrandrum *Curtis.*
Sandy seashores—frequent. Fl. April—June.
Down—Greencastle; *Ir. Flor.* Holywood; *Flor. Ulst.* Common on shores at Mountstewart, Donaghadee, and from Strangford to Newcastle; S.A.S. Ballyholme; T.H.C. Sandy ground near Giant's Ring; C. Dickson.
Antrim—Giants' Causeway, and Ballintoy; *Templeton,* 1799. Shores at the Gobbins, and Brown's Bay Islandmagee, also at Redbay, and north coast of Antrim, and frequent on shores of Lough Neagh; S.A.S.
Derry—At the City of Londonderry; *Templeton.* Sandy warrens by the coast; D.M. Magilligan; T.H.C.

5. C. arvense *Linn.*
Sandy banks—very rare. Fl. May—July (*Cyb. Hib.*).
Down—About Greencastle; *Ir. Flor.*
Antrim—Plentiful on a sandy bank in Massereene Park at Antrim; D. Redmond.

SPERGULARIA *Persoon.*

1. S. rubra (*Linn.*) *Pers.*
Sandy margins of lakes and rivers—very rare. Fl. June and July.
Antrim—Sparingly on the shores of Lough Beg southeast of Toome; S.A.S.
Derry—By the Moyola River above Ballynascreen; D.M. Shore of Lough Neagh southwest of Toome; S.A.S.

2. S. rupestris *Lebel* (S. RUPICOLA *Kind.*).
Seashores—especially where rocky—common. Fl. July and August.
The conditions on the Derry shore are not so suitable, and we have no note of this species in that county.

3. S. salina *Presl.*
Muddy and wet sandy seashores—frequent. Fl. July and Aug.
Down—Shores of Belfast and Strangford loughs; S.A.S.
Antrim—Belfast Lough, and by the lighthouse at Islandmagee; S.A.S.
No doubt more widely spread, but the distribution of this plant has not yet been worked out.

4. S. media *Persoon* (S. MARGINATA D.C. S. MARINA *Wahl.*). SEA SPURREY.
Salt marshes, and muddy seashores—common around the coast. Fl. mid. June till mid. Aug.

MALVACEÆ.

SPERGULA Linn. SPURREY.

1. S. arvensis *Linn.* CORN SPURREY.

Amongst crops in fields, and in sandy waste ground—very common. Fl. mid. May till mid. Aug.

SCLERANTHUS Linn. KNAWEL.

1. S. annuus *Linn.*

Sandy waste ground, and borders of sandy or gravelly fields—frequent. Fl. mid. May till mid. Sept.

Down—"In great plenty"; *Harris's Down*, 1744. Kilkeel; *Ir. Flor.* Strandtown, Belmont, and Knock; *Flor. Belf.* Giant's Ring; T.H.C. Rostrevor, Newcastle, Dundrum, Castlewellan, Ballynahinch, and Hillsborough; S.A.S.

Antrim—Malone; *Flor. Belf.* Lough shore southeast of Toome, and on Rathlin Island; S.A.S.

Derry—Frequent in many places, as Rosse's Bay; D.M. Draperstown; S.A.S.

ORDER XII. **MALVACEÆ.**

MALVA Linn. MALLOW.

1. M. moschata *Linn.* MUSK MALLOW.

Roadsides, and borders of sandy fields—rare, and probably often a garden escape. Fl. July and Aug.

Down—Greyabbey; *Harris's Down*, 1744. Fields near Gilford; *Templeton*, 1795. Near Holywood; *Flor. Hib.*

Antrim—Near Templepatrick; *Templeton*. Side of Crumlin river near its mouth, with white flowers; *F. Whitla*, (still there, but we saw only pink flowers). Whitehouse; *W. Millen*. Ram's Island; *Belf. Nat. Field Club*, 1875. About Bellaghy; S.A.S.

Derry—Frequent near gardens, but rare truly wild; D.M. Ballyronan, and Moneymore; *Flor. Ulst.*

2. M. sylvestris *Linn.* COMMON MALLOW.

Roadsides, and waste ground—common throughout the district. Fl. mid. June till mid. Aug.

HYPERICACEÆ.

3. M. rotundifolia *Linn.*

Roadsides, and sandy seashores—rare. Fl. July—Sept. Very rare inland, and diminishing in frequency to the northwest. Native unquestionably.

Down—Common in the Ards; *Templeton*, 1798. Rostrevor, and Strangford; *Flor. Ulst.* St. John's Point, and on the shore opposite Gunn's Island, and thence, occasionally, to Ardglass, also inland on the roadside at Saul near Downpatrick; S.A.S. Abundant at Killough; T.H.C.

Antrim—Roadside halfway between Belfast and Milewater (loc. obsolete), and a half mile south of Carrickfergus; *Templeton*. Rathlin Island; *Miss Gage (Flor. Ulst.)*—not seen there by S.A.S. Magheramorne; R.Ll.P.

LAVATERA *Linn.*

1. L. arborea *Linn.* TREE MALLOW.

Rocky seashores—occasionally. Fl. mid. June till mid. Aug.

Known to Templeton in 1800, who noted it at Carrick-a-Rede near Ballycastle, Co. Antrim. The frequency of tree mallow in cottage gardens round the coast, taken in connection with its scarcity in similar gardens at a distance from the sea, points to the conclusion that it was an original native of our maritime rocks, from whence gardens about the shore were stocked.

ORDER XIII. **HYPERICACEÆ**.

HYPERICUM *Linn.* ST. JOHN'S WORT.

1. H. androsæmum *Linn.* TUTSAN.

Woods, and damp shady banks—general, but rarely abundant. Fl. July—Sept.

Down—Holywood, and Dundonald; *Flor. Belf.* Crawfordsburn, and Ballynahinch; *Flor. Ulst.* Rostrevor, glens in Mourne Mountains, Downpatrick, Killyleagh, Greyabbey, Braniel Hill, and Hillsborough Park; S.A.S. Tullycairne, and Gillhall; *C. Dickson.* Newtownards, Castlereagh, and Belvoir Park; T.H.C. Newcastle; R.Ll.P.

Antrim—Colin Glen, Whiterock, and Whiteabbey; *Flor. Belf.* Carnmoney, Castle Dobbs, Cranfield, and Rathlin; *Flor. Ulst.* Cave Hill, Crow Glen, Glynn, Glendun, and Glenariffe; S.A.S.

HYPERICACEÆ.

By Crumlin River, and at Woodburn; T.H.C. Tannaghmore near Antrim; *D. Redmond.*
Derry—Benevenagh, and frequent through the county; D.M. Castledawson; S.A.S.

2. H. tetrapterum *Fries* (H. QUADRANGULUM *Sm.*).
Wet pastures, ditch banks, and damp places—common everywhere. Fl. July and Aug.

3. H. perforatum *Linn.*
Shrubby and shady places—frequent. Fl. mid. July till end of Aug. Occurs in Derry at 1200 feet; D.M.
Down—Holywood; *Flor. Belf.* Tollymore Park; *Flor. Ulst.* Newcastle, Cloughey, Groomsport, Malliagh Hill, and Belmont; S.A.S. Dundrum, and Newtownards; T.H.C. Ballynahinch; R.Ll.P.
Antrim—Stranmillis, and banks of Lagan; *Flor. Belf.* Wolfhill, and Shane's Castle; *Flor. Ulst.* Banks of Crumlin River; T.H.C. Cave Hill, Portmore, Ballinderry, and Massereene Park; S.A.S.
Derry—Benevenagh, and other places; D.M. Toome; S.A.S.

4. H. quadrangulum *Linn.* (in part) (H. DUBIUM *Leers.*).
Hedge banks—very rare. Fl. mid. July till mid. Sept. The var. *maculatum* only occurs.

Var. β H. MACULATUM *Bab.*
Antrim—Allan's meadow 200 yards north of second lock, Lagan canal; *Templeton,* 1797. Ditch bank at right angles to the tow-path between third lock and Shaw's Bridge; S.A.S. Sparingly at intervals between second and third locks; T.H.C.
The Newcastle plant of *Flor. Ulst.* is H. *perforatum,* and the Antrim locality unreliable.

5. H. humifusum *Linn.* TRAILING ST. JOHN'S WORT.
Sandy, and gravelly banks—frequent. Fl. July and Aug. Rises to 1000 feet in Derry.
Down—Knock, Strandtown, Bangor, and Holywood Hill; *Flor. Belf.* Orlock Point; T.H.C. Rostrevor, Greencastle, Dundrum, Newcastle, Greyabbey, Comber, Newtownbreda, and Dundonald; S.A.S. Comber; R.Ll.P.
Antrim—By the Lagan canal near Lisburn; *Prof. Cunningham.* Lisburn, Broughshane, and Portrush; S.A.S. Glendun, Ballycastle, and Cushendall; R.Ll.P.
Derry—Common through the county; D.M. Castledawson, Magherafelt, Dungiven, and frequent about Derry; S.A.S.

GERANIACEÆ.

6. H. hirsutum *Linn.*
Pasture field—very rare, and apparently now lost. Fl. July and Aug.
Antrim—" Found at Macedon Point, near the Whitehouse, two or three years ago, by Mr. R. Tennant. I had looked for it in vain until in 1809, when I found it in flower, and settled my doubt"; *Templeton.*
The plant grew, in small quantity, in the upper corner of the field by the shore beyond Macedon coastguard station, close to the demesne wall, It was found there until about 1874. The field, which had lain so long in pasture, has been broken up, and converted into a kitchen garden, and since then *H. hirsutum* has not appeared in its old *habitat*, nor was it found when carefully searched for in the adjoining fields.

7. H. pulchrum *Linn.*
Dry banks, and rocky heathy places, ascending to 1500 feet on Carntogher in Derry—common. Fl. July and Aug.

8. H. elodes *Hudson.* MARSH ST. JOHN'S WORT.
In very wet bogs, especially in hilly situations—not common, and now becoming rare. Fl. July and Aug.
Down—Plentiful in Newtownards Park, Ballygowan moss, Annahilt bog, and half-way between Ballynahinch and Clough; *Templeton,* 1793. Greyabbey bog; *Ir. Flor.* Conlig Hill, and near Newcastle; *Flor. Ulst.* Slievenabrock above Newcastle, and abundant on west side of Conlig Hill; S.A.S.
Antrim—Plentiful on wet heath near the lighthouse Rathlin Island; S.A.S.
Derry—Plentiful in a bog joining Enagh Lough, and only there; D.M. Seen there recently by S.A.S.

ORDER XIV. **GERANIACEÆ.**

GERANIUM *Linn.* CRANESBILL.

1. G. sylvaticum *Linn.*
Damp shady places—very rare. Fl. June and July (*Cyb. Hib.*).
Down—In a small den above Holywood; *Flor. Ulst.* (*spec. in herb. C.C.B.*).
Antrim—On the bank of the river in Glenarm Park, July, 1800, and in the little Deer Park, Glenarm, plentifully; *Templeton.* Subsequently observed in the above station by D.M.; still more recently by T.H.C., who noted it as in Glenarm Park, on the

left bank of the river, at the great waterfall; also in 1885 by R.Ll.P. who says east side of Linford branch.

The above are the only reliable stations for this plant. The record of Flora Belfastiensis was founded on a single chance or stray plant, which has long since disappeared. Templeton's stations of Murlough, and Giants' Causeway were errors. The Antrim record of Flora of Ulster, and also those of Irish Flora require confirmation. Prof. Murphy's Dunluce plant was *G. pratense*.

2. G. pratense *Linn.* FLOWER OF DUNLUCE (loc.).

Dry rocks, and sandy wastes—rare, but usually conspicuously abundant where it does occur. Fl. mid. June till mid. Aug.

Antrim—About Ballintoy, Dunluce Castle, and all the north parts of Antrim; *Templeton.* Abundantly at Whitepark, Ballintoy 1836; D.M. in *Ord. Surv. Derry.* Port Ballintrae; *G. C. Hyndman.* On blown sand at Port Bradden; *R. Tate (Cyb. Hib. Suppl.).* On limestone near Carrick-a-Rede; T.H.C. Glenarm deerpark, and glen above Milltown at Cairncastle, plentiful; R.Ll.P.

First found in Ireland by D.M. in 1836.

3. G. sanguineum *Linn.* BLOODY CRANESBILL.

Rocky ground—very rare. Fl. June and July. (*Cyb. Hib.*).

Down—Bushy places by the shore near Crawfordsburn; *Flor. Ulst.* Since verified by Prof. R. O. Cunningham.

Antrim—Coast at Portrush; *Flor. Ulst.*

4. G. perenne *Huds.* (G. PYRENAICUM *Linn.*).

Hedge banks—Fl. June and July. Well established and has no appearance of an introduction.

Down—Hedge side beyond Groomsport; *Miss Maffett, (Flor. Belf.).* Plentiful on roadside about one mile south of Portavo House; S.A.S. Field near Strangford; T.H.C. Hedges between Downpatrick and Saul; *C. Dickson.* Ditch bank about one mile east of Groomsport; R.Ll.P.

5. G. molle *Linn.*

Roadsides, banks and borders of fields—common. Fl. May—Aug.

6. G. dissectum *Linn.*

Waysides, fields, and waste ground—common. Fl. mid. May till mid. Aug.

7. G. lucidum *Linn.* SHINING CRANESBILL.

Bare stony places, and occasionally on walls—frequent. Fl. mid. May till mid. July.

GERANIACEÆ.

Antrim—In plenty on rocks and among stones in the southwest corner of the Belfast Deer Park, 1793, and on Sallagh Braes, 1808; *Templeton.* Colin Glen, and Cave Hill tramway; *Flor. Belf.* Knockagh, and Whitehead; *Flor. Ulst.* Glenoe; S.A.S.
Derry—Benevenagh; R.Ll.P.

8. G. robertianum *Linn.* HERB ROBERT.

Hedge banks, walls, and waste ground—very common. Fl. May—Sept.

ERODIUM *L' Heritier.* STORKSBILL.

1. E. cicutarium *L' Herit.*
Sandy shores—frequent on the coast, but very rare inland. Fl. mid. May till mid. Sept.
Down—Sands at Kirkiston, and Donaghadee; *Templeton,* 1793. Between Greencastle and Kilkeel; *Ir. Flor.* Ballyholme, and Holywood; *Flor. Belf.* On the coast from Strangford to Newcastle; S.A.S.
Antrim—Carrickfergus; *Flor. Belf.* (*sub E. moschatum*). Redbay, and Portrush; S.A.S. Shane's Castle; T.H.C. Gravelly waste ground at the mouth of the Sixmile River Antrim; *Rev. W. S. Smith.* Ballycastle, and Carnlough; R.Ll.P.
Derry—Abundant from the Foyle to the Bann; D.M.

Var. E. PIMPINELLÆFOLIUM *Sibth.*
Antrim—Cushendall, and Ballycastle; C.C.B., Sept. 1852.

2. E. moschatum *L' Herit.*
Sandy and gravelly shores, and on roadsides—very rare. Fl. June and July (*Cyb. Hib.*).
Down—Near Portaferry; *Templeton,* 1806. Abundant at Kilkeel Bay; *Ir. Flor.* (? *Eds.*). Roadside near Dundrum; T.H.C.
Antrim—Hedge bank by the roadside at Gortnacross; D.M. Ballycastle; R.Ll.P.
Many errors have been made with respect to this plant, and the published records are most unsatisfactory. It does not now occur in Templeton's station at Dunmurry. Only the preceding species is found at Carrickfergus, and it is to be inferred that a form of this was mistaken for *E. moschatum* in that locality. Millen's stations, both in Flor. Belf. and Flor. Ulst. also refer to the preceding species, and the Glenarm, and Cave Hill stations, of Flor. Ulst. are not reliable. The Donaghadee, and Magilligan records of Stewart, in Cyb. Hib. are errors likewise, and those stations must be assigned to *E. cicutarium.* The Dundrum, and Ballycastle localities are the only ones in which we can state, with certainty, that *E. moschatum* grows at present.

3. E. maritimum *L' Herit.*
Sandy shores, and waste ground near the sea—rare. Fl. June and July.
Down—Sands about Newcastle, and plentiful on the sandy shores about Kirkiston; *Templeton.* Shore opposite Gunn's Island, and at Portlabar north of St. John's Point, also on roadside halfway between Rathmullen and Tyrella; S.A.S.
Antrim—Glenarm, and Ballycastle; *Flor. Ulst.* Abundant in a field below Whiteabbey; *Cyb. Hib.*
Derry—Castlerock; *Flor. Ulst.*

ORDER XV. **OXALIDACEÆ.**

OXALIS *Linn.* WOOD SORREL.

1. O. acetosella *Linn.*
Woods, and damp shady banks—common. Fl. April and May. Ascends to 2200 feet in Derry; D.M.

ORDER XVI. **LINACEÆ.**

LINUM *Linn.* FLAX.

1. L. catharticum *Linn.* PURGING FLAX.
Heaths, dry banks, upland pastures, and sandhills—common, and abundant. Fl. mid. June till mid. Sept.

RADIOLA *Gmelin.*

1. R. linoides (*Linn.*) *Gmel.* (R. MILLEGRANA *Sm.*) ALLSEED.
Gravelly and sandy shores, and gravelly waste ground—rare, and usually by the coast. Fl. June—Aug.
Down—North and east side of Holywood warren, 1797, and plentiful about Ballygowan Lough, 1801; *Templeton.* In boggy stations along the sides of rivers about Kilkeel, and on the sands at Murlough; *Wade Rar.* Wet sandy ground by the riverside below Kilkeel; *Ir. Flor.* Bangor, Newtownards; *Flor. Ulst.* Sandy shore north of Greencastle, Magherascouse,

and gravel pits half a mile from Comber on the road to Newtownards; S.A.S. Conlig; R.Ll.P.
Antrim—Gravelly place on shore of Ushet Lake in the island of Rathlin —sparingly; S.A.S. *Proc. R.I.A.* 1882.
Derry—On wastes; *Sampson's Surv. Derry.* Plentiful in sandy warrens from Portrush to Downhill; D.M.

ORDER XVII. **CELASTRACEÆ.**

EUONYMUS *Linn.* SPINDLE TREE.

1. E. europæus *Linn.*
Rocky places, and in thickets by the margins of rivers and lakes—frequent. Fl. June and first half of July.
Down—Holywood; *Templeton*, 1810. Rockport; *Flor. Belf.* Hedge near Comber; T.H.C. Abundant by the wooded shore south of Rostrevor, and by the west bank of the Quoile two to three miles below Downpatrick, also in Dundonald Glen; S.A.S. Tollymore Park, Rademon, and abundant in hedges in townland of Tullycairne; *C. Dickson.*
Antrim—Belfast Deer Park; *Templeton*, 1797 (still there, *Eds.*). Shore of Lough Neagh at Shane's Castle; *Flor. Ulst.* Southern end of Lough Beg near Toome; *B.N.F.C.*, 1867. Shore of Lough Neagh at Cranfield Point, and on the Knockagh; S.A.S. Roadside near Whitehouse; T.H.C.
Derry—By the Bann below Kilrea, and by the sides of small rivers in the parish of Arboe; D.M.

ORDER XVIII. **RHAMNACEÆ.**

RHAMNUS *Linn.*

1. R. catharticus *Linn.* BUCKTHORN.
Bushy places by lake shores—very rare. Fl. May and June; (*Cyb. Hib.*).
Antrim—A young seedling in Portmore wood; *Templeton*, 1794. Lough Neagh by the outlet of the Lagan canal, and near Toome; *Cyb. Hib.* Creagh Island in Lough Beg; D.M.
Derry—In a hedge near Salterstown; D.M.

2. R. frangula *Linn.* BLACK ALDER.

In boggy places, and by lake shores—very rare. Fl. May and June; (*Cyb. Hib.*).

Down—Side of a lake (Bow Lake, *Eds.*) in the townland of Creevytenant near Ballynahinch, and plenty among other shrubs in a bog near Ballygowan bridge; *Templeton.*

Antrim—Creagh Island in Lough Beg; D.M. Portglenone, and shores of Lough Neagh near Shane's Castle, and near Toome; *Cyb. Hib.*

The two species of *Rhamnus* just enumerated are plants which are extinct, or nearly so, through drainage. We are not aware that any one has seen either, in our district, in recent years.

ORDER XIX. **LEGUMINOSÆ.**

ULEX *Linn.* WHIN, FURZE, GORSE.

1. U. europæus *Linn.*

Stony heaths, and ditch banks in moory lands—common. Fl. mid. Feby. till end of May.

Mr. Templeton was of opinion that this is an introduced plant. He says —" *U. nanus*, or dwarf whin, appears to be the native plant, the other is evidently a naturalised plant rather tender for the climate, being in severe winters much hurt by frost, which the other is not" *Temp. M.S.* *U. europæus* does not grow at any great elevation. It has a very definite altitude, the upper limit ranging from 700, to about 1000 feet. Mr. Templeton states that it was unknown in Rathlin Island until 1794 or 1795, when it was sown there by Mr. Gage.

The so called Irish whin—(U. EUROPÆUS β MINOR *Hk.* U. STRICTUS *Mackay*) was found in the grounds of Mountstewart, Co. Down, by Mr. Murray of Comber, who brought it into his nursery. Mr. John White, who saw it there, took plants to Dublin, and was the first to make it known. It is a mere sport, or inconstant variety, and we believe cannot now be found in the wild state.

2. U. Gallii *Planch.*

Rocky heaths—frequent in Down, but not common in Antrim or Derry. Fl. Aug.—Oct. Ranges from sea level to 1700 feet on Slieve Croob.

Down—Newtownards Park; *Templeton*, 1793. Near Newry, Newcastle, Strangford, and Ballynahinch; *Flor. Ulst.* Slievenagriddle Hill near Saul; *B.N.F.C.*, 1867. Knockbracken, and abundant

on the Mourne Mountains, and Slieve Croob; S.A.S. Conn's Hill, and elsewhere on Holywood Hills; R.Ll.P.
Antrim—Glenravel near Clough; *Flor. Ulst.* Abundant on the bare heaths of Rathlin Island; S.A.S. *Proc. R.I.A.*, 1884.
Derry—Frequent in the Barony of Loughinshollen; D.M.

SAROTHAMNUS *Linn.*

1. S. scoparius *Linn.* BROOM.

Heaths, and ditch banks in hilly districts—common, but rarely abundant. Fl. May—July.

ONONIS *Linn.* REST HARROW.

1. O. repens *Linn.* (O. ARVENSIS *auct.*).

Sandy seashores—frequent, rare inland. Fl. June—Aug.
Down—Sandy shores at Kirkiston, and at Clough; *Templeton*, 1793. Newcastle, and Portaferry; *Flor. Ulst.* Orlock Point; T.H.C. Ballywalter; *Miss Reid.* Sandy shores at Cloughey, and Millisle, and abundant on sandy shores, and the bordering fields, from Strangford to Newcastle; S.A.S.
Antrim—Seashore at Redbay; *Templeton.* Plentiful on sandy shores at Ballycastle, Cushendun, and Cushendall, also in dry sandy pastures in Massereene Park; S.A.S.
It is singular that this plant has not been noted in the sandy warrens of the Derry coast.

2. O. spinosa *Linn.* (O. CAMPESTRIS *Koch*).

Dry gravelly pastures—very rare.
Down—Sparingly in pasture field by the east side of Lough Erne three miles northwest of Ballynahinch; S.A.S.
Not found in fruit, but from the characters of flower and leaf, together with the habit and general aspect, there seems little reason to doubt the identification.

MEDICAGO *Linn.* MEDICK.

1. M. lupulina *Linn.* BLACK MEDICK.

Roadsides, pastures, and sandy waste ground—frequent. Fl. mid. May till mid. Aug.
Down—Ballymaghan, and Cultra; *Flor. Belf.* Crawfordsburn; T.H.C. Newcastle, Downpatrick, Strangford, Craigavad, Groomsport, Donaghadee, Comber, and Newtownbreda; S.A.S.

Antrim—Curran of Larne; *Templeton*, 1794 (still there). Whiteabbey;
    Flor. Belf. Roadside near Crumlin; T.H.C. Lisburn, Colin
    Glen, Kilroot, Magheramorne, and Bushmills; S.A.S.
Derry—Magilligan, and by Lough Neagh; D.M.

TRIFOLIUM *Linn.* CLOVER.

1. T. pratense *Linn.* PURPLE CLOVER.

Pastures, waste places, and occasionally on rock cliffs—common. Fl. mid. May till mid. Sept.

This clover being sown so extensively, renders it probable that the wild specimens found in the lowlands, are all descended from cultivated plants. There is, however, on the mountains, a very coarse hairy form, which seems distinct, and is probably truly native.

2. T. medium *Linn.*

Stony ground, and short pastures, especially in hilly districts—very rare in Down—frequent in Antrim and Derry. Fl. mid. June till mid. Sept. Ascends to 1000 feet in Derry; D.M. A plant of the basalt especially.
Down—Ditch bank opposite the glen gate Portavo; T.H.C. Newcastle; R.Ll.P.
Antrim—Islandmagee, Whiteabbey, and very common about Doagh;
    Templeton, 1794. Elevated ground between Larne and Glenarm; *Flor. Hib.* Black Mountain, Cave Hill, Carnmoney Hill, and Knockagh; *Flor. Ulst.* Ballypallady, Muckamore, Larne, Glenarm Deer Park, and very luxuriant on road from Craigbilly to Slemish; S.A.S. Glenariffe and Carnlough; R.Ll.P.
Derry—Base of Umbra rocks, Glen of the Roe above Dungiven, and plentifully along the basaltic range; D.M. Ballyronan; S.A.S.

3. T. arvense *Linn.* HARE'S-FOOT TREFOIL.

Sandy ground—rare. Fl. July and Aug.

As is the case with many other annuals of sandy ground, the precise stations of this plant are very inconstant.
Down—Railway banks near Bloomfield; *Flor. Belf.* Rough Island two
    miles east of Comber; S.A.S. Sands at Ringsallin Point east of Dundrum; T.H.C.
Antrim—Church Bay in Rathlin, and on slipped ground at Blackhead;
    Templeton, 1794. (not seen by S.A.S. in Rathlin—1882). Ulsterville near Belfast; *Miss Maffett.* Coasts of Antrim; *Cyb. Hib.* Sandpits at Stranmillis; S.A.S. Portrush sandhills; T.H.C.
Derry—Frequent in the sandy fields of Magilligan, and by the shores of
    Lough Neagh; D.M. Abundant in Magilligan; *B.N.F.C.*, 1863.

LEGUMINOSÆ.

4. T. repens *Hudson*. WHITE CLOVER, SHAMROCK.
Pasture fields, grassy heaths, etc.—very common. Fl. June—Sept.

5. T. striatum *Linn*.
Gravelly seashores—very rare, and not seen recently. Fl. May and June (*Cyb. Hib.*).
Antrim—" On sloping gravelly banks on the shore at Rhanbuoy (*Rahanbuidhe* yellow strand)"; *Templeton*, 1811. Near the seaside at Whitehead; *Wm. Millen*, also *Prof. Dickie*, in *Flor. Ulst.* Whitehead; *G. C. Hyndman*, 1866 (*spec. in herb. Hyndman*).
Mr. Templeton was not altogether certain that his plant was correctly named, he says—"I have some doubts that this may prove to be *T. maritimum*" (*Temp. M.S.*). Mr. F. Whitla by a note in the margin of his copy of *Flor. Hib.* says that the Ranboy plant was *T. maritimum*, but as this species has not been found in Ireland, and as *T. striatum* was assuredly met with within a few miles of Templeton's station, it seems certain that the name above cited is right.

6. T. procumbens *Linn*. HOP TREFOIL.
Pastures, and wastes in dry sandy ground—frequent. Fl. mid. June till mid. Aug.
Down—Ballymaghan, and Groomsport; *Flor. Belf.* Roadsides Newcastle to Castlewellan, and thence to Clough, also Downpatrick, Castlereagh Hill, and road from Bangor to Crawfordsburn; S.A.S.
Antrim—Banks of Larne river, 1794, also Rathlin (*abundant now in Rathlin*; S.A.S.), Blackhead, walls of Dunluce Castle, and of the ruin in Islandmagee, 1808; *Templeton*. Railway bank at Randalstown; *Flor. Ulst.* Redbay, and shore close to the Gobbins; S.A.S. Railway bank at Larne; T.H.C.
Derry—Sandy fields by the Roe opposite Bellarena, and at Magilligan; D.M. Downhill; *B.N.F.C.*, 1863.

7. T. dubium *Sibth*. (T. MINUS *Sm.*, T. PROCUMBENS *Huds.*).
Pastures, ditch banks, and waste ground—very common. Fl. mid. May till mid. Aug.

LOTUS *Linn*. BIRD'S-FOOT TREFOIL.

1. L. corniculatus *Linn*.
Roadsides, and banks—common. Fl. mid. May till end of Aug. Ascends to 1500 feet in Derry (D.M.).
Var. γ L. CRASSIFOLIUS *Persoon*.
Stony and gravelly places by the seashore.

LEGUMINOSÆ.

Down—At the Kinnegar Holywood; *Flor. Belf.* Railway bank between Sydenham and Tillysburn ; S.A.S.
Antrim—Shore at Kilroot ; S.A.S.

2. L. pilosus *Beeke* (L. MAJOR *Sm.*).
Marshy ground, and damp ditch banks—frequent. Fl. June—Aug.
Down—Knock, and Belmont; *Flor. Belf.* Banbridge, Ballynahinch, Conlig, Comber, Newtownards, Drumbo, and Moneyrea; S.A.S. Orlock Point ; T.H.C. Holywood ; R.Ll.P.
Antrim—Whiteabbey ; *Templeton,* 1806. Carrick Junction, and near the Forth River; *Flor. Belf.* Whiterock, Cave Hill, and Macedon Point; T.H.C. Knockagh ; S.A.S.
Derry—Limavady, and other places ; D.M. Magilligan ; *Hart, Proc. R.I.A.,* 1884.

ANTHYLLUS *Linn.* KIDNEY VETCH.

1. A. vulneraria *Linn.* LADY'S FINGER.
Rocks, and dry gravelly banks—common, especially abundant on the coast. Fl. mid. May till early in Aug.

VICIA *Linn.* VETCH, TARE.

1. V. hirsuta (*Linn.*) *Koch.*
Borders of fields, and in waste sandy ground—frequent. Fl. June—Sept.

2. V. sylvatica *Linn.* WOOD VETCH.
Woods, and shady rocks—rare. Fl. mid. June till mid. Aug.
Down—In the wood of Rostrevor ; *Templeton,* 1793. Drumbo; *Flor. Ulst.* Charlesville Glen ; *Cyb. Hib.* Railway bank near Clandeboye ; R.Ll.P.
Antrim—Plenty about Glenarm, and on sea cliffs near Larne ; *Templeton,* 1808. Crow Glen, Trooper's Lane, and Woodburn ; *Flor. Belf.* Ram's Island, Cave Hill, Knockagh, Whitehead, and Ballycastle; *Flor. Ulst.* Side of Lough Neagh at Langford Lodge, and abundant on the rocky coast of Antrim ; *Cyb. Hib.* Springfield Glen, and shore south of the Gobbins; S.A.S. Kilroot, and Garron Head ; T.H.C. Tor Head, etc. ; R.Ll.P.
Derry—Tircrevan, and sides of small lakes above Kilrea ; D.M. Magilligan ; *Flor. Ulst.* South side of Benevenagh ; *Cyb. Hib.*

3. V. orobus (*Linn.*) D.C.
On rocks—very rare. Fl. July and Aug.

LEGUMINOSÆ.

Down—"*Orobus sylvaticus nostras*, found near Rostrevor, Dr. Sherard"; *Raii. Syn. ed. II.* 1696.

This does not seem to have been seen at Rostrevor by any botanist subsequent to Sherard, and may have been an error, but the fact that it grows now in Antrim renders it probable that the determination of Sherard was correct.

Antrim—Basaltic rocks of Sallagh Braes; *Guide to Belf. and Adj. Cos., by Belf. Nat. Field Club*, 1874, also *Proc. Belf. Nat. Field Club*, 1882-83. One very large bushy plant, with a profusion of flowers, seen by S.A.S. in July, 1873, and seen in fruit by Belf. Nat. Field Club in 1882 (still there).

4. V. cracca *Linn.*
Hedges, and ditch banks—very common. Fl. mid. June till mid. Aug.

5. V. sepium *Linn.* BUSH VETCH.
Hedge banks, and wooded places—common. Fl. May—July.

6. V. angustifolia *Roth.*
Sandy, and gravelly waste ground, and borders of sandy fields—frequent. Fl. May and June. Known to Templeton in 1797, who pronounced it "most certainly indigenous."
Down—Scrabo Hill; *Flor. Ulst.* Donaghadee, Groomsport, Ballyalloley, Castlereagh, and at Ballyoran quarries; S.A.S. Newcastle, Dundrum sandhills, and at Ballyholme; T.H.C. Holywood; R.Ll.P.
Antrim—By Lough Neagh; *Cyb. Hib.* Whitewell quarries, and by the lake at Selshan, and Derrymore; S.A.S. Lake shore at Antrim; T.H.C.
Derry—Abundant all along the coast; D.M. Lough Neagh shore; *Cyb. Hib.*

7. V. lathyroides *Linn.*
Sandy seashores—rare. Fl. mid. April till end of May.
Down—Sandy shore at Ballyholme; *G. C. Hyndman*, 1829. On the warren at Donaghadee, and frequent on sandy shores from thence to Ballywalter; S.A.S.
Antrim—Gravelly bank on the shore at Three-mile-water Point near Carrickfergus; *Templeton*, 1808. Ballycastle; *Flor. Ulst.* (still there 1882).
Derry—Coast at Magilligan—rare; D.M.
First found in Ireland by Mr. Templeton.

LATHYRUS *Linn.* VETCHLING.

1. L. pratensis *Linn.*
Meadows, damp pastures, and ditch banks—common. Fl. June—Aug.

2. L. palustris *Linn.*

Marshy places—very rare. Fl. June and July (*Cyb. Hib.*).

Antrim—In a moist meadow a little way north of where the Lagan canal enters Lough Neagh; *Templeton*, and subsequently D.M. Drains by Lough Neagh at Selshan; D.M. in *Cyb. Hib.* Shore of Lough Neagh near Ellis's cut; *G. C. Hyndman* (*spec. in herb. Hyndman*).

This plant was at one time abundant in the above localities, but has not been seen there for perhaps 40 or 50 years, and owing to the lower level of the lake, and to drainage of marshes is probably now lost in Antrim. It has, however, been recently found by *Rev. H. W. Lett* on islets in Closet River, in Co. Armagh. This station, though outside our district, is less than six miles southwest of Selshan. Rev. George Robinson has also found this plant on Scawdy Island, which is further west.

3. L. macrrorhizus *Wimm.* (OROBUS TUBEROSUS *Linn.*) NAPPERTY (*loc.*).

Heaths, and dry banks—frequent throughout the district. Fl. May and June.

Var. β L. TENUIFOLIUS *Roth*.

Dry stony banks—rare.

Down—Hedge side at Purdysburn, and dry bank at Knock; S.A.S. Hedge bank on old road from Newtownbreda to Castlereagh; T.H.C. Dundonald, Holywood reservoir, and Craigauntlet; R Ll.P.

ORDER XX. **ROSACEÆ.**

PRUNUS *Linn.*

1. P. communis *Huds.* (P. SPINOSA *Linn.*) SLOE, BLACKTHORN.

Glens, hedges, and stony waste ground—common. Fl. April and May.

2. P. padus *Linn.* BIRD CHERRY.

Thickets in damp rocky places—rare, and decreasing. Fl. mid. May till end of June.

Down—Margin of the Lagan at Tullycairne near Dromore; *C. Dickson*, and T.H.C.

Comber was printed in *Flor. Ulst.* for Cumber, which is in Co. Derry.

Antrim—In Glenarm Park below the junction of the two rivers, and plentiful in the great deer park, and in shrubby grounds below

Glenarm, also on limestone rocks at Sallagh Braes near Larne ;
Templeton. Seen at Sallagh by the *Belf. Nat. Field Club* in
1874. Low glens in Glenshesk ; *Ir. Flor.* Glenravel ; *Flor.
Ulst.* Glenariffe; *Cyb. Hib.* Carnlough glen; *G. C. Hyndman*
(seen there recently by the *Eds.*).

Derry—On the banks of the Faughan about a mile below Cumber;
Templeton. By the Roe between Dungiven and Limavady,
and side of the Faughan below Clady, also by the Agivey
above Garvagh ; D.M.

3. P. avium *Linn.* WILD CHERRY.

In glens, and hedges—frequent. Fl. May and early part of June.
Natural in glens, planted in hedges.

Down—Hedges at Ballymaghan, Trench, Gransha, and in glen at Knock ;
Flor. Belf. Dundonald glen ; S.A.S.

Antrim—"Among alder, and other native trees on the margin of Lough
Neagh, in many places, also pretty common in hedges, Fruit
black, sweet, and eaten by birds" ; *Templeton.* Glen of
Crumlin River ; T.H.C.

Derry—Apparently wild by the side of Curley Burn, and in Ballynascreen
near Draperstown ; D.M.

SPIRÆA *Linn.*

1. S. ulmaria *Linn.* MEADOW SWEET.

Roadsides, meadows, and in damp places—very common. Fl. July
and Aug.

SANGUISORBIA *Linn.*

1. S. officinalis *Linn.* GREATER BURNET.

Sloping banks—very rare. Fl. June—Aug. (*Cyb. Hib.*).

Down—Pastures, Donaghadee ; *Miss Maffett, in Flor. Belf.* About half
a mile from Donaghadee on the inland side of the road to
Millisle ; T.H.C.

Antrim—On small gravelly mounds in townland of Maddykeel, parish of
Finvoy near Rasharkin ; *Cyb. Hib.* Near Carnlough ; (*W.
Hancock*), *A. G. More, in Jour. Bot.* April, 1873. On a
sloping bank by the road to Garron Point about 3 miles from
Carnlough ; *C. Dickson.*

Derry—On a dry bank sloping to the Bann one mile below Agivey ;
Cyb. Hib.

AGRIMONIA *Linn.* AGRIMONY.

1. A. eupatoria *Linn.*

Roadsides, dry banks, and stony pastures—rare on the silicious grits of

Down, but frequent in the Trappean districts of Antrim and Derry. Fl. mid. June till early in Sept.

Down—Links at Newcastle, also Holywood; *Flor. Ulst.* Benderg Bay near Ardglass; T.H.C. Shore between Comber and Newtownards, and at Killough; S.A.S.

Antrim—Belfast Hills; *Flor. Belf.* Colin Glen, Cave Hill, Knockagh, and Rathlin; *Flor. Ulst.* Ballinderry, Muckamore, by the canal near Moira, Carnmoney Hill, Macedon, Carrickfergus, Whitehead, Ballycarry, Glenarm, and Ballintoy; S.A.S. Crumlin, and Islandmagee; T.H.C. Cairncastle, Drumnasole, and Glenariffe; R.Ll.P.

Derry—Abundant about Dungiven; D.M. Toome; S.A.S.

2. A. odorata *Mill.*

On damp banks—very rare. Fl. July and till mid. Sept.

Down—In some quantity on roadside at Haw Hill about one mile northeast of Comber; S.A.S.

Antrim—Rather abundant on the rocky and bushy shore of Lough Neagh near Shane's Castle, and on flat ground to the west of the castle; *Rev. W. W. Newbould, and Prof. Babington.* Banks of Crumlin River half a mile above waterfoot bridge; S.A.S. By the river in Glenarm Park, in considerable abundance above the waterfall; T.H.C.

First found in Ireland by Rev. W. W. Newbould, and Prof. Babington in 1852.

ALCHEMILLA *Linn.*

1. A. vulgaris *Linn.* LADY'S MANTLE.

By streams, and on short damp pastures—common, especially on the hills. Fl. May—Sept. Ascends to 2000 feet on mountains in Derry; D.M.

  Var. β A. MINOR *Huds.*, (A. HYBRIDA *Pers.*, A. MONTANA *Willd.*).

Mountain pastures at elevations of near 1000 feet—rare.

Antrim—South side of Cave Hill; (*J. Ball*) *Cyb. Hib.* Heathy pastures on Black Mountain at about 700 feet, and plentiful on Cave Hill below the cliffs; S.A.S.

Derry—Rocky, and grassy slopes on Benevenagh, at about 1000 feet; S.A.S.

The distribution of this montane var. has not yet been worked out, and the above notes do not afford sufficient data to define either its horizontal, or vertical limits.

2. A. arvensis (*Linn.*) *Lamk.* PARSLEY PIERT.

Borders of fields in sandy lands—common. Fl. mid. May till mid.

ROSACEÆ.

Sept. Occurs at 1200 feet on Spalga near Rostrevor.

POTENTILLA *Linn.* CINQUEFOIL.

1. P. anserina *Linn.* SILVER WEED, Mashcorns (loc.).
Roadsides, and pastures—very common in damp, stony or gravelly places.
Fl. mid. May till mid. Aug.

2. P. reptans *Linn.*
Dry banks, and gravelly waste ground—frequent. Fl. July and Aug.
Down—Abundant about Warrenpoint, and Killough ; S.A.S. Ardglass, and Crawfordsburn ; T.H.C.
Antrim—Abundant on gravel around the shores of Lough Neagh; *Templeton*, 1797. Cave Hill, Colin Glen, and banks of the Lagan ; *Flor. Ulst.* By the Forth River, also shore of Lough Neagh near Toome, and roadsides between Larne and Glenarm ; S.A.S.
Derry—Abundant through the county ; D.M. Castledawson ; S.A.S.

3. P. tormentilla *Sibth.* TORMENTING ROOT (loc.).
Pastures, heaths, and wastes—very common. Fl. June—Sept. Found on our highest mountain summits.
    Var. β P. PROCUMBENS *Sibth.* (P. NEMORALIS *Nestl.* TORMENTILLA REPTANS *Linn.*).
Sandy and gravelly banks—rare ?
Down—Railway bank between Belfast and Connswater ; S.A.S.
Derry—Not unfrequent by the side of Lough Neagh ; D.M.

4. P. fragariastrum (*Linn.*) *Ehr.* BARREN STRAWBERRY.
Hedge banks, and dry waste ground—common. Fl. Feby.—April.
Not easy to define the flowering time, as it varies much according to the mildness or severity of the season.

COMARUM *Linn.*

1. C. palustre *Linn.* MARSH CINQUEFOIL.
Marshes, and bog drains—frequent throughout the district. Fl. mid. June till end of July.

FRAGARIA *Linn.* STRAWBERRY.

1. F. vesca *Linn.* WILD STRAWBERRY.
Hedge banks, and grassy heaths—common, especially abundant in rocky and shady places on the basaltic hills. Fl. mid. April till mid. July.
With white fruit on railway bank at Marino near Holywood ; R.Ll.P.

ROSACEÆ.

RUBUS Linn. BRAMBLE, BLACKBERRY.

It will be understood that the brambles here enumerated have very unequal systematic values. The names adopted are those of Babington's Manual, 8th edition; corrected in accordance with his Notes in Journal of Botany for 1886.
The determination of fruticose *Rubi*, by ordinary botanists, is so unreliable that we pass over all notes of such, which have not the sanction of special authorities; those recorded here have been either named, or confirmed by Prof. Babington.

1. R. idæus *Linn.* WILD RASPBERRY.
Hedge banks, and thickets—common, but especially abundant on the lower slopes of the hills. Fl. mid. June till mid. July. Ascends to 1600 feet on the northeast side of Slieve Gallion; D.M.

2. R. suberectus *Anders.*
Derry—Limavady deerpark, at the upper end; D.M.

3. R. fissus *Lindley.*
Down—Marshy margin of the northeast corner of Carrickmannan Lake near Saintfield; S.A.S.
Derry—By the Foyle near Londonderry; *Cyb. Hib.*

4. R. plicatus *W. & N.*
Derry—Clady, and Kilrea; D.M.

5. R. rhamnifolius *W. & N.*
Down—Railway bank at Connswater; S.A.S.
Antrim—Blackhead, and Massereene Park; S.A.S.

6. R. rusticanus *Merc.* (R. DISCOLOR *auct. Brit.*).
Waste places, and neglected hedge banks—the most abundant of our brambles. Fl. July—Sept.

7. R. pubescens *Wirt.* (THYRSOIDEUS *auct. Brit.*).
Very rare.
Derry—Near Benevenagh; *Cyb. Hib.*

8. R. leucostachys *Smith.*
Not rare?, but distribution imperfectly known.
Down—Dundonald; *R. Tate.*
Antrim—Lagan-side at second lock; S.A.S.
Derry—In County Derry; *Cyb. Hib.*

ROSACEÆ.

9. R. pyramidalis *Kaltenberg.*
Down—Newtownards; S.A.S.
Antrim—By the Lagan canal near Belfast; S.A.S.

10. R. Grabowskii *Genevier.* (R. MONTANUS *Wirtgen?*).
Antrim—At second lock Lagan canal; S.A.S.

11. R. Salteri *Babington.*
Down—Wood at Shrigley; S.A.S.
Derry—Frequent in Derry (*R. calvatus*); *Cyb. Hib.*

12. R. carpinifolius *W. & N.*
Down—On road from Bangor to Clandeboye, and by the Causeway Water in south Down; S.A.S.
Antrim—By the second lock Lagan canal; T.H.C. Cave Hill above Whitewell, Glenshesk, and Rathlin Island ("*curious form*" C.C.B.); S.A.S. Cushendun; *Rev. S. A. Brenan.*

13. R. villicaulis *W. & N.*
Not rare, but range imperfectly known.
Down—Roadside Castlewellan, and hedge near Castle Espie, and by the Causeway Water; S.A.S.
Antrim—By canal near Belfast, and on the Cave Hill, also on rocks in Glenarm Park; S.A.S. Bushmills; *Cyb. Hib.*

14. R. macrophyllus *Weihe.*
In some of its forms frequent throughout the district. Fl. mid. June till end of July.

Var. α R. UMBROSUS *Arrh.*
Down—Newtownbreda; T.H.C.
Antrim—Macedon Point, and Cave Hill quarries; S.A.S.

Var. β R. MACROPHYLLUS *W. & N.*
Down—Castlewellan, Ballynahinch, and Newtownards; S.A.S.
Antrim—Black Mountain; *Cyb. Hib.* By the canal near Belfast; *R. Tate.*
Derry—At Londonderry; D.M. Coleraine ("*true*" C.C.B.), and by the Bann opposite Coleraine ("a peculiar form" C.C.B.); S.A.S.

Var. γ R. SCHLECHTENDALII *W. & N.*
Down—Hedges near Castle Espie; S.A.S.

Var. δ R. AMPLIFICATUS *Lees.*
Down—Woods at Rademon; S.A.S.
Antrim—Carnmoney; *Flor. Belf.* By small stream on Black Mountain; S.A.S.

15. R. mucronatus *Blox.*
Antrim—Rocky heath at Whiterock near Belfast; S.A.S.

16. R. radula *Weihe.*
Down—Railway bank at Connswater; S.A.S.
Antrim—By stream near the foot of Black Mountain, on the east side; S.A.S.
   Var. γ R. DENTICULULATUS *Bab.*
Derry—Kilrea; *Cyb. Hib.*

17. R. Koehleri *Weihe.*
Hedges, and waste ground—common. Occurs, in some of its forms, abundantly. Fl. mid. June till end of Aug.
   Var. α VERUS.
Down—Hedge at Knock; S.A.S.
Antrim—Carnmoney; *Cyb. Hib.*
   Var. β R. INFESTUS *Bab.*
Down—Roadside near Clough, and waste ground on Castlereagh Hill; S.A.S.
   Var. γ R. PALLIDUS *Weihe.*
Down—Hedge at Sydenham; T.H.C. Hedges at Cregagh; S.A.S.

18. R. Lejeunii *Weihe.*
Antrim—In a hedge near Dunadry on the road to Templepatrick; T.H.C.

19. R. flexuosus *Weihe.*
Derry—By the road from Garvagh to Kilrea; *Cyb. Hib.*

20. R. humifusus *Weihe.*
Derry—By the Foyle above Londonderry; *Cyb. Hib.*

21. R. foliosus *Weihe.*
Down—By stream in Tollymore Park; S.A.S.
This is the form considered by Bloxam to be typical *foliosus*.

22. R. corylifolius *Smith.*
Hedges, and waste places—common. Fl. mid. June till end of July.
   Var. α R. SUBLUSTRIS *Lees.*
Down—Side of stream at Cultra; S.A.S. Hedge near Sydenham; T.H.C.

Antrim—By the shore at Blackhead, and on Rathlin Island, and lake shore at Selshan ; S.A.S.
Derry—Faughanvale, Clondermot, and Templemore ; D.M.

Var. β R. CONJUNGENS *Bab.*
Antrim—Lagan-side, and Carnmoney Hill; *R. Tate.* Woodburn, and Bushmills ; S.A.S.

Var. γ R. FASCICULATUS *Mull.*
Down—Brett's Glen ; *Cyb. Hib.* Hedges at Comber, and Newtownards, and waste ground near Downpatrick ; S.A.S.
Derry—By the Ballinderry river near Ardtrea ; S.A.S.

23. R. scabrosus *Mull.*
Antrim—By the Lagan canal near Belfast ; S.A.S.

24. R. cæsious *Linn.* DEWBERRY.
Dry places—very rare.
Antrim—By Lough Neagh ; C.C.B. in *Cyb. Hib.* A specimen found by S.A.S. in Massereene Park seems to belong here, but too imperfect to decide without doubt.

Var. γ R. LIGERINUS *Genevier.*
Antrim—Hedge bank at Drumadarragh near Doagh ; S.A.S.

25. R. saxatilis *Linn.*
Rocky ledges and screes in the mountains—frequent, especially on the basaltic range, at elevations up to 1000 feet. Fl. mid. June till end of July.
Down—Glen near Newtownards ; *Ir. Flor.* Slieve Martin ; *H. C. Hart.*
Antrim—Near the head of Colin Glen, and similar places ; *Templeton,* 1793. Rocky mountains near Glenarm ; *Ir. Flor.* Cave Hill, Carr's Glen, and Black Mountain ; *Flor. Ulst.* Wolfhill ; *W. Millen.* Glens in the North of Antrim ; *Cyb. Hib.* Crow Glen, Woodburn, and Sallagh Braes ; S.A.S. Glenariffe ; *Bab. Brit. Rubi.* Carnlough, and Ardclinis ; R.Ll.P.
Derry—Erigal banks above Garvagh, stony places about Swatragh, and abundant about the base of Benevenagh ; D.M. Umbra rocks ; *J. Ball.* Cliffs on Slieve Gallion ; S.A.S.

DRYAS *Linn.*

1. D. octopetala *Linn.* MOUNTAIN AVENS.
Basaltic cliffs—very rare, one station only. Fl. June and July.
Derry—"Northern face of Magilligan rocks where mould has stopped in its fall from the top, in seed Aug. 17th, 1796"; *Templeton.* Abundant on Benevenagh at 1000 to 1100 feet—northern

aspect only; D.M., 1835. On Benevenagh rocks; *B.N.F.C.*, 1863. Benevenagh at 1100 feet; T.H.C., 1878.

In Flora Hibernica Mr. Templeton is erroneously credited with finding this plant in Antrim.

GEUM *Linn.* AVENS.

1. G. urbanum *Linn.* WOOD AVENS, HERB BENNET.
Under hedges, and on borders of woods and thickets—common. Fl. mid. May till mid. July.

2. G. intermedium *Ehr.*
Woods, and damp shady glens—frequent. Fl. June and July.
Antrim—Carr's Glen; *Templeton*, 1806. By the Forth River, and at Cushendall; *Cyb. Hib.* Glenarm, Bruslee, Springfield Glen, and abundant in upper Colin Glen; S.A.S. Shane's Castle woods; T.H.C.
No doubt often passed over as a form of the following species, which it resembles most. If a hybrid, remarkably constant, having held its ground in Carr's Glen over 80 years.

3. G. rivale *Linn.* WATER AVENS.
Banks of streams, and wet places in woods and shady glens—common. Fl. mid. May till mid. July.

ROSA *Linn.* ROSE, BRIAR.

1. R. spinosissima *Linn.* BURNET ROSE, SCOTCH ROSE.
Dry rocky heaths, and sandy seashores—frequent throughout the district, but more abundant on the coast. Fl. mid. June till mid. July. Ascends to 1000 feet on the Derry mountains; D.M.

Var. R. CIPHIANA *Sibbald*, which has petals pink, or streaked with red —a beautiful little rose—has been found in Antrim, at Larne; *Templeton*, Rathlin; S.A.S., and in Derry at Castlerock; R.Ll.P.

2. R. hibernica *Smith.* IRISH ROSE.
Hedge banks, and rocks—very rare. Fl. mid. June till mid. July.
Down—Left of the road going to Holywood, south of the Whinny Hill,
(*This was the old road from Belfast to Holywood*), and on the east side of the Lagan about a quarter of a mile below Stranmillis; *Templeton*, 1795. Not now found in the latter locality, and only to be met with in a hedge on the right side of the present road to Holywood.
Antrim—Sparingly on rocks in the little Deerpark at Glenarm (var. *glabra*); *Cyb. Hib.*

ROSACEÆ.

Derry—In a glen at Magilligan; *Templeton,* and subsequently; D.M.
Rocks between Umbra and Benevenagh, and basaltic cliffs on
Benevenagh (var. *glabra*); *Cyb. Hib.*
First found in Ireland by Mr. Templeton, at the Holywood station.

3. R. involuta *Smith.*
On rocks and banks in glens, and by roadsides—rare, and only in the
north. Fl. latter end of June till end of July. Some of the varieties in
which this rose occurs are difficult to discriminate. Typical form not
found.

Var. R. SABINI *Woods.*
Antrim—Glenarm, Garron Head, Glenariffe, and Tor Head; *Cyb. Hib.*
Hedge at Ticloy in the Braid; *B.N.F.C.*, 1871. Roadside
at Racavan near Slemish; S.A.S.
Derry—Benone, Umbra, and townland of Craigs; D.M. Plentiful in
Bennedy Glen near Dungiven; *Cyb. Hib.*

Var. R. DONIANA *Woods.*
Antrim—Near Glenarm; *Templeton,* 1814. Glenariffe, and Garron Head;
Cyb. Hib.
Derry—On the old beach mark (raised beach) at Benone, Magilligan,
D.M.

Var. WILSONI *Borrer.*
Derry—Sparingly on Umbra rocks; D.M. in *Cyb. Hib.*
The following varieties are recorded in Baker's Monograph of the
British Roses:—

Var. GRACILESCENS *Baker.* Co. Antrim; D.M.
Var. ROBERTSONI *Baker.* Co. Derry; D.M.
Var. LÆVIGATA *Baker.* Cos. Antrim and Derry; D.M.
Var. MOOREI *Baker.* Near the sea at Tamlaghtard, Co.
Derry; D.M.
The stations assigned to *R. Sabini* in *Flor. Belf.* are erroneous.

4. R. mollissima *Willd.* (R. MOLLIS *Smith,* R. VILLOSA *Linn.* in
part).
Bushy places, especially along the slopes of the basaltic hills—rare.
Fl. mid. June till mid. July. Fr. formed in July—ripe early in Sept.
Down—Sparingly by the Causeway Water west of Kilkeel; S.A.S.
Antrim—Cave Hill, and near Glenarm, also between Garron Head and
Glenariffe; *Cyb. Hib.* Undercliff at the Knockagh, and on
Cave Hill a little to the south of, and higher up than the
Whitewell quarries, also at Crumlin waterfoot; S.A.S.
Hedge at Glenavy, T.H.C.
Derry—Foot of Umbra rocks, and in Whitewater Glen near Draperstown
D.M. Bennedy Glen; *Cyb. Hib.*

D

ROSACEÆ.

5. R. tomentosa *Smith.*
Hedges, thickets, and by streams—common. Fl. mid. June till mid. July.
Var. *subglobosa* occurs, and is sometimes taken for the preceding species.

6. R. rubiginosa *Linn.* SWEETBRIAR, EGLANTINE.
Hedges, and bushy places by streams—rare. Fl. mid. June till mid. July.
Stray bushes planted in hedges, or escaped from cultivation, are of frequent occurrence, but such are omitted here. We consider the plant to be native in most of the localities which follow. Mr. Templeton thought our plant to be different from the sweetbriar sold in nurseries.
Down—Roadside one mile south of Ballylesson; *Templeton*, 1802. Hedges at Portavo, Knockbracken, and on road from Greyabbey to Kirkcubbin; S.A.S.
Antrim—Carr's Glen; *Templeton.* Chalk rocks above Larne; *Cyb. Hib.* Thickets near Oldstone; *Flor. Ulst.* Plentiful in hedges from Lisburn to Ballinderry, and hedges from Crumlin waterfoot to Langford Lodge (seems planted here), also in woods at Muckamore, and abundant on quarry spoil-bank at Magheramorne, and in old limestone quarry at Gleno; S.A.S.
Derry—Side of the river Roe where the Curlyburn joins it, roadside between Swatragh and Kilrea, and side of Agivey River near Ballydevitt; D.M.

7. R. micrantha *Smith.*
Hedges and thickets—very rare. Fl. June and July (*Cyb. Hib.*).
Antrim—Cushendun; *Rev. S. A. Brenan, spec.*
This extends the range in Ireland of the small flowered sweetbriar over 200 miles to the northeast.

8. R. canina *Linn.* DOG ROSE.
Hedges, copses, and waste places—very common. Fl. mid. June till mid. July.
Several forms of this variable species occur. Vars. *pruinosa, corifolia,* and *marginata* were gathered in Derry by Dr. Moore. These were identified by Mr. Baker, and published in his Monograph. In the absence of similar authoritative determination it is safer to omit the names of other forms reported to occur in the district.

9. R. arvensis *Hudson.*
Hedges, and by river banks—rare. Fl. mid. June till mid. July.
Down—In Tollymore Park, and near Ballynahinch, also between Tandragee and Gilford; *Templeton*, 1795. Hedge at Islanderry near Dromore, and one plant by wall of Clandeboye demesne; T.H.C. Annalong; *H. C. Hart.*

Antrim—North side of Glenariffe near the great waterfall; *Templeton*.
Ballycastle; *Flor. Ulst*. Along the Crumlin River from Cidercourt to the waterfoot, and in hedges adjacent; S.A.S.
Plentiful by Glenavy River; T.H.C. Glendun; *C. Dickson*.
Cushendun, and Cushendall; R.Ll.P.
Derry—Near Garvagh, and in roadside hedges from Moneymore to Ballyronan, and shore of the lough near Salterstown; D.M.

CRATÆGUS *Linn*. HAWTHORN.

1. C. oxyacantha *Linn*.
Mountain cliffs, waste ground, fields, and hedges—very common. Fl. mid. May till end of June. Our plant is the variety MONOGYNA *Sibth*.

PYRUS *Linn*.

1. P. malus *Linn*. CRAB APPLE.
Thickets—very rare in the wild state, but specimens of the cultivated apple are frequent in hedgerows. Fl. May.
Down—Holywood Glen, three or four bushes remaining, May 1797;
Templeton. Knock Glen (*acerba*); S.A.S. Castlereagh and Ballymaghan glens; *Flor. Belf*.
Antrim—Very plentiful among natural wood on margin of Lough Neagh;
Templeton. Colin Glen, and Lisburn; *Flor. Belf*. Hedges at Stranmillis, and by Forth River (*mitis*), scarcely native; S.A.S.
Derry—General in woods and hedges through the country; D.M.

2. P. aucuparia (*Linn*.) *Gaertn*. MOUNTAIN ASH, ROWAN TREE, QUICKEN TREE.
Rocks, glens, and mountainous woods—frequent. Fl. mid. May till end of June. Ascends to 1900 feet on Donard; *Hart*. 1400 on Slemish; *Templeton*. Often planted about houses.
Down—Mourne Mountains; *Templeton*, 1793. Slieve Donard; *Flor. Ulst*.
Dundonald and Castlereagh hills; S.A.S.
Antrim—Slemish; *Templeton*. Colin Glen, and Black Mountain; *Flor. Belf*. Carnlough Glen; *Flor. Ulst*. Sallagh Braes; T.H.C.
Cave Hill cliffs, and in Glendun; S.A.S. Straidkilly, and Pollan Burn; R.Ll.P.
Derry—Benevenagh, Lignapeiste, and frequent in the glens in the mountains; D.M. Castledawson, Dungiven, and frequent in the glens of the Sperrin Mountains; S.A.S.

3. P. aria (*Linn*.) *Smith*. WHITE BEAM TREE.
Rock cliffs at elevations under 1000 feet—almost extinct. Fl. May and June (*Cyb. Hib*.).

Down—"Brought from off the Mourne Mountains to Lord Clanbrassil";
Templeton (not found in Down subsequently).
Antrim—"Several trees in a thriving condition, from the little Deerpark at Glenarm, seen July 15th, 1808"; *Templeton.* Inaccessible rocks, hanging over the sea, in the little Deerpark at Glenarm, and in low glens, Co. Antrim; *Ir. Flor.* 1833. Steep cliffs south of Glenarm; *Flor. Ulst.* Near Ballycastle, and not unfrequent along the rocky coast of Antrim; *Cyb. Hib.* "A plant, which I consider this, on inaccessible basalt cliffs southeast of Carrick-a-Rede"; T.H.C.
Derry—Sparingly on Umbra rocks, and between them and Benevenagh; D.M.

The plants of Glenarm and Ballycastle are referred to var. RUPICOLA *Syme*, and it is doubtful whether restricted *P. aria* occurred in the district.

ORDER XXI. **LYTHRACEÆ.**

LYTHRUM *Linn.*

1. L. salicaria *Linn.* PURPLE LOOSETRIFE.
Marshes, and banks of streams—common; very abundant on the sandy coast of Derry. Fl. mid. July till mid. Sept.

PEPLIS *Linn.* WATER PURSLANE.

1. P. portula *Linn.*
Marshy places in sandy or rocky ground—frequent. Fl. July and Aug.
Down—Kinnegar, and Ballyalloley; *Flor. Belf.* Margin of Corbet Lake near Banbridge, Newcastle warren, roadside near Moneyrea, and by road from Comber to Strangford Lough; S.A.S. Downpatrick, and Ballynahinch; R.Ll.P.
Antrim—Carrickfergus commons; *Flor. Ulst.* Rathlin Island; S.A.S. Antrim; *D. Redmond.* Summit of Fairhead; T.H.C. Cushendall, and Cushendun; R.Ll.P.
Derry—Common through the county; D.M. Moneymore; *Flor. Ulst.* Shores of Lough Beg, and between Moneymore and Ballyronan; S.A.S.

ONAGRACEÆ.

ORDER XXII. ONAGRACEÆ.

EPILOBIUM *Linn.* WILLOW HERB.

1. E. angustifolium *Linn.* ROSE BAY. BLOOMING SALLOW (loc.).
Damp glens, and rocky ledges on the mountains—rare. Fl. July and Aug. Ascends to 1100 feet. Our plant is var. MACROCARPUM *Steph.* It prefers limestone or basalt rocks.

Down—[The statement, in *Cyb. Hib.*, that this plant is frequent among rocks, and in thickets in Co. Down is erroneous. It occurs in a railway cutting between Comber and Newtonards, and is frequent near cottages, but only where planted, or escaped from cultivation. "Sydenham"; *Flor. Belf.* The above remarks apply to this also.]

Antrim—Cave Hill on rocks beyond the caves, and near the head of the north branch of the Sixmilewater, also Skerry rocks, and rocks about Murlough; *Templeton*, 1793. Rocks between Tor Point and Fairhead; *Ir. Flor.* Banks of Woodburn stream near the source; *F. Whitla.* Abundant at Woodburn waterfall; *G. C. Hyndman.* Ballycastle; *Flor. Ulst.* Rocks at Cushendall; *B.N.F.C.*, 1866. Carrick-a-Rede; *Miss Reid.* Steep cliffs at Sallagh Braes, and rocks at Redbay; S.A.S. Cushendun and Caramurphy glens; R.Ll.P. (Wolfhill; *Flor. Belf.* and *Flor. Ulst.*—plant not wild).

Derry—Benbradagh, and braes near Carntogher; *Templeton*. Not rare along the basaltic range from near Umbra to Clontygearagh; D.M.

2. E. hirsutum *Linn.* GREAT WILLOW HERB.
Marshes and borders of streams and ditches—common. Fl. mid. July till mid. Aug.

3. E. parviflorum *Schreb.*
Marshy ground, and by ditches and streams—common. Fl. mid. June till end of Aug.

4. E. montanum *Linn.*
Roadsides, and shaded banks—common. Fl. mid. June till mid. Aug.
Var. γ E. HUMILE.
Antrim—Dripping bank on Black Mountain near Belfast; D.M., 1846 (C.C.B. in *herb.*).

5. E. obscurum *Schreb.*

Roadsides, walls, and dry banks—common. Fl. mid. June till end of Aug.

6. E. palustre *Linn.*

Marshes, and wet peat bogs—frequent. Fl. July and Aug.

Down—Cotton moss, and Ballygowan moss; *Flor. Belf.* Wolf Island bog, Clandeboye demesne, and bogs near Saintfield; S.A.S. Wet rocks near Orlock, and on the road to Groomsport; T.H.C.

Antrim—Rathlin Island, and peat bogs near Bushmills; S.A.S.

Derry—Fens in the old channel of the Roe; *Samp's. Surv.* Frequent in boggy places; D.M.

[E. ROSEUM *Linn.* We don't accept this as a native plant, though we have the following references for it. Mr. Templeton noted it as in his orchard in 1820. Dr. Robt. Templeton, as abundant by the towing path between the first and second locks of the Lagan canal. Mr. F. Whitla, in a MS. note, says there is every reason to suppose it to be very common. *Flor. Belf.* gives it as occurring in glens in Castlereagh Hill, and *Flor. Ulst.* as in Holywood Glen, fields near Ulster Railway terminus, and waste ground near Belfast Workhouse, the two last localities being practically one. We have failed to find the plant, either by the canal, or at Malone, and fear a wrong identification in those instances. The records for Castlereagh, and Holywood are certainly errors of determination; so also Mr. Whitla's note. That it was found in the locality near the railway station seems certain, there being a specimen, in *herb. Babington*, marked, Belfast, 1846; W. Thompson. This we are confident was Mr. Orr's plant (*Flor. Ulst.*). It is not now to be found either in the waste ground or fields as stated, and must, in this case, be reduced to the rank of a casual only.]

CIRCÆA *Tourn.* ENCHANTER'S NIGHTSHADE.

1. C. lutetiana *Linn.*

Woods, and shady banks—common. Fl. mid. June till mid. Sept.

2. C. alpina *Linn.*

Shady and rocky places on the hills, and in glens—rare. Fl. mid. July till mid. Sept.

Down—Holywood, and Clandeboye—sparingly; R.Ll.P.

Antrim—In a stone ditch on the roadside about halfway to Hannahstown, and also in Glenariffe; *Templeton*, 1804. Mountains about Glenarm; *Ir. Flor.* Colin Glen; *Flor. Hib.* Cave Hill; *Flor. Ulst.* Cave Hill above the Whitewell quarries, and in Redhall Glen, and Glendun; S.A.S.

Derry—Muff Glen, Faughanvale, and Desertmartin; D.M.

HALORAGACEÆ.

The Cliftonville plant of *Flor. Ulst.* was a small form of sp. 1, and it is most probable that the Shane's Castle record is also erroneous.

Var. β C. INTERMEDIA *Ehr.*

Antrim and Derry—Under bushes on the east side of the Cave Hill; *J. Ball.* In subalpine woody glens in the counties of Antrim and Derry; *Cyb. Hib.* Shady places by the lower end of the path to the caves on the Cave Hill; S.A.S. (! C.C.B.).

ORDER XXIII. **HALORAGACEÆ**.

MYRIOPHYLLUM *Linn.* WATER MILFOIL.

1. M. verticillatum *Linn.*

Ditches—very rare, only one station known. Fl. June and July (*Cyb. Hib.*).

Antrim—Bog drains at Portmore; *F. Whitla.* Drains near Lough Beg (Portmore lough); *Cyb. Hib.* Sparingly in a muddy drain on the south side of Lough Beg (Portmore); S.A.S.

2. M. spicatum *Linn.*

Lakes, dams, and rivers—frequent. Fl. mid. June till end of Aug.

Down—Drains by Saintfield lakes, and in the Quoile at Downpatrick; S.A.S. Ditches at Connswater and Victoria Park; T.H.C.
Antrim—By Lough Neagh at Portmore, dam at Whitehouse, and lake on Fairhead; S.A.S.

No doubt occurs in Derry, but has not there been distinguished from the next sp. The *M. spicatum* of Flor. Belf. includes this, and the following.

3. M. alterniflorum *D. C.*

Lakes, rivers, and ditches—common. Fl. mid. June till end of Aug.

HIPPURIS *Linn.* MARESTAIL.

1. H. vulgaris *Linn.*

Wet marshy places, and stagnant ditches—frequent. Fl. June and July.

Down—Holywood bog, and Ballyalloley marsh; *Flor. Belf.* Marshes at Victoria Park, Rockport, and between Comber and Newtownards, also at Monlough, and by the canal near Hills-

borough; S.A.S. Warrenpoint; *Rev. C. H. Waddell.*
Downpatrick; *C. Dickson.* Tullycairne and Drumkee bogs;
T.H.C. Conlig, R.Ll.P.
Antrim—In many ditches; *Templeton*, 1797. King's moss; *Flor. Belf.*
Bog meadows, Ballyeaston, Castle Dobbs, and Whitehead;
Flor. Ulst. Shore of Lough Neagh at Selshan, also at Duneane,
and Carrickfergus; S.A.S.
Derry—By Enagh Lough, also at Magilligan, and by the Bann above
Kilrea, not infrequent in the county; D.M.

ORDER XXIV. **CRASSULACEÆ.**

SEDUM *Linn.* STONECROP.

1. S. rhodiola D.C. (RHODIOLA ROSEA *Linn.*). ROSEROOT.
Damp shady cliffs—rare. Fl. June and July.
Down—On black rocks (Silurian grit) of Shanslieve, and Slieve-na-Glough
in the Mourne range; R.Ll.P.
Antrim—South end of Slemish, and north of Rathlin; *Templeton.* Still
plentiful on steep cliffs at the northwest of the island; S.A.S.
1882. Tor Head, and Fairhead; *Cyb. Hib.* Grayman's path
Fairhead; *B.N.F.C.*, 1870. Glenariffe; R.Ll.P.
First found in Ireland by Mr. Templeton. Cave Hill station of *Cyb.
Hib.*—erroneous, the plant was never there in recent times.

2. S. telephium *Linn.* ORPINE LIVE-LONG.
Ditches, and wet ground—rare, and probably only as a garden escape.
Fl. July and Aug. (*Cyb. Hib.*).
Down—Hedges between Newry and Loughbrickland; *Ir. Flor.* Ballyholme, and Orlock; *Flor. Belf.* Near Ballynahinch, and
Mealough Hill; *Flor. Belf.* Roadside ditch near Stormount
Castle, also near Comber rail. station, and strand between
Newtownards and Mountstewart; S.A.S. Roadside near
Castlereagh; T.H.C.
Antrim—Roadside from Colin Glen to Hannahstown; *Flor. Belf.* Old
quarries in parish of Duneane; *Cyb. Hib.* Galgorm; *R. Tate.*
Shane's Castle woods; T.H.C.
Not sufficiently investigated, but our plant is probably Var. S. FABARIA
Koch.

3. S. anglicum *Hudson.*
Flat rocky surfaces—common by the coast, rare inland, but occurs at

Saintfield, and by the shores of Lough Neagh. Not found by S.A.S. in the Sperrin Mountains. Fl. mid. June till mid. Aug. Rises to 1200 feet on Benevenagh; D.M.

4. S. acre *Linn.* WALL PEPPER.
Dry crumbling rocks, walls, roofs, and sandy wastes—common, but less general than No. 3. Fl. mid. June till end of July.

5. S. reflexum *Linn.*
Walls, roofs, and dry banks—naturalised. Fl. July and Aug.
Down—Old walls at Greyabbey; *Ir. Flor.* Stone fence on old road from Newtownbreda to Castlereagh; S.A.S.
Antrim—Houses and walls at Antrim, and on the shore at Shane's Castle; *Templeton*, 1810. Between Belfast and Carrickfergus; *Ir. Flor.* Carrickfergus; *F. Whitla.* Ballygomartin; *Flor. Belf.* Bellahill; *Flor. Ulst.* Portmore ruins; S.A.S. Railway bank south of Antrim; *D. Redmond.*
Derry—In abundance on the rocks by the Roe about two miles above Limavady; *Templeton*, 1813. On the Churchyard wall at Muff, and abundant by the Roe at O'Cahan's rocks; D.M.

6. S. rupestre *Huds.*
Very rare, and only naturalised.
Antrim—Rocks at Shane's Castle; *Flor. Ulst.*
Derry—Gravelly banks in a field by the southeast side of Lough Foyle near to entrance of the old canal at Ballykelly—plants very small; D.M.

COTYLEDON *Linn.*

1. C. umbilicus *Linn.* NAVELWORT.
Walls, and clefts of dry rocks—common, and especially abundant on the grits and slates of Co. Down. Fl. mid. June till mid. Aug. Rises to 1000 feet at Rostrevor.

ORDER XXV. **SAXIFRAGACEÆ.**

SAXIFRAGA *Linn.* SAXIFRAGE.

1. S. stellaris *Linn.*
Wet rocks on lofty mountains—rare, seems to avoid basalt, and lime-

stone rocks. Fl. mid. June till mid. Aug. Ranges from 600 to over 2000 feet altitude.

Down—Poolagaragh south of Tollymore, 1793, and common on the Mourne Mountains; *Templeton.* 1799. Shady cliffs on Slieve-na-Gloch, and many more of the Mourne Mountains; *Wade Rar.* Still frequent on the Mourne range at elevations above 1000 feet.

Derry—Rivulets on the sides of Sawel, and abundant on the top of Dart, also on Clontygeragh ; D.M.

2. S. hirculus *Linn.*

Moory peat bogs—very rare. Fl. July and early half of Aug.

Antrim—In considerable abundance in an elevated moor near Dunloy, parish of Rasharkin, a little west of some low rocks called "Cohinnen" and "Lough Rocks"; D.M. in *Cyb. Hib.* (*spec. in herb. Babington dated* 1841). Rediscovered, by Rev. S. A. Brenan, growing sparingly on swampy ground near Lough Naroon, townland of Glenbuck, Rasharkin, July 31st, 1884. *Jour. Bot. vol.* 22, 1884. Mr. Brenan's loc. is very near, but not precisely the original station. Boggy, and elevated plateau behind Garron Point; R.Ll.P., July 8th, 1884. *Jour. Bot. vol.* 22, 1884. Mr. Praeger describes the plant as plentiful here. The locality is about 14 miles east of the original *habitat*.

3. S. aizoides *Linn.*

Damp shady rocks—very rare. Fl. July and Aug.

[Down—"Rocks by the cataract at Donard Lodge"; *Flor. Ulst.*]

This is a plant not readily overlooked, and usually abundant where it occurs. It was not noticed by Mr. Templeton, or any of his contemporaries, nor yet by any recent botanist, and we conclude that it was not native on Donard, but existed for a brief time as an introduced, ornamental plant. This conclusion is fortified by the mineralogical instincts of this species, as displayed in its Irish distribution. It seems to be confined, in Ireland, almost, if not altogether, to limestone and its near ally (for phytological effects) trap rock, and therefore not likely to be found on grit.

Antrim—Murlough; *Miss Hincks* in *Flor. Ulst.* Plentifully at Murlough Bay near Fairhead; D.M. in *Flor. Ulst.* In considerable quantity in woody places amongst rocks by streams, and on wet rocks by several streamlets on the north side of Tor Head ; T.H.C.

4. S. hypnoides *Linn.* Mossy Saxifrage.

Ledges of rocks, and more especially abundant on debris and under-cliff—frequent, but confined to limestone and basalt, not found on silicious rocks. Fl. May and June, in shaded places later. Ranges from 200 to 1300 feet.

Antrim—Cave Hill, and Sallagh Braes; *Templeton*, 1797. Rocks about Giants' Causeway; *Wade Rar*. Deerpark Glenarm, and mountain pastures between Tor Point and Ballycastle; *Ir. Flor.* Basaltic rocks of Knockagh; *Flor. Belf.* Murlough, and Rathlin Island; *Flor. Ulst.* Black Mountain, and Woodburn Glen; S.A.S. Garron Head; R.Ll.P.

Var. S. QUINQUEFIDA *Haworth*. (S. SPONHEMICA *Gmel.*, S. AFFINIS *Don*).

Apparently rare, but distribution not fully known; habitat similar to the preceding.

Antrim—On a moist rock by the Clough road near Cushendall (*S. platypetala*); *Templeton*. Cave Hill; *J. Ball*, 1837. Garron Head; C.C.B., 1852.

Derry—With the type, occasionally; D.M., 1835. Benevenagh; T.H.C., 1878. Plentiful by the base of Benevenagh cliffs; S.A.S.

5. S. tridactylites *Linn.*

Walls, dry rocks, and coast sandhills—rare. Fl. mid. April till mid. June.

Antrim—Common on old walls about Lisburn; *Templeton*, 1800. Still found sparingly on these old walls. On a bridge northeast side of Lisburn; *G. C. Hyndman*, 1827.

Derry—Frequent on sandy warrens from the Foyle to Portstewart; D.M. Basaltic rocks at Downhill; *Flor. Hib.*

6. S. granulata *Linn.*

Under trees, and on dry banks—rare. Fl. May and June.

Down—Abundant in Belvoir Park; *B.N.F.C.*, 1870; Castlereagh Hill; S.A.S. Doubtfully native in both of the above localities.

Antrim—About one mile south of Carnlough; *Cyb. Hib.*

Derry—Only found about Springhill near Moneymore, where it is abundant in copses; D.M.

7. S. oppositifolia *Linn.*

Rock cliffs—very rare. Fl. April and May (*Cyb. Hib.*).

Derry—"Only found very sparingly on basaltic rocks on north side of Benevenagh, at 1100 feet"; D.M., 1835, also *H. C. Hart*, 1882.

CHRYSOSPLENIUM *Linn.* GOLDEN SAXIFRAGE.

1. C. oppositifolium *Linn.*

By streams, and on wet dripping rocks—abundant, ascending to 2000 feet on Dart Mountain. Fl. mid. March till end of May.

PARNASSIA *Linn.* GRASS OF PARNASSUS.

1. P. palustris *Linn.*
Wet sandy, or rocky banks—absent or rare in the south of the district, but becoming frequent in the north. Fl. July and Aug.
Antrim—North end of Islandmagee, west of Ballintoy, and moist places about Ballycastle; *Templeton*, 1794. Woodburn Glen; *Flor. Belf.* (not found there now). Brown's Bay in Islandmagee; *B.N.F.C.*, 1869. Rathlin Island; *Flor. Ulst.* Carrick-a-Rede; S.A.S. Tor Head, and Murlough Bay; T.H.C. Glenariffe; R.Ll.P.
Derry—Moist bottoms of Magilligan and Dunboe; *Samp's. Surv.*, 1814, also D.M., 1835, and *B.N.F.C.*, 1863. Desertcreat; *Rev. W. T. Whan.*

ORDER XXVI. **UMBELLIFERÆ.**

HYDROCOTYLE *Linn.* PENNYWORT.

1. H. vulgaris *Linn.* WHITEROT.
Marshes, and wet pastures—common. Fl. June and July.

SANICULA *Linn.* SANICLE.

1. S. europæa *Linn.* WOOD SANICLE.
Woods, and damp bushy places—common. Fl. mid. May till mid. July.

ERYNGIUM *Linn.* ERYNGO.

1. E. maritimum *Linn.* SEA HOLLY.
Sandy seashores—rare. Fl. mid. July till mid. Sept.
Down—Sandy shore at Kirkiston, and shore opposite Portavo; *Templeton*, 1793. Between Greencastle and Kilkeel; *Ir. Flor.* A single plant at Ballyholme Bay; *Flor. Belf.* Sandhills at Dundrum and Newcastle; *Flor. Ulst.* Benderg Bay near Ardglass, and abundant on sandhills at Ringsallin Point; T.H.C. Greencastle, and Ballykinler, also shore opposite Gunn's Island, and abundant on the sandy coast north of Cloughey in the Ards; S.A.S.
Antrim—Sands at Bush waterfoot; *Templeton.* Ballycastle; *Flor. Ulst.*

UMBELLIFERÆ.

Redbay; *B.N.F.C.*, 1866. Cushendall; S.A.S. Cushendun; R.Ll.P.

Derry—Sandy warrens between the Black rock and Portrush; D.M.

CICUTA *Linn.* WATER HEMLOCK.

1. C. virosa *Linn.* COWBANE.

Ditches, and very wet marshes—rare. Fl. mid. June till mid. Aug.

Down—Legacurry northeast of Hillsborough, in the mill dam, and down the river; *Templeton.* Drumkee bog Tullycairne; T.H.C.

Antrim—In the Lagan above Lisburn, and drains at Portmore; *Templeton*, 1794. Portmore, *B.N.F.C.*, 1872. Rathlin Island; *Flor. Ulst.* (Not found by S.A.S. in 1882, and not likely to occur in the rocky swamps of Rathlin.) Abundant near Lough Neagh, and Lough Beg, and occasionally by the Bann to Coleraine; *Cyb. Hib.*

Derry—By the west bank of the Bann occasionally; D.M.

APIUM *Linn.*

1. A. graveolens *Linn.* WILD CELERY.

Salt marshes—rare. Fl. mid. June till end of July.

Down—About the ropewalk at upper end of Belfast Bay; *Templeton*, 1793. Shore of Strangford Lough at Cherryvalley, and shore at Groomsport; *Flor. Belf.* Shores of Dundrum Bay; *Flor. Ulst.* Killough Bay; S.A.S.

Antrim—North of Carrickfergus at Eden; *Flor. Belf.* East of Larne by the Lough shore; *Flor. Ulst.*

Derry—At Magilligan, and by the Foyle at Brookhall; D.M.

2. A. nodiflorum (*Linn.*) *Reich.* (HELOSCIADUM *Koch*).

Drains, and stagnant ditches—common. Fl. early June—Aug.

Var β H. REPENS *Linn.*

Down—In marshy places by the riverside at Kilkeel; *Wade Rar.*

Antrim—On the south side near the mouth of the large drain at Portmore; *Templeton.* This was, most probably, the form noted by Dr. Moore, and referred to *H. Moorei.*

3. A. inundatum *Reich.* (HELOSCIADUM *Koch*).

Ditches, pools, and sluggish streams—frequent. Fl. mid. June till mid. Sept.

Down—Lough Leagh, and ditch by second lock of the Lagan canal; *Templeton*, 1797. Holywood, Ballyalloley, and near Bangor; *Flor. Belf.* Greyabbey, and Slieve Croob; *Flor. Ulst.* Dro-

more, Downpatrick, Newcastle, Saintfield, Monlough, and
Moira; S.A.S. Conlig, and Dundonald; R.Ll.P.
Antrim—Loughmourne; *Flor. Belf.* Bog.meadows, and Rathlin; *Flor.
Ulst.* By the canal, occasionally, from Lisburn to Lough
Neagh, also near Toome; S.A.S. Shane's Castle; R.Ll.P.
Derry—Near the race course Derry, by Enagh Lough, and by Lough
Neagh; D.M. Marshy shore of Lough Beg; T.H.C. Benevenagh; S.A.S.

Var. MOOREI *Syme.*

Down—Amongst grass on the marshy margin of a rill that flows into the
Quoile between Inch Abbey and the wood below it, also very
fine under railway bridge near the same place; S.A.S.
Antrim—In drains near Portmore; D.M. Drains at Selshan; S.A.S.

ÆGOPODIUM *Linn.* GOUTWEED.

1. Æ. podagraria *Linn.* FARMERS' PLAGUE.
Hedge banks, and dry shady places—abundant. Fl. June and July.
May have been introduced at a distant date, but now among the most
conspicuous wild plants of the district.

CARUM *Linn.*

1. C. verticillatum (*Linn.*) *Koch.*
Wet pastures—very rare. Fl. July and Aug.
Antrim—In the rushy bog near Orange Grove (Cranmore); *Templeton.*
By the Lagan near Belfast; *Cyb. Hib.* Crumlin waterfoot;
B.N.F.C. (Leaves only, and may not be right).
Derry—Bannside below Coleraine; *Templeton,* 1797. Abundant by the
Bann, both above and below the bridge; D.M., 1835.
Plentiful, on both sides of the river, below the bridge at
Coleraine; S.A.S., 1885.

The County Down station of *Flor. Belf.* is probably an error, and the
Bog meadows station of *Flor. Ulst.* is not mentioned in the Templeton
MS. The plant does not now grow near Belfast, and as the Crumlin
station is doubtful it is probable that Coleraine is the only locality which
can at present be quoted with certainty.

BUNIUM *Linn.*

1. B. flexuosum *With.* PIG NUT.
Woods and bushy places, and in short pastures—frequent. Fl. mid.
May till mid. July.

UMBELLIFERÆ.

PIMPINELLA *Linn.*

1. P. saxifraga *Linn.* BURNET SAXIFRAGE.

Rocky pastures on the hills, and by sandy seashores, and dry roadside banks—frequent. Fl. July—Sept.

There are two, perhaps three, distinct looking forms of this species. The neat elegant plant of rocky pastures is very different from the large (2-3 feet) coarse form of the sandhills and roadsides. This latter has more the aspect of P. MAGNA *Linn.* and some luxuriant examples correspond to P. DISSECTA *Retz.*, which Prof. Babington places under that species.

Down—Newcastle, and Portaferry; *Flor. Ulst.* Frequent about Ballylesson and Drumbo, also abundant by the coast from Strangford to Newcastle, and on Mourne Mountains; S.A.S. A very large form on roadsides near Clough, and on sandhills at Ballykinler; S.A.S. Roadsides near Downpatrick, and Ballynahinch (Var. DISSECTIFOLIA *Wallr.*, HIRCINA *Lees.*); S.A.S.

Antrim—Rocks of the Cave Hill; *Templeton*, 1797. Slopes of the Belfast Hills; *Flor. Belf.* Knockagh; *Flor. Ulst.* Hedges by the Lagan near Belfast; T.H.C. Lisburn, Castlerobin, and Carnmoney Hill; S.A.S.

Derry—Abundant about Kilrea, and Castledawson; D.M. Bellaghy; S.A.S.

SIUM *Linn.* WATER PARSNEP.

1. S. latifolium *Linn.*

Marshes and ditches—rare. Fl. July and Aug.

[Down—The Saintfield record of *Flor. Belf.* was probably erroneous; young leaves, only, were found.]

Antrim—In the large drain at Portmore Park; *Templeton*, 1794. By Lough Neagh near Ballinderry; *G. C. Hyndman.* West end of Glenarm Park; *Flor. Ulst.* Marsh at Selshan; *Cyb. Hib.*

Derry—Marsh at Culmore (Donegal strictly), and between Portstewart and the Bann, also by shore of Lough Neagh; D.M.

2. S. erectum *Huds.* (S. ANGUSTIFOLIUM *Linn.*).

Ditches and marshes—rare. Fl. July—Sept.

Down—Several places in Ballynahinch river; *Templeton*, 1813 (with a ?). Roadside near Downpatrick; *Flor. Ulst.* Ditches by the Strand Lake Killough; T.H.C.

Antrim—Ditches by Lough Neagh; *Cyb. Hib.* Ditch by roadside above Millbay in Islandmagee; S.A.S. Specimens with stems six feet long were seen in this place.

Derry—By Lough Neagh; *Cyb. Hib.* Marshy margins of the Bann at Coleraine; S.A.S.

UMBELLIFERÆ.

ŒNANTHE *Linn.* WATER DROPWORT.

1. Œ. fistulosa *Linn.*
Marshes, ditches, and wet meadows—rare. Fl. July and Aug.
Down—Abundant in several places by the Newry canal from Newry
 northwards; S.A.S.
Antrim—Portmore wood; *Templeton*, 1798. Between Lisburn and Lough
 Neagh; *G. C. Hyndman*, 1830. The Carrickfergus and Rath-
 lin localities of *Flor. Ulst.* are most likely errors.
Derry—Frequent by the Bann, and in wet meadows near Bellarena and
 other places; D.M. Abundant on the marshy margin of the
 Bann at Coleraine; S.A.S.

2. Œ. Lachenalii *Gmel.*
Salt marshes—rare. Fl. July and Aug.
Down—Holywood warren; *Templeton*, 1810. Victoria Park, Kinnegar
 at Holywood, also Bangor, and Groomsport; *Flor. Belf.*
 Plentiful by the shore for two or three miles on each side of
 Ardglass, and at upper end of the bay at Clough, also at
 Sydenham sparingly; S.A.S.
Antrim—By the Lagan canal below the first lock; *Templeton*, and *Flor.
 Belf.*
Derry—Mouth of the Roe, and abundant by the Bannfoot; D.M.

3. Œ. crocata *Linn.*
Ditches, and banks of rivers and streams—common. Fl. mid. June
till end of Aug.

4. Œ. phellandrium *Lamk.* HORSEBANE.
Muddy ditches—frequent. Fl. mid. June till mid. Aug.
Down—Newry canal, and by the Lagan above and below Moira; S.A.S.
 Saul dam near Downpatrick; *C. Dickson.*
Antrim—By the Lagan canal near Belfast; *Templeton*, 1797 (still there).
 Portmore; *B.N.F.C.*, 1872. Selshan bogs, and by the river
 Bann; S.A.S. Crumlin waterfoot; T.H.C.
Derry—Frequent by the Bann, and Lough Beg; D.M. Abundant by
 the Bann near Portglenone; S.A.S.

ÆTHUSA *Linn.*

1. Æ. cynapium *Linn.* FOOL'S PARSLEY.
Fields, and waste ground in sandy soil—not common. Fl. July and
Aug.
Down—Holywood, and Sydenham; *Flor. Belf.* Cultra, Stormount, and
 Knocknagoney; S.A.S. Clandeboye; R.Ll.P.

UMBELLIFERÆ. 65

Antrim—Malone, and Hannahstown; *Flor. Belf.* Carrickfergus, and Ballycastle; *Flor. Ulst.* Lisburn, and Ballinderry; S.A.S. Dunadry, Cushendall, and Glenariffe; R.Ll.P.
Derry—Magilligan, and Garvagh; D.M. Bellaghy; S.A.S.

FŒNICULUM *Adanson.* FENNEL.

1. F. officinale (*Linn.*) *All.*
Rocky, and gravelly seashores—very rare. Fl. July and early Aug.
Down—Damp rocky beaches on both sides of Killough Bay, away from houses, and having quite the appearance of a native; S.A.S.
Antrim—Curran of Larne (introduced); D.M. Not now found on the Curran.
The plants found at Ballyholme (*Flor. Belf.*) were escapes from cultivation.

LIGUSTICUM *Linn.* LOVAGE.

1. L. scoticum *Linn.* SCOTTISH LOVAGE.
Rocky seashores—very rare. Fl. July and Aug. (*Cyb. Hib.*).
Down—On the rocks about Donaghadee, and the Copeland Isles; *Templeton*, 1793. Mew Island off Donaghadee; T.H.C.
Antrim—On the back of a ditch in Ballinleg south of Ballycastle; *Ir. Flor.* Garron Head; *Cyb. Hib.*
Derry—Between Portrush and Portstewart in several places where the sea spray washes over the rocky coast; D.M.
First found in Ireland by Mr. Templeton.
The Carrickfergus plant of *Flor. Belf.* was a state of *Smyrnium*.

SILAUS *Besser.*

1. S. pratensis (*Linn.*) *Besser.* MEADOW SAXIFRAGE.
Gravelly banks—very rare. Fl. June—Aug. (*Cyb. Hib.*).
Derry—On a gravelly field by the side of the Foyle, in the townland of Tully above Londonderry; D.M., 1835.
Not known elsewhere in Ireland, and may have been only a casual. The county Antrim records are believed to be erroneous.

ANGELICA *Linn.*

1. A. sylvestris *Linn.* WILD ANGELICA.
Ditches, sides of streams, and in marshy pastures—common, but seldom abundant. Fl. July and Aug.

E

UMBELLIFERÆ.

HERACLEUM *Linn.*

1. H. sphondylium *Linn.* HOGWEED.
Woods, and damp shady laces—very common; occurs also on rock cliffs at elevations up to 1200 et. Fl. mid. June till mid. Sept.

DAUCUS *Linn.*

1. D. carota *Linn.* WILD CARROT.
Ditch banks, and waste ground—very common. Fl. July—Sept.

TORILIS *Adanson.*

1. T. anthriscus (*Hudson*) *Gaert.* HEDGE PARSLEY.
Hedges, banks, and waste places—common. Fl. July and Aug.

2. T. nodosa (*Hudson*) *Gaert.*
Dry, stony or gravelly banks—frequent on some parts of the Down coast—rare elsewhere. Fl. mid. June till mid. Aug.

Down—Frequent on exposed roadside stony hedge-banks, in dry places, from Mill-quarter Bay north of Ardglass, to Clough on the southwest, a line of about 20 miles, being specially plentiful about Killough, and from Ringboy to Rathmullen, also on waste ground at Cloughey in the Ards, and occasionally to near Ballyhalbert ; S.A.S.

Antrim—Gravelly banks on the Curran of Larne; *Templeto*n, 1797. Carnlough ; *Cyb. Hib.* Dunseverick; *R. Tate.* Roadside near Portbradden, and Giant's Causeway ; S.A.S.

Derry—Ballyronan ; *Flor. Ulst.*

SCANDIX *Linn.* SHEPHERD'S NEEDLE.

1. S. pecten-veneris *Linn.* VENUS'S COMB.
Borders of cultivated fields on light sandy soil—widely distributed, but not common. Fl. June—Aug.

Down—Sandy fields at Ballynafeigh, and frequent in Lecale ; *Templeton.* Comber, Stormount ; *Flor. Belf.* Holywood ; *Flor. Ulst.* Belmont, Groomsport, Ballywalter, Ballyhalbert, Mountstewart, Ballynahinch, and frequent in fields by the coast from Strangford to Dundrum ; S.A.S. Orlock Point; T.H.C. Ballygowan, and Dundonald ; R.Ll.P.

Antrim—Malone, and Derriaghy; *Templeton*, 1797. Glenarm; *F. Whitla.* Limestone quarry near Lisburn ; *G. C. Hyndman.* Antrim ; *Flor. Ulst.* Portmore, and Ballintoy; S.A.S. Islandmagee ; T.H.C. Cushendall ; R.Ll.P.

Derry—Sides of the Foyle, also lower slopes of Benevenagh, and many other places ; D.M.

CHÆROPHYLLUM *Linn.* CHERVIL

1. C. sylvestre *Linn.* (ANTHRISCUS *Hoff.*) WILD CHERVIL.
Hedge banks, and shady places—very common. Fl. mid. April till end of June.

2. C. anthriscus *Lamk.* (ANTHRISCUS VULGARIS *Pers.*).
Dry banks, and roadsides in sandy, or calcareous soil—not common. Fl. May and June.
Down—By the strand north of Mountstewart ; *B.N.F.C.*, 1864. Shore at Portaferry; *Flor. Ulst.* Roadside from Rostrevor to Greencastle, and frequent by roadsides on the Ards coast from Donaghadee to Cloughey ; S.A.S. By the warren at Donaghadee ; T.H.C. Dundrum ; *C. Dickson.*
Antrim—Gravelly banks on the Curran of Larne ; *Templeton*, 1794. Ballintoy; *Cyb. Hib.* Waste ground behind Carnlough; R.Ll.P.
Derry—Abundant at Magilligan ; D.M., 1835, and T.H.C., 1878.

3. C. temulum *Linn.* ROUGH CHERVIL.
Hedge banks—very rare. Fl. June and July (*Cyb. Hib.*).
Antrim—Ballycastle (*Miss Hincks*) ; *Flor. Ulst.* Plentiful on roadside for about 200 yards south of Derrymore national school, and on bushy banks of Lough Neagh at same place ; S.A.S.
Derry—On the shores of Lough Neagh near Ballyronan ; D.M. in *Cyb. Hib.*
The station at Malone, in *Cyb. Hib.* was an error, and the note in *Flor. Belf.* is also erroneous.

MYRRHIS *Linn.*

1. M. odorata *Linn.* SWEET CICELY.
Roadsides, and waste places ; often found near houses as an escape ; certainly naturalised in some places, but probably nowhere native. Fl. mid. May till mid. July.
Down—Hillsborough, Newtownards, and Knock ; S.A.S. Roadside between Comber and Castle Espie ; T.H.C. Dundonald; R.Ll.P.
Antrim—Orchards, and waste places, probably only a naturalised plant ; *Templeton.* Pastures at Brookmount ; *Flor. Belf.* Colin Glen, and Ballycarry; *Flor. Ulst.* About Lisburn, Duneane, Potter's Walls, and Glenarm ; S.A.S. Abundant about Cairncastle ; R.Ll.P.

Derry—Not unfrequent about houses, but apparently wild at foot of Umbra rocks; D.M. Fields at Magilligan ; T.H.C. Ballyronan, and Moneymore; *Flor. Ulst.* Roadside near Dungiven, and plentiful on road above Bellarena station; S.A.S. By the river Roe ; *H. C. Hart.*

CONIUM Linn. HEMLOCK.

1. C. maculatum *Linn.*

Hedges, ditch banks, and waste ground—common on the light sandy soils, but rare or absent on the clay lands. Fl. mid. June till mid. Aug.

Down—General throughout the county, localities need not be specified.

Antrim—At Eden below Carrickfergus ; *Flor. Belf.* Rathlin ; *Miss Gage.* Glenavy, and of frequent occurrence on roadsides, and waste ground by the seacoast, in great abundance on Rathlin, seemingly replacing *Chærophyllum sylvestre* on that island ; S.A.S.

Derry—Abundant on waysides, and in waste places; D.M. Shores of Lough Neagh near Toome; *Flor. Ulst.* Plentiful about Dungiven, and Bellarena ; S.A.S.

SMYRNIUM *Linn.*

1. S. olusatrum *Linn.* ALEXANDERS.

Damp rocky, and gravelly places, and hedge banks—frequent, especially by the coast. Fl. May and June.

Down—About maritime villages, and rocks on the shore ; *Templeton,* 1797. Holywood, and Bangor ; *Flor. Belf.* Greyabbey, and Crawfordsburn ; *Flor. Ulst.* Donaghadee, and frequent by the Ards coast ; S.A.S.

Antrim—Lisburn, Whitehouse, and Carrickfergus ; *Flor. Belf.* Woodburn, Larne, and Rathlin ; *Flor. Ulst.* Carnmoney, Castle Dobbs, and rocky shore of Lough Neagh between Conns Point and the Three Islands, also frequent by the Antrim coast ; S.A.S.

Derry—Plentiful by the Roe at O'Cahan's Rocks, and below Limavady; D.M. Moneymore, and Ballyronan ; *Flor. Ulst.*

ORDER XXVII. **HEDERACEÆ**.

HEDERA *Linn.* IVY.

1. H. helix *Linn.*
Rocks, walls, woods, and hedges—common. Fl. mid. Oct. till end of Nov.

ORDER XXVIII. **CAPRIFOLIACEÆ.**

ADOXA *Linn.*

A. moschatellina *Linn.* MOSCHATEL.
Rocky, and shady banks—very rare. Fl. mid. March till end of April.
Antrim—" Found on the shaded banks of the Milewater River about a quarter mile above where the tide flows. First observed there by Dr. J. L. Drummond, and seen by myself in April, 1820"; *Templeton.* Glen of Jennymount (Milewater stream) ; *Drummond*, and *Whitla.* Seen growing sparingly in the above stations, which are practically one, by T.H.C. in 1878. In the southwest side of the Deerpark, Cavehill, near the quarries; J. (Canon) Grainger ; in *Flor. Belf.*
Though the plant still grows in the Belvoir Park station, of Flora of Ulster, nevertheless that may be dismissed as a suspicious locality, but there is every reason to believe it native on the Cave Hill, and in its first observed *habitat* by the Milewater stream. The Botanic Gardens were not formed until 1828, and doubtless it was from Jennymount, or the Cave Hill that the *Adoxa* was brought into them, and most probably into the grounds of Belvoir also. This plant may be looked on as one of those which are losing their hold on the ground they once occupied in greater force, and soon destined to be extinct as an Irish plant. It now hides away in one little spot at the Milewater, and under the friendly shelter of one or two fallen rocks in the Deerpark, and the first alterations, or improvements effected in these places will be its death warrant.

SAMBUCUS *Linn.* ELDER.

1. S. ebulus *Linn.* DWARF ELDER, DANE'S BLOOD.
Waste ground, dry banks, and fields—rather frequent. Fl. mid. July till end of Aug.
Down—" Plentiful a quarter mile east of Magheralin, whole fields being covered with it"; *Harris's Down*, 1744. At St. Colman's well near Magheralin ; *Rev. H. W. Lett*, 1884. Fields below Holywood ; *Templeton*, 1793. Ruins of Greyabbey ; *Ir. Flor.* Side of road near Lambeg ; *G. C. Hyndman.* Warrenpoint, and road east of Tyrella, also near Bangor; *Flor. Ulst.*

Cliffs south of Killough ; S.A.S. Near the graveyard at Inch Abbey ; *B.N.F.C.*, 1867.
Antrim—' Abundant in fields, and ditches in the lands of Ballymena "; *Ir. Flor.* Near Lisburn ; *Flor. Hib.* Antrim, Hannahstown, and Lylehill; *Flor. Ulst.* On the fort at Rathmore ; *B.N.F.C.*, 1884. Culraney, and near Cushendun ; R.Ll.P.
Derry—Near Porstewart; *Templeton.* Plentiful on roadsides in the parish of Agivey, and near Ballyronan ; D.M.

2. S. nigra *Linn.* ELDER.

Hedges, thickets, and low glens—common, and especially abundant to the north and northwest. Fl. June and July.

VIBURNUM *Linn.*

1. V. opulus *Linn.* GUELDER ROSE.

Glens, and damp bushy places—frequent, less common in hedges. Fl. June and July.
Down—Newtownards, and Dunleidy; *Ir. Flor.* Ballygowan, and Ballymaghan ; *Flor. Belf.* Newcastle ; *Flor. Ulst.* Glens at Holywood, Belmont, and Castlereagh ; S.A S. Knock Glen ; T.H.C. Rademon woods, and steep banks of river at Tullycairne ; *C. Dickson.* Dundonald Glen, and Downpatrick ; R.Ll.P.
Antrim—Steep rocks in the great Deer Park at Glenarm; *Ir. Flor.* Upper end of Colin Glen, and banks of Lough Neagh at Antrim ; *Flor. Hib.* Glens at Springfield, and Woodburn ; *Flor. Belf.* Wolfhill ; *Flor. Ulst.* Roadsides at Glenavy, Crumlin waterfoot, Andersonstown, and Carmavy, also on rocks at upper end of Crow Glen ; S.A.S. By Inver River ; R.Ll.P.
Derry—Frequent in bushy places throughout the county ; D.M. Magherafelt ; *Flor. Ulst.* Roadsides near Toome ; S.A.S.

LONICERA *Linn.*

1. L. periclymenum *Linn.* HONEYSUCKLE.

Hedges, and dry shady and bushy places—common. Fl. mid. June till end of Aug.

ORDER XXIX. **RUBIACEÆ.**

SHERARDIA *Dillenius.*

RUBIACEÆ. 71

1. S. arvensis *Linn.* FIELD MADDER.

Sandy, and gravelly fields, and waste ground—frequent. Fl. mid. May till mid. Sept.

Down—Dundonald, and Strandtown; *Flor. Belf.* Newcastle, Bryansford. Conlig, Crawfordsburn, Holywood, and Ballylesson; *Flor. Ulst.* Frequent from Strangford to Dundrum, also at Greyabbey, Donaghadee, and Hillsborough; S.A.S. Bangor; T.H.C.

Antrim—Malone, and Ballydrain; *Templeton,* 1797. Cave Hill tramway; *Flor. Belf.* Lisburn, Glenavy, Cranfield, Donegore, Carnmoney, and Ballycarry; S.A.S. Islandmagee; T.H.C.

Derry—Frequent in the barony of Loughinshollen, and other places; D.M. Ballyronan; *Flor. Ulst.* Portstewart, and Draperstown; S.A.S.

ASPERULA *Linn.*

1. A. odorata *Linn.* SWEET WOODRUFF.

Woods, and damp places amongst bushes—frequent. Fl. May—July.

GALIUM *Linn.* BEDSTRAW.

1. G. boreale *Linn.*

Damp rocky places, principally river banks and lake shores—frequent, but not general. Fl. mid. June till mid. Aug.

Antrim—Stony margin of Lough Neagh a little south of Portmore, also river at Randalstown, Glenwherry River, and Glenarm River; *Templeton,* 1793. By the lough shore at Antrim; *Flor. Hib.* Islands in Lough Neagh, and lake shore near Toome; *Flor. Ulst.* Abundant on Lurigethan Mountain; *Cyb. Hib.* Shore of Lough Neagh a little north of Lagan canal, and at Selshan, also on rocky lake shore west of Cranfield Point; S.A.S. By the river at Cushendun; *Rev. S. A. Brenan.* Carnlough River, Pollan Burn, and rocks on Runabay Head; R.Ll.P.

Derry—Abundant along the shores of Lough Neagh, and Lough Beg, also on Clontygearagh Mountain; D.M. Magilligan; *E. Murphy* in *Mag. Nat. Hist.* vol. I., 1829. Ballyronan, and Toome; *Flor. Ulst.*

2. G. cruciatum (*Linn.*) *Withering.* CROSSWORT.

Banks, and waste ground—very rare—Downpatrick only. Fl. May and June (*Bab.*).

Down—"Among the rubbish of the cathedral of Downpatrick"; *Harris's Down,* 1744. " . . . where I have observed it for the last 30 years"; *F. Whitla* (marginal note in copy of Flora

Hibernica, no date). Field adjoining the marshes, and on the side of the rath; *Prof. Hodges*, 1842. Field near the cathedral; *Rev. W. E. Mulgan*, 1868.

3. G. aparine *Linn.* GOOSE GRASS, ROBIN-RUN-THE-HEDGE (*loc.*).
Hedge banks, and waste ground—very common. Fl. mid. May till mid. Sept.

4. G. mollugo *Linn.*
Grassy lawns—very rare. Fl. July and Aug. (*Cyb. Hib.*).
Down—Abundant on the lawn at Aghaderg Glebe near Loughbrickland; *Rev. H. W. Lett.*
Antrim —In Mr. Tennant's lawn at Mount Vernon (Belfast); *Templeton*, 1797. Lawn at Glenarm Castle; *Dr. J. S. Holden, F.G.S.*, and *Flor. Ulst.*

5. G. verum *Linn.* LADY'S BEDSTRAW.
Heaths, dry banks, and coast sandhills—very common at all elevations up to 1200 feet. Fl. July—Sept.

6. G. saxatile *Linn.*
Heaths, peat bogs, and stony banks—very common from sea level to the tops of our highest mountains. Fl. June –Aug.

7. G. palustre *Linn.* MARSH BEDSTRAW.
Marshes, and ditches—very common. Fl. mid. June till mid Sept.

ORDER XXX. **VALERIANACEÆ.**

VALERIANA *Linn.* VALERIAN.

1. V. officinalis *Linn.*
Ditches of clear water, and margins of streams—frequent. Fl. mid. June till mid. Aug.

Plants like V. SAMBUCIFOLIA *Mikan* occur, but not, as yet, indentified with sufficient certainty.

VALERIANELLA *Tournefort.*

1. V. olitoria (*Linn.*) *Moench.* CORN SALAD.
Waste ground, and borders of fields in sandy or gravelly land—frequent. Fl. mid. April till end of July.

DIPSACEÆ.

Down—Sandy ditches about Knocknagoney, and about Moira; *Templeton.* Crawfordsburn, and Groomsport; *Flor. Belf.* Newcastle, Newtownards, and Holywood; *Flor. Ulst.* Old walls, and rocks at Dundrum Castle; T.H.C. Frequent by the shore of the Ards from Ballyhalbert to Kearney's Point, and from Newtownards to Mountstewart; S.A.S. Dundonald; R.Ll.P.
Antrim—Woodburn; *Flor. Belf.* Knockagh, and Whitehead, also park and cliffs at Glenarm; *Flor. Ulst.* Shane's Castle; *B.N.F.C.*, 1863. Blackhead, Islandmagee, Sallagh Braes, and abundant on the sands at Portrush and Bushfoot; S.A.S.
Derry—Common, especially along the sandy seacoast, and by the shores of Lough Neagh; D.M. Shores of the lough from Ballyronan to Toome; *Flor. Ulst.*

2. V. dentata *Willd.*
Cultivated ground, and borders of sandy fields—rare. Fl. June and July.
Down—Bangor, and Groomsport; *Flor. Belf.* Strangford; T.H.C.
Antrim—Sandy places on the Curran of Larne; *Templeton*, 1794. Carnlough; *Prof. R. Tate.* Rathlin Island; *Cyb. Hib.* Fields at north end of Islandmagee: S.A.S.
Derry—By the side of Lough Foyle; *Cyb. Hib.*

ORDER XXXI. **DIPSACEÆ.**

DIPSACUS *Tournefort.*

1. D. sylvestris *Hudson.* WILD TEASEL.
Gravelly waste ground—very rare. Fl. July and Aug.
Down—Abundant on a bank by the shore near the coastguard station at Greencastle; S.A.S.
The teasel is often met with as an escape, but in the above station it seems really native.

SCABIOSA *Linn.* SCABIOUS.

1. S. succisa *Linn.* DEVIL'S BIT.
Grassy heaths, upland pastures, and ditch banks—very common. Fl. mid. July till mid. Sept.

2. S. (KNAUTIA) arvensis *Linn.* FIELD SCABIOUS.
Dry banks, and sandy or gravelly fields—local, but in some cases

abundant. Fl. July and Aug.

Down—Holywood road near Knocknagoney ; *Templeton*, 1797. Marino; *Flor. Belf.* Abundant between Downpatrick and Strangford ; *Flor. Ulst.* In profusion in fields by the coast from Strangford to Dundrum, also at Rostrevor, Castlewellan, Groomsport, Newtownards, and Dundonald ; S.A.S. Conlig ; T.H.C. Carrowreagh ; R.Ll.P.

Antrim—Banks by Lagan canal, and at Carrickfergus ; *Flor. Belf.* Magheramorne, Curran of Larne, and Rathlin ; *Flor. Ulst.* Abundant on railway banks from Doagh to Cookstown Junction, and at Glenavy ; S.A.S. At Balmoral, and by Crumlin River ; T.H.C.

Derry—Abundant about Moneymore ; D.M. Ballyronan ; *Flor. Ulst.*

Order XXXII. COMPOSITÆ.

EUPATORIUM *Linn.*

1. E. cannabinum *Linn.* HEMP AGRIMONY.

Wet sandy, or rocky places—rather rare. Fl. July and Aug.

Down—Hollow places in rocks west of Ardglass, and ditch near Lambeg; *Templeton*, 1793. Damp shady rocks south of Killough ; T.H.C. Rocks near an old sea cave south of Newcastle ; *C. Dickson.*

Antrim—Plentiful on shores of Lough Neagh, especially near Cranfield ; *Templeton.* Rathlin Island ; *Miss Gage.* Shane's Castle, and near Toome ; *Flor. Ulst.* Portmore, and Knocklayd ; S.A.S. Near Redbay ; *B.N.F.C.*, 1875. Glenarm Park, and Cushendall ; R.Ll.P.

Derry—Not general through the county, but plentiful in Magilligan, and Bennedy Glen ; D.M.

PETASITES *Tournefort.*

1. P. vulgaris (*Linn.*) *Desf.* BUTTER BUR.

Wet gravelly places by streams—common. Fl. latter end of Feb. till mid. May.

TUSSILAGO *Tournefort.*

1. T. farfara *Linn.* COLTSFOOT.

Wet clay fields and banks—very common, especially abundant on the stiff Triassic marls. Fl. mid. Feby. till mid. May.

COMPOSITÆ.

ASTER *Linn.* STARWORT.

1. A. tripoleum *Linn.*
Muddy salt marshes, and wet maritime rocks—common around the coast. Fl. mid. July till mid. Sept.

ERIGERON *Linn.*

1. E. acris *Linn.* BLUE FLEABANE.
Coast sandhills—very rare. Fl. July and Aug. (*Cyb. Hib.*).
Down—Ballykinler near Dundrum—one small patch on the sand dunes; R.Ll.P.
The station on Cave Hill, reported in Flor. Belf., was no doubt an error, the plant is certainly not there. *P. dysenterica* was probably the plant intended.

BELLIS *Linn.* DAISY.

1. B. perennis *Linn.*
Pastures, banks, and heaths—very common everywhere, and at all elevations up to 1500 feet. Fl. Feb. till Nov.

SOLIDAGO *Linn.* GOLDEN ROD.

1. S. virgaurea *Linn.*
Ledges of rocks—widely distributed, but not very common. Fl. latter end of July till mid. Sept. Occurs at elevations of 300 to close on 2000 feet.
Down—Plentiful in the mountains of Mourne; *Harris's Down*, 1744. Tollymore Park, and rocks at Blue Lake in Mourne Mountains; *Flor. Ulst.* Slieve Donard; S.A.S. By the Shimna, and Kilkeel rivers, also on Slieve Bernagh, and Bignian mountains; *C. Dickson.* Rostrevor, and Kilbroney; R.Ll.P.
Antrim—Rocks of Slemish; *Templeton*, 1793. Knockagh; *Flor. Ulst.* Glenariffe; *B.N.F.C.*, 1875. Sallagh Braes, and Garron Point; S.A.S. Carnlough Glen; R.Ll.P.
Derry—Slieve Gallion rocks; *Templeton.* Common on the north side of the basaltic mountains; D.M. Benbradagh, banks of the Roe, and in Lignapeiste Glen; S.A.S. Sawel; *H. C. Hart.*
  Var. γ S. CAMBRICA *Hudson.*
Down—Banks of the river in Tollymore Park; *Wade Rar.*
This may be correct, but we have not seen any specimens that could be considered as identical with Hudson's plant, and most likely Wade's plant was merely a dwarf form of the type.

COMPOSITÆ.

INULA *Linn.*

1. I. helenium *Linn.* ELECAMPANE.

Hedge banks, and bushy places—rare, most probably not native, but certainly quite naturalised. Fl. July and Aug.

Antrim—Banks of a stream two to three miles south of Glenarm; *Templeton*, 1793. Upper Ballysillan west of Belfast; *F. Whitla.* Colin Mountain, Whiterock. Whitehead, and Rathlin; *Flor. Ulst.* Apparently wild on limestone rocks between Larne and Glenarm; *Cyb. Hib.* Braid Valley below Skerry; *B.N.F.C.*, 1871. Glendun; T.H.C. Colin Glen, sparingly, but plentiful by shore of Lough Neagh at Cranfield; S.A.S.

Derry—Naturalised in several places; D.M.

Occurs in many other places as a garden escape.

PULICARIA *Gaert.* FLEABANE.

1. P. dysenterica *(Linn.) Gaert.*

Damp places by roadsides, and in waste ground—common. Fl. Aug. and Sept.

FILAGO *Linn.* CUDWEED.

1. F. germanica *(Huds.) Linn.*

Dry banks, and sandy fields—not common. Fl. July and Aug.

Down—Victoria Park, Ballyhackamore, and Cherryvalley near Comber; *Flor. Belf.* Newcastle sandhills, and on the Kinnegar at Holywood; *Flor. Ulst.* Sandy fields at Newtownbreda, and Knock, also at Rostrevor, and roadsides from Newcastle to Castlewellan, and Clough; S.A.S. Conlig Hill, and sandhills of Ballykinler; T.H.C.

Antrim—Sandy wastes at Toome, and at Redbay, not common in Antrim; S.A.S. Shane's Castle, Cushendun, and Ballycastle; R.Ll.P.

Derry—Common in Magilligan; D.M. Coleraine, Draperstown, and Dungiven; S.A.S.

2. F. minima *(Hudson) Fries.*

Sandy fields, and wastes—frequent, abundant on coast sandhills, and on the shores of Lough Neagh. Fl. mid. June till end of Aug.

GNAPHALIUM *Linn.* CUDWEED.

1. G. uliginosum *Linn.* MARSH CUDWEED.

Wet places by roadsides, and in sandy land—common. Fl. July and Aug.

COMPOSITÆ.

2. G. sylvaticum *Linn.*

Damp places on sandy coasts, and shady spots, at low elevations, in the hills—frequent, but not usually in much abundance. Fl. July and Aug.

Down—Dundonald; *G. C. Hyndman.* Comber, and Conlig; *Flor. Belf.* Newcastle, and Holywood; *Flor. Ulst.* Plentiful about Dundrum, and on the Mealough Hill; S.A.S. Near Gillhall, and at Kilkeel; *C. Dickson.* Sandpit at Giant's Ring; T.H.C. Mourne Mountains; R.Ll.P.

Antrim—Carrickfergus commons; *Flor. Belf.* Ballintoy; *Prof. R. Tate.* Shane's Castle; *B.N.F.C.*, 1866. Portglenone, Glenshesk, Knocklayd, and Bushfoot; S.A.S. Tor Head; T.H.C. Glendun; R.Ll.P.

Derry—Wood at Enagh Lough, and many other places; D.M. Moneymore; *Flor. Ulst.* Near the mouth of the Bann; *Cyb. Hib.* Castledawson; S.A.S. Benbradagh, and by the Roe; *H. C. Hart.*

ANTENNARIA *R. Brown.*

1. A. dioica (*Linn.*) *Br.* CAT'S FOOT.

Dry stony heaths, and coast sandhills—frequent. Fl. June and July. From sea level to 2500 feet.

Down—Newtownards heath; *Templeton*, 1793. Newcastle; *Flor. Ulst.* Near the summit of Slieve Donard, and very luxuriant on Conlig Hill; S.A.S.

Antrim—Carnmoney Hill; *Templeton.* Fairhead; *F. Whitla.* Cave Hill, and commons of Carrickfergus; *Flor. Belf.* Near summit of Divis, also on Wolfhill, and heath above Giant's Causeway; S.A.S. Sallagh Braes; T.H.C. Portrush sandhills; *W. H. Phillips.*

Derry—Slieve Gallion at 1500 feet, and abundant on heathy ground between Kilrea and Swatragh; D.M. Downhill; *G. C. Hyndman.* Abundant on sandhills at Magilligan; R.Ll.P. Bog at Ballyronan; S.A.S. Mullaghmore; *H. C. Hart.*

ACHILLEA *Linn.*

1. A. ptarmica *Linn.* SNEEZEWORT.

Hedge banks, and neglected fields—common. Fl. mid. July till mid. Sept.

2. A. millefolium *Linn.* YARROW.

Dry banks, and sandy or gravelly pastures and wastes. Fl. mid. June till mid. Oct. Ascends to 1000 feet in Derry; D.M.

ANTHEMIS *Linn.*

1. A. nobilis *Linn.* CHAMOMILE.

Damp sandy or gravelly wastes and pastures—very rare. Fl. July—Sept.

Down—Foot of Tullybranagan mountain west of Newcastle; (*John White*) *Wade Rar.* Commons near Comber; *Flor. Ulst.* Old gravel pits on the Newtownards road about one mile northeast of Comber; R.Ll.P.

Antrim—Banks of Lough Neagh; *Cyb. Hib.*

Derry—West side of Enagh Lough, and roadsides between Moneymore and Ballyronan; D.M. Roadside between Toome and Ballyronan, and pastures by Lough Beg; *Flor. Ulst.*

Perhaps truly wild in some of the above stations, but may in others be merely a garden escape.

MATRICARIA *Linn.*

1. M. inodora *Linn.* MICHAELMAS DAISY.

Roadsides, waste ground, and borders of fields—very common. Fl. June—Sept.

Var. β M. SALINA *Babington.*

Rocky and gravelly places on the seashore—frequent around the coast.

Down—Near the quay at Killyleagh, and other places within reach of the salt water; *Templeton.*

Antrim—Frequent along the coast; *Flor. Ulst.* Larne, Cushendun, and Rathlin; S.A.S.

Derry—Abundant on the coast from Umbra to near Portrush; D.M.

2. M. chamomilla *Linn.* WILD CHAMOMILE.

Cultivated fields—very rare, and perhaps sometimes imported with seed. Fl. July and Aug.

Down—In the Ards near Kirkestown; *Harris's Down*, 1744. Field near Comber; *Templeton*, 1800. Holywood; R.Ll.P.

Antrim—Near Dunmurry, and in a wheat field at Malone; *Templeton*, 1804. Field at Mountcollier, and field at Kilroot; *Flor. Ulst.* In great abundance in an oat-field on the west side of Islandmagee; S.A.S., 1880. Larne, and Cairncastle; R.Ll.P. Waste ground—Bog meadows near the distillery; *Richard Hanna.*

CHRYSANTHEMUM *Linn.*

1. C. leucanthemum *Linn.* OXEYE DAISY, ESPIBAWN, and GILLGOWANS (*loc.*).

COMPOSITÆ.

Pastures, and grassy heaths—very common. Fl. mid. June till mid. Aug.

2. C. segetum *Linn.* CORN MARIGOLD.
Cultivated fields—very common, in many places a pest. Fl. June—Sept.

ARTEMISIA *Linn.* WORMWOOD.

1. A. vulgaris *Linn.* MUGWORT.
Sandy banks, and waste ground—common; very abundant about Londonderry, Magilligan, and the northern part of the Antrim coast, also in the Downpatrick and Newcastle district.

TANACETUM *Linn.* TANSY.

1. T. vulgare *Linn.*
Ditch banks, and by streams and gravelly lake shores—frequent, and well naturalised in many places, but probably nowhere native. Fl. Aug. and early part of Sept.

Down—Seaforde; *W. Thompson.* Near Killyleagh; *G. C. Hyndman.* Bangor; *Flor. Belf.* Newcastle; *Flor. Ulst.* Roadside between Greencastle and Rostrevor, also plentiful in two or three places on road from Comber to Newtownards; S.A.S. Between Newcastle and Kilkeel; *C. Dickson.*

Antrim—Sandy ditch banks at south end of Lisburn, a doubtful native; *Templeton,* 1793. Islandmagee; *Flor. Belf.* Carrickfergus, Curran of Larne, Ballycastle, and Rathlin; *Flor. Ulst.* Oldforge; *Miss Maffett.* Whitehead; *B.N.F.C.*, 1863. Plentiful by the shore at Crumlin waterfoot, and by the river occasionally to Crumlin; S.A.S. Banks of river at Cushendall waterfoot, and near Dunseverick; T.H.C.

Derry—On the embankments of the Roe, and sandy ditches at Dumblane; *Samp's. Surv.,* 1814. By the Faughan near Clady, and several other parts; D.M. Downhill; *B.N.F.C.*, 1863.

SENECIO *Linn.*

1. S. vulgaris *Linn.* GROUNDSEL.
Roadsides, waste ground, and among the green crops in cultivated fields—very common in lowland districts, rare on mountains, but ascends to the upper limits of cultivation. Flowers throughout the year, except in severe winters.

2. S. sylvaticus *Linn.*
Waste ground and borders of fields, in sandy or gravelly land, and by

80   COMPOSITÆ.

sandy shores—frequent. Fl. June—Sept.
Down—Found about lakes and bogs in the neighbourhood of Ballynahinch
    ("*S. lividus*"); *Templeton*, 1813.   Holywood, Groomsport,
    Ballygowan, and Comber ; *Flor. Belf.* Shore at Crawfords-
    burn ; T.H.C.   Newcastle sandhills, and roadsides about
    Tyrella, also by canal between Lisburn and Hillsborough,
    and at Moneyreagh ; S.A.S.
Antrim—Stranmillis ;   *Templeton*, 1797.   Near Larne ; *Flor. Hib.*
    Whiterock, and Cavehill ; *Flor. Ulst.* Roadsides near Toome;
    T.H.C.   Roadside between Dervock and Ballycastle, also
    Rathlin Island ; S.A.S. Cushendall ; R.Ll.P.
Derry—Near the natural wood of Ballykelly—rare ; D.M. Bellaghy ;
    S.A.S.

3. S. jacobæa *Linn.* RAGWEED, BENWEED.

Roadsides, pastures, and wastes—very common. Fl. mid. June till mid. Oct.

4. S. aquaticus *Hudson.*

Marshy ground, damp pastures, and by ditches and streams— common. Fl. June—Aug.

BIDENS *Linn.* BUR MARIGOLD.

1. B. tripartita *Linn.*

Boggy places, and stagnant ditches—rather rare. Fl. Aug. and Sept.
Down—Ballygowan moss ; *Flor. Belf.* Bog holes at Conlig ; *Flor. Ulst.*
    Abundant on shore of Corbet Lake near Banbridge, and by the
    Quoile near Downpatrick, also margins of Lough Aghry, and
    Carrickmannan Lake ; S.A.S.
Antrim—Watery places at Stranmillis ; *Templeton*, 1797. Abundant on
    shores of Portmore Lake, also at Portglenone ; S.A.S. An-
    trim ; *D. Redmond.*
Derry—Frequent in Magilligan, and by the side of Enagh Lough ; D.M.
    Shore at Toome, and margins of the Bann ; S.A.S.

2. B. cernua *Linn.*

In bogs, and in marshy places by lakes, local, but not rare. Fl. Aug. and Sept.
Down.--Abundant between Clough and Tollymore ; *Templeton*, 1797.
    Ballygowan moss, and marsh at Ballyallolly ; *Flor. Belf.*
    Abundant, and luxuriant in several places by the Newry canal,
    and by the Quoile at Downpatrick, also in bogs, and on lake
    margins about Saintfield and Ballynahinch, and shore of Lough
    Neagh at Kilmore ; S.A.S. Cotton moss ; R.Ll.P.
Antrim—Magheralave near Lisburn, and bog between Toome and Lough

COMPOSITÆ. 81

Neagh; *G. C. Hyndman.* Ditches at Lambeg; *Flor. Ulst.* By the Bann at Portglenone; S.A.S. Antrim; *D. Redmond.* Derry—Enagh Lough, and many parts of the county; D.M.

The form with radiant flowers (*B. radiata*), has been found at Ballygowan in Down, and Portmore in Antrim.

ARCTIUM *Linn.* BURDOCK.

The distribution of the segregate forms of this genus has not been fully, and certainly ascertained. When the plants are well developed they are sufficiently distinct, but only such can be relied on, and all notes based on imperfect, or immature specimens must be discarded, and also notes by botanists who have not specially studied the group, unless fortified by satisfactory examples. The records which follow are meagre, but they have been fiated by high authority.

[A. MAJUS *Schkur.* A plant collected by S.A.S., in Rathlin, was considered by Prof. Babington to belong here, but as only one specimen was gathered, and that not as good as might be desired, it is advisable to leave it for future research. The note on this species in *Flor. Belf.* is of no authority.]

1. A. nemorosum *Lej.*
Gravelly and sandy waste ground—frequent? Fl. latter end of July, mature in Aug.
Down—Near Donaghadee, and gravelly pastures on Rough Island, above Newtownards; S.A.S.
Antrim—Quarry spoil-bank at Magheramorne, very fine; S.A.S.
Derry—Roadside at foot of Benevenagh east of Bellarena station; S.A.S.

2. A. intermedium *Lange.*
Roadsides, and waste ground—frequent? Fl. July and Aug.
Down—There is little doubt of its occurrence in the county, but authoratative specimens have not been localised, and special stations cannot be quoted at present.
Antrim—Coast of Antrim; *Prof. R. Tate.* Roadsides at Antrim, and common on Rathlin Island; S.A.S. About Belfast, and other places in Antrim; *Cyb. Hib.*
Derry—Uncertain.
[A. MINUS *Schkur.* Plants which seem to be this form have been collected, but they are not sufficiently mature, and may belong to the preceding species. The plant of *Flor. Belf.* was not certain.]

CENTAUREA *Linn.*

1. C. nigra *Linn.* KNAPWEED.

F

Hedge banks, pastures, and heaths—very common. Fl. mid. July till end of Sept.

The radiant form occurs at Castle Espie in Down. We are not certain as regards Dr. Moore's Derry plant whether it was the var. decipiens *Thuill.* or simply the form of *C. nigra* with radiant flowers.

2. C. cyanus *Linn.* CORN BLUEBOTTLE.

In fields of flax—frequent, occasionally in corn. Fl. July and early part of Aug. An alien introduced with seed, but now well established.

CARDUUS *Linn.* THISTLE.

1. C. crispus *Linn.*
Var. β C. ACANTHOIDES *Linn.*

Dry gravelly fields, and wastes—very rare, and apparently sometimes as a casual only, or perhaps erroneously determined. Fl. July and Aug.

Down—Roadside at Milltown; *Flor. Ulst.* (not found there now). Pasture field at Quarry Hill north of Moira; S.A.S., 1865 (one plant only, and not seen since).

Antrim—Road from Lisburn to Lambeg; *Templeton.* Rubbish heaps at York Street Terminus; *W. Millen*, (now lost). Antrim; *Flor. Ulst.* Curran of Larne; *Cyb. Hib.* Not seen recently in either of the two last named stations, and the Rathlin record of Dr. Marshall was an error.

Derry—Not rare in Magilligan; D.M.

2. C. tenuiflorus *Curtis.*

On sandy seashores—frequent, inland very rare. Fl. mid. June till mid. Aug. This thistle occurs in profusion on the sandy coasts of Down and Derry. It is less common in Antrim, but grows in plenty at Ballycastle, Portrush, etc. Mr. Praeger finds it in great abundance and luxuriance at the iron mines by the Inver River, Glenariffe. This is the only inland station that we know of.

3. C. lanceolatus *Linn.* SPEAR THISTLE.

Roadsides, waste ground, and borders of fields—very common. Fl. July and Aug.

4. C. arvensis *Curtis.*

Waste ground, and sandy pasture fields—common. Fl. mid. June till mid. Sept.

5. C. palustris *Linn.*

Ditch banks, meadows, and damp pastures—very common. Fl. mid. June till mid. Oct.

COMPOSITÆ. 83

6. C. pratensis *Hudson.*
Meadows, and wet pastures—rare. Fl. mid. June till end of Aug.
Down—Between Holywood and Bangor ; *Flor. Ulst.* (? *Eds.*).
Antrim—Ballycastle ; *Flor. Ulst.* Sparingly on the south side of the lake at Portmore, and plentiful on grassy heath above Giants' Causeway ; S.A.S.
Derry—Abundant along the base of the basaltic mountains near Dungiven; D.M.
  Var. β C. FORSTERI *Smith.*
Derry—Garvagh demesne, and wet meadows by the Bann below Coleraine ; D.M.

LAPSANA *Linn.*

1. L. communis *Linn.* NIPPLEWORT.
Hedge banks, and waste ground—common. Fl. July and Aug.

CICHORIUM *Linn.* CHICORY.

1. C. intybus *Linn.*
Gravelly banks, and waste ground—very rare, possibly native. Fl. July and Aug.
Down—On the mountains of Mourne; *Harris's Down*, 1744. Embankment from Connswater to Queen's Island ; *H. Robinson, M.R.I.A.* Holywood Hills ; R.Ll.P.
Antrim—Field two miles north of Templepatrick ; *Templeton.* Muldersly's Hill at south end of Islandmagee ; *Flor. Belf.* Squire's Hill, and Jennymount ; *Flor. Ulst.*
Derry—Salters Castle ; *Flor. Ulst.*

HYPOCHÆRIS *Linn.* CATSEAR.

1. H. radicata *Linn.*
Roadsides, borders of fields, and rock cliffs—very common. Fl. mid. June till mid. Oct. Ascends to 1500 feet.

LEONTODON *Linn.* HAWKBIT.

1. L. hirtus *Linn.* (THRINCIA D.C.).
Pasture fields—very rare. Fl. July and Aug.
Down—Sides of new roads at Ballymaghan ; *Flor. Belf.* Near Newtownbreda; *Cyb. Hib.*
Antrim—On the Lagan side above the first lock ; *Templeton,* 1797. In

F 2

an old pasture field by the shore below the coastguard station at Macedon; S.A.S., 1869, but now much reduced, owing to a portion of the field having been broken up, and drained.

The Ballymaghan station of Tate's Flora, and that of Templeton at the Lagan were originally assigned to *Apargia hispida*; there can scarcely be a doubt, however, that the plant in each case was the present species.

2. L. hispidum *Linn.* (APARGIA *Willd.*).

Sandy, and gravelly ground—very rare. Fl. June—Aug. (*Cyb. Hib.*). Derry—Magilligan; D.M.

The localities near Belfast ascribed to this plant in Flora Belfastiensis are transferred to the preceding species.

3. L. autumnalis *Linn.*

Damp pastures, ditch banks, and waysides—common. Fl. mid. July till mid. Sept.

This plant varies exceedingly; forms occur with foliage, scape, and involucre quite glabrous, whilst in others they are extremely hairy. Between these there are intermediates of every degree. The var. A. TARAXACI *Willd.* is reported from mountains in the three counties.

4. L. taraxacum *Linn.* DANDELION.

Hedge banks, pastures, and waste places—very common. Fl. April—June, and flowers again in the autumn until cut down by frost.

Var. γ L. ERYTHROSPERMUM D.C.

Antrim—Roadside in Glendun; *Rev. S. A. Brenan.*

Var. δ L. PALUSTRE *Smith.*

In marshes, and by streams in hilly districts.

Down—Glens at Rostrevor, and Drumbo, also in boggy places at Mountstewart, and Moneyreagh; S.A.S.

Antrim—Rathlin Island; *Flor. Ulst.* Not rare on wet moors, and sides of mountains; *Cyb. Hib.* Black Mountain, and Cave Hill; S.A.S.

PICRIS *Linn.*

1. P. echioides *Linn.* OX-TONGUE.

Gravelly banks, and fields—very rare. Fl. latter end of July till mid. Sept.

Antrim—Amongst briars on a bank by the shore near the north end of Islandmagee, and abundant in a field by the shore between Castle Chichester and Blackhead; S.A.S.

SONCHUS *Linn.* SOWTHISTLE.

COMPOSITÆ.

1. S. oleraceus *Linn.*
Roadsides, and waste ground—very common. Fl. June—Sept.

2. S. asper *Hoffman.*
Ditch banks, and waste places —common. Fl. June—Sept.

3. S. arvensis *Linn.* CORN SOWTHISTLE.
Corn fields—common. Fl. mid. June till mid. Sept. A maritime form occurs, occasionally, on sandy or gravelly seashores.

CREPIS *Linn.* HAWKSBEARD.

1. C. virens *Linn.*
Meadows, damp pastures, and by streams—common. Fl. June—Aug.

2. C. paludosa (*Linn.*) *Moench.*
Rocky glens, and banks of mountain streams—frequent at elevations from 100 to 1400 feet. Fl. July till mid. Sept.
Down—Newtownards Glen; *Mack. Cat.* Crawfordsburn; *Flor. Ulst.* Slieve Donard; T.H.C. Tollymore, and glens at Drumbo, Holywood, and Cregagh; S.A.S.
Antrim—Moist rocks in Colin Glen; *Templeton*, 1797. Black Mountain; *Flor. Belf.* Lough Mourne; *Flor. Ulst.* In all the glens of the Belfast hills, also Drumadarragh Hill, Knockagh, Woodburn, Glynn, Magheramorne, and Glenariffe; S.A.S. A few plants by the Lagan canal between first and second locks; T.H.C. Donegore; R.Ll.P.
Derry—Common by river sides; D.M. Benbradagh; *H. C. Hart.*

HIERACIUM *Linn.* HAWKWEED.

1. H. pilosella *Linn.*
Pastures, dry banks, and rocky heaths—common. Fl. June—Aug.

2. H. anglicum *Fries.*
Rock cliffs—frequent on the basalt, but rare on the silicious rocks of Down and Derry. Fl. July and Aug.
Down—Cliffs of Slieve Bignian; T.H.C. Tollymore Park; *Hart, Jour. Bot.* Feby. 1886.
Antrim—Basaltic rocks at second cave of the Cave Hill; *Flor. Belf.* Garron Head, and Glenariffe; *Ord. Surv. Derry*, 1837. Limestone rocks at Garron Head, and other places on the coast of Antrim northwards; S.A.S. Cushendall, and coast of Antrim; *Cyb. Hib.* Knockagh; *B.N.F.C.*, 1875. Rocks

at Sallagh Braes, and Retreat Castle in Glenballyeamon;
T.H.C.
Derry—Magilligan; *Back. Brit. Hier.* Sparingly on Benevenagh; D.M.
Basaltic rocks of Benbradagh, Mullaghmore, and Slievegallion; S.A.S.
Var. β H. DECIPIENS *Syme.*
Antrim—Abundant on the cliffs of the Cave Hill between the caves and MacArt's fort; S.A.S. Probably elsewhere in the district.
Var. γ H. ACUTIFOLIUM *Back.*
Basaltic cliffs of Sallagh Braes at about 400 feet; S.A.S.

3. H. iricum *Fries.*
Basaltic rocks—rare. Fl. July and Aug. Ranges from 100 feet in Rathlin to 1100 in Derry.
Antrim—Garron Head; *Bab. Manual.* Glenariffe; *Cyb. Hib.* Cave Hill, a few plants only, on cliffs south of the caves, and frequent on rocks at the southeast of Rathlin Island; S.A.S.
Derry—Sparingly on the basaltic cliffs of Benevenagh; S.A.S.

4. H. pallidum *Fries.*
Basaltic rocks, at 1000 to 1200 feet, on limestone in Rathlin at 150 feet—very rare. Fl. July and Aug.
Antrim—Steep rocks south of the caves of the Cave Hill, 1877, and sparingly on limestone cliffs at southwest of Rathlin; S.A.S.
Derry—Basalt rocks of Mullaghmore, 1885, and steep cliffs of Benevenagh; S.A.S.

5. H. cinerascens *Jord.* (H. LASIOPHYLLUM *Bab.*).
Ledges of rock at 200 to 1400 feet, and on coast sandhills in one instance—very rare. Fl. mid. June till end of July.
Down—Spalga Mountain northeast of Rostrevor; S.A.S., 1872. One considerable patch of plants on the sandhills of Ballykinler; R.Ll.P., 1886.
Antrim—Steep cliffs south of Glenarm; *Flor. Ulst.* Glenariffe; *Cyb. Hib.* and subsequently; T.H.C. Cave Hill, Sallagh Braes, and Ballycastle; S.A.S.
Derry—Benbradagh; S.A.S., 1869.
First found in Ireland by Dr. David Moore.

6. H. argenteum *Fries.*
Rocky banks of stream—extremely rare. Fl. July and Aug. (*Bab. Man.*).
Down—By the Causeway Water at Broughnamady northeast of Rostrevor; *H. C. Hart, Jour. Bot.* Feb. 1886.
First found in Ireland by Mr. Hart.

COMPOSITÆ. 87

7. H. murorum *Linn.*

Ledges of rock—frequent, especially in the basaltic mountains. Fl. mid. June till mid. Sept. Ranges from 100 to 1400 feet.

Down—Newtownards Glen; *Templeton*, 1797. Summit of Spalga Mountain northeast of Rostrevor; S.A.S.
Antrim—Common on rocks of the Cave Hill; *Templeton*, 1809, (still found there). Glenarm, Cairncastle, and Cushendall; *Ir. Flor.* Knockagh; *Flor. Ulst.* Tor Head and rocks on the Antrim coast; *Cyb. Hib.* Woodburn; T.H.C. Sallagh Braes, Glenariffe, Fairhead, Ballycastle, and Ballintoy; S.A.S.
Derry—Abundant on the basaltic range of Derry; D.M. Benevenagh, with rounded, cordate leaves, resembling the var. *rotundata* of Backhouse; S.A.S., 1885.

8. H. cæsium *Fries.*

Basaltic cliffs—very rare. Fl. July and Aug. (*Bab. Man.*).
Antrim—*Fairhead; Prof. Babington* in *Cyb. Hib.*
Derry—Very sparingly on the cliffs of Benevenagh at 1100 feet; S.A.S.

9. H. flocculosum *Backhouse.*

Wet shady rocks—extremely rare. Fl. July and Aug. (*Bab. Man.*).
Down—Sparingly on rocky banks of the Spinkwee River above Tollymore Park, at about 1500 feet; S.A.S. Not found elsewhere in Ireland.

10. H. vulgatum *Fries.* (H. SYLVATICUM *Smith*).

Damp rocks, and sometimes on walls in moist shady places—not rare. Fl. mid. June till mid. Sept. Ranges from 50 to 1400 feet above sea level.

Down—Along the river bank at Bryansford (Tollymore); *Wade Rar.* Slieve Donard, and Conlig Hill; S.A.S.
Antrim—Woody mountains between Larne and Glenarm; *Ir. Flor.* Frequent in Antrim; *Cyb. Hib.* Among rocks on the shore of Lough Neagh at Cranfield; S.A.S. Cliffs at Garron Head; *B.N.F.C.*, 1866. On cliffs at Sallagh Braes, Glenarm Park, Glenariffe, and Glendun; T.H.C.
Derry—On rocks at the "Dog's Leap" near Limavady; *Templeton*, 1809. Abundant on Benbradagh, and Benevenagh; D.M. Abundant in Derry; *Cyb. Hib.* Lignapeiste Glen; *B.N.F.C.*, 1876. Frequent in Bennedy Glen, and at Dungiven old church, and all rocky river banks about Dungiven; S.A.S.

11. H. gothicum *Fries.*

Rocky glens—very rare. Fl. July—Sept. (*Bab. Man.*).
Antrim—Near Carrickfergus; *Back. Brit. Hier.* Glenballyeamon; *Cyb. Hib.* Rocky banks at lower end of Glenariffe; S.A.S. Cushendun; *Rev. S. A. Brenan.*

COMPOSITÆ.

12. H. prenanthoides *Villars.*

Damp shady places in mountainous districts—extremely rare. Fl. July and Aug. (*Bab. Man.*).

Derry—Near Garvagh, and meadows near Donald's Hill in the parish of Bovevagh ; *Cyb. Hib.*

There have been many other notices of the occurrence of this hawkweed in our district, but inasmuch as the identifications were in most part mere guesses we do not further refer to them. *H. crocatum* seems to have been most often the plant really seen.

13. H. strictum *Fries.*

Rocky places—very rare. Fl. July and Aug. (*Bab. Man.*).

Antrim—Garron Head ; *Back. Brit. Hier.*

Derry—Garvagh ; *Cyb. Hib.*

The plant of Tollymore Park (*Cyb. Hib.*) proves to be H. CORYMBOSUM *Fries.*

14. H. umbellatum *Linn.*

Damp rocks—very rare. Fl. July and Aug. (*Bab. Man.*).

Antrim—Glenariffe ; *Cyb. Hib.* "Harper's" (Harbour) Island in Lough Neagh (var. FILIFOLIUM *Backhouse*) ; *Back. Brit. Hier.* (Not found there now).

Derry—At 1000 feet on Benevenagh ; *Cyb. Hib.*

15. H. crocatum *Fries.*

Damp rocks, and banks of streams—not rare. Fl. mid. June till mid. Sept. Occurs at elevations from 100 to 1100 feet, and is sometimes very luxuriant. One plant noticed on Cave Hill had 118 flower heads.

Down—By the stream in Rademon demesne (two plants, only, seen); *B.N.F.C.*, 1885. Tollymore Park ; *H. C. Hart, Jour. Bot.* Feb. 1886.

Antrim—River bank in Glenballyeamon, and by the waterfall in Woodburn Glen ; *J. Backhouse.* (Seen recently in both of the above stations ; T.H.C.). Cave Hill, Garron Head, Glenariffe, and not infrequent in the Glens of Antrim ; *Cyb. Hib.* Abundant on the damp rocks of the Cave Hill above Whitewell, and sparingly on Carnmoney Hill, and on Lough rocks above Dunloy in Rasharkin ; S.A.S.

Derry—Benevenagh ; *Cyb. Hib.* By Owenrigh River near Dungiven ; *H. C. Hart, Jour. Bot.* Feb. 1886.

16. H. corymbosum *Fries.*

Damp rocks, and rocky banks of rivers—rare. Fl. mid. July till mid. Sept. Ranges from 100 to 1000 feet altitude.

CAMPANULACEÆ. 89

Down—Banks of Spinkwee River at upper end of Tollymore Park;
S.A.S.
Antrim—Glenariffe; *Back. Brit. Hier.* Garron Head; *Cyb. Hib.* Rocky
banks of the river Bush a short distance above Bushmills, and
rocks of Cave Hill above Whitewell; S.A.S.
Derry—Copses near Garvagh; *Cyb. Hib.* Tircrevan; *Rev. S. A. Brenan.*

17. H. boreale *Fries.*
Shady rocks—very rare. Fl. Aug. and Sept. (*Bab. Man.*).
Down—Moist rocks near the middle bridge in Tollymore Park; *Templeton*, 1793. Slievenamady; *Wade Rar.*
Antrim—Below the rocks at Murlough, and on rocks about the upper fall at Glenariffe; *Templeton.* Cairncastle, and rocky mountains between Larne and Glenarm; *Ir. Flor.* Cave Hill; *Dr. Mateer.* Glenariffe; C.C.B. There are specimens from the two last named localities in *herb.* Babington.
Derry—By the Agivey below Garvagh, and in Tircrevan glen near Magilligan; D.M.

Order XXXIII. **CAMPANULACEÆ.**

LOBELIA *Linn.*

1. L. Dortmanna *Linn.*
Margins of lakes—very rare, and apparently decreasing. Fl. July and Aug. Ascends to 1500 feet on Mourne Mountains.
Down—In Lough Neagh at the townland of Kilmore near Moira; *Ray's Synopsis*, 1696. Blue Lake in the Mourne Mountains; *F. Whitla.* Pond by the Kilkeel River near the "miner's hole"; S.A.S. Castlewellan Lake, and in a lake south of Strangford; *Flor. Ulst.* Bignian Lake; *H. C. Hart.*
Antrim—In Lough Neagh at Portmore; *Templeton.* In the lough at Selshan; *G. C. Hyndman.* In Lough Neagh; *Rev. S. A. Brenan*, 1865.
Found in Lough Neagh, still more recently, but further south and outside our bounds, by *Rev. H. W. Lett.*
Derry—In Lough Beg; D.M.

JASIONE *Linn.*

1. J. montana *Linn.* Sheep's Scabious.

Dry rocks, and heaths, and on sandy seashores—common. Fl. July and Aug. Ranges from sea level, to 1000 feet on the hills.

CAMPANULA *Linn.* BELLFLOWER.

1. C. rotundifolia *Linn.* HAIRBELL.
Rocks, and dry banks—common, very abundant on the basaltic hills. Fl. July and Aug. Altitudinal range—sea level to over 2000 feet.

ORDER XXXIV. **ERICACEÆ.**

ARCTOSTAPHYLOS *Adanson.*

1. A. uva-ursi (*Linn.*) *Sprengel.* BEARBERRY.
Rocky heaths—very rare. Fl. May (*Cyb. Hib.*).
[Down—"Top of Slieve Donard"; *Flor. Ulst.* Probably an error— not seen by any other botanist on that well-searched summit.]
Antrim—At Fairhead, plentifully; *Templeton*, 1814. Fairhead; *Dr. McDonnell*, annot. in Belfast Museum copy of *Fl. Brit.* By the northern lake on Fairhead; *G. C. Hyndman*, 1837, spec. in *herb.* Hyndman.

ANDROMEDA *Linn.*

1. A. polifolia *Linn.* WILD ROSEMARY.
Heathy parts of turf bogs—very rare, and destined, ere long, to become extinct in this district. Fl. mid. April till mid. July.
Down—"About a quarter of a mile in the first bog on the left side of the road from Greyabbey to Donaghadee"; *Templeton*, 1794. Cotton moss near Donaghadee; *Flor. Belf.* Sparingly in Wolf Island bog north of Greyabbey; S.A.S.

CALLUNA *Salisb.*

1. C. erica D.C. (*C.* VULGARIS *Salisb.*). LING.
Heaths, and peat bogs—common at all elevations, from sea level to summit of Slieve Donard. Fl. July and Aug.

ERICA *Linn.* HEATH, HEATHER.

1. E. tetralix *Linn.*

ERICACEÆ.

Peat bogs, and wet mountain heaths—common. Fl. mid. June till mid. Aug. Ascends to 1800 feet in Down and Derry; *Hart.*

2. E. cinerea *Linn.*
Heaths, peat bogs, and coast sandhills—very common at all elevations up to 2200 feet.

VACCINIUM *Linn.*

1. V. myrtillus *Linn.* BILBERRY, BLAEBERRY (*loc.*).
Heaths, peat bogs, rocks, and mountainous woods and thickets—very common. Fl. mid. April till mid. June. Rises to 2200 feet in Derry, and 2700 in Down.

2. V. vitis-idæa *Linn.* COWBERRY.
Rocks, and stony heaths on the mountains at elevations from 500 to 2400 feet—rare. Fl. mid. June till mid. Aug.

Down—Bog in townland of Creevytenant near Ballynahinch; *Templeton,* 1796. Cliffs of Scalleree west of Bencrom, and on Slievenamady; *Wade Rar.* Top of Slieve Bignian; *G. C. Hyndman,* 1825 (spec.). Slieve Donard; *F. Whitla.* Slieve Comedagh, and Slieve Bernagh; *Hart, Proc. R.I.A.,* 1884.

Antrim—Top of Divis; *Templeton.* (Not found there now). Cliffs at west end of Cave Hill, and on the top Slievenanee; *Flor. Ulst.* Some error as to Cave Hill, the plant cannot be found there, and there are no cliffs at the west end. Trap rocks two miles east of Rasharkin, and plentiful on moor at Lough Naroon, not far from the same place; S.A.S. Rocks on Slievenanee at 1500 feet; T.H.C.

Derry—At 2000 feet on Dart Mountain; *Cyb. Hib.* Mullaghmore, and Benbradagh; *Hart, Proc.*; *R.I.A.,* 1884.

3. V. oxycoccos *Linn.* CRANBERRY.
Wet peat bogs at low elevations—becoming rare. Fl. June and July.

Down—Bog near Greyabbey; *Ir. Flor.* (now worked out). Cotton moss; *Flor. Belf.* Birky moss, bog near Bangor, and bog between Newtownards and Donaghadee; *Flor. Ulst.* Wolf Island bog; *B.N.F.C.,* 1864.

Derry—Plentiful in bogs by Lough Neagh between the Moyola and Ballyronan, and among the lakes above Kilrea; D.M. Bogs about Moneymore; *Flor. Ulst.*

PYROLA *Tournefort.* WINTER-GREEN.

1. P. media *Swartz.*

Rocky and heathy places on the hills—rare. Fl. mid. June till late in July.

Down—On the sides of the hill of Scrabo near Newtown (probably this sp.); *Harris's Down*, 1744. Northwest side of Newtownards Park, and shrubby places at Ballygowan bog (not there now); *Templeton*, 1797. Conlig Hill; *Flor. Ulst.*

Antrim—Heathy ground, called Highwood, at the head of the Sixmilewater, 1794, and Fairhead, 1815; *Templeton* (still at Fairhead). Crow Glen; *F. Whitla.* Black Mountain; *Flor. Belf.* Wolfhill; *Flor. Ulst.* (The Colin Glen plant of *Flor. Ulst.* is *P. minor*). Glenarm; *Cyb. Hib.* Glen at Cairncastle, and plentiful in the upper cover of Glenarm Deerpark; R.Ll.P. Clinty Hill near Ballymena; *D. Redmond.*

Derry—Side of a glen in Ballynascreen mountains; *Templeton.* In the parishes of Banagher, and Ballynascreen, also at Glenedra, Altahoan, and on Errigal banks above Garvagh; D.M.

2. P. minor *Linn.*

On rocks in shady and bushy places in the hills, ranging from about 400, to 1000 feet—rare. Fl. mid. June till latter end of July.

Down—Tollymore Park; *C. Dickson.*

Antrim—Among heath on the north side of Slemish 1797, and rocks in Glenariffe,* 1809; *Templeton.* Crow Glen; *Flor. Ulst.* On Divis above Crow Glen; *B.N.F.C.*, 1878. Near Ballymoney; *Rev. S. A. Brenan.* Colin Glen, Woodburn, and Glendun; *Cyb. Hib.* Basaltic cliffs of Cave Hill above Whitewell, plentiful in one spot; S.A.S.

Derry—Woods at Lissan, and Castledawson, also in Altahoan Glen, and on Errigal banks; D.M. Mountain glen near Garvagh (Errigal); *Flor. Hib.* By the fall in Lignapeiste Glen near Banagher; *Flor. Ulst.*

First found in Ireland by Mr. Templeton.

* Not Glenarm, as printed in Flora Hibernica, and Flora of Ulster.

3. P. secunda *Linn.*

Basaltic rocks—extremely rare. Fl. July (*Cyb. Hib.*).

Antrim—Glenarm, and Sallagh Braes, also on Agnew's Hill near Larne; D.M. in *Cyb. Hib.*

Derry—In considerable quantity on Errigal banks, facing the north side of the Agivey River above Garvagh; D.M. The exact spot is between the lint mill, and Errigal bridge.

It is singular that this plant which was first discovered in Ireland, by Dr. Moore, more than 50 years since, has not been noticed by any other botanist.

Order XXXV. AQUIFOLIACEÆ.

ILEX Linn. HOLLY.

1. I. aquifolium *Linn.*
Glens, mountain woods, and hedges in hilly districts—frequent. Fl. May and June.

Order XXXVI. OLEACEÆ.

LIGUSTRUM *Linn.*

1. L. vulgare *Linn.* PRIVET.
Hedges, and by streams—common, but usually planted. Fl. mid. June till end of July.

The privet grows spontaneously in a few localities, and was noted, by Mr. Templeton, as established in a number of places at the close of last century, but not considered, by him, as native.

FRAXINUS *Linn.*

1. F. excelsior *Linn.* ASH.
Woods, and glens—frequent. Fl. May and early part of June.

Indigenous by some mountain streams, and perhaps, also, in some of the old woods, but difficult to distinguish the native from the introduced localities.

Order XXXVII. GENTIANACEÆ.

ERYTHRÆA *Renealm.* CENTAURY.

1. E. pulchella (*Swartz*) *Hooker.*
Sandy shores—very rare. Fl. July and Aug.
Down—[Rocky ground below Bangor; *Templeton*, 1804.] Shore of Strangford Lough near Newtownards; *Flor. Ulst.* Sparingly by the Quoile below Downpatrick; S.A.S., 1885.

The Downpatrick plant is right, and the Newtownards plant may be right, but cannot now be vouched. The Rathlin record must be deleted, and Templeton's Bangor plant is most probably wrong also, *E. pulchella* being a plant of the pure sand, and not petrophilous.

2. E. centaurium *(Curtis) Persoon.*
Dry sandy, or rocky pastures—common. Fl. July and Aug.

GENTIANA *Linn.* GENTIAN.

1. G. amarella *Linn.*
Dry rocks—very rare. Fl. July and Aug. *(Cyb. Hib.).*
Antrim—With double flowers on Kenbane Head ; *Templeton.*
Derry—Observed growing sparingly, at 700 feet, on the basaltic debris of Keady Hill ; D.M.

The erroneous records of Cave Hill, and other places in Antrim are due to negligent transcription of the original notes into Flora Hibernica, copied thence into subsequent Floras. Mr. Templeton only noted *G campestris* in the above mentioned localities.

2. G. campestris *Linn.*
Coast sandhills, sandy shores, and bare mountain pastures—frequent. Fl. Aug. and Sept. Altitudinal range—from sea level to 1400 feet.
Down—Northeast side of Slieve Donard, and at Armor's Hole south of Newcastle, also on Holywood warren ; *Templeton,* 1796. Links at Newcastle; *Flor. Ulst.* Rostrevor ; *Rev. Geo. Robinson.* By the shore between Bangor and Crawfordsburn ; *Prof. Cunningham.* Abundant on Conlig Hill ; T.H.C. Bloody bridge south of Newcastle, and Carngaver Hill; S.A.S.
Antrim—About the Cave Hill, and many other places ; *Templeton.* Knockagh, and Glenarm; *Flor. Ulst.* Shores of Lough Neagh between Crumlin and Antrim, also Rathlin Island, sands at Ballycastle, and frequent on grassy mountain slopes around the Antrim coast; S.A.S. At Skerry in the Braid ; *B.N.F.C.,* 1871. Headlands near the Giants' Causeway ; *Miss Reid.* Knockdhu, Glenarm Park, Torhead, Ballintoy, and Carrick-a-rede ; T.H.C.
Derry—General throughout the county ; with double flowers on Benevenagh, and with white flowers on same mountain. D.M. Benbradagh ; S.A.S. Sandhills of Castlerock ; R.Ll.P.

A large flowered form is often mistaken for the preceding species.

MENYANTHES *Tournefort.* BUCKBEAN.

1. M. trifoliata *Linn.* BOGBEAN (*loc.*).
Bogs, and marshy places on heaths—common at various elevations up to more than 1000 feet. Fl. mid. May till end of June.

ORDER XXXVIII. **CONVOLVULACEÆ**.

CONVOLVULUS *Linn.* BINDWEED.

1. C. arvensis *Linn.*
Banks, and waste ground in sandy districts—not common. Fl. July and Aug.
Down—Near Holywood; *Flor. Ulst.* Warrenpoint, and sandy fields between Rathmullan and Tyrella, also on sandy shore at Ballyholme; S.A.S. Comber; R.Ll.P.
Antrim—Dry ditch on the roadside at the Curran of Larne; *Templeton*, 1794, and *B.N.F.C.*, 1869. Banks of Lagan near the first lock; *Flor. Belf.* Templepatrick; *Flor. Ulst.* Glenarm; *Dr. J. S. Holden.* Whitehead, and Magheramorne; T.H.C. Railway bank at Glenmore; *J. H. Davies.* By the Six-mile River at Shane's Castle; R.Ll.P.
Derry—Sparingly on a dry bank by the Foyle above Londonderry; D.M.

2. C. sepium *Linn.* GREAT BINDWEED.
Hedges, and dry banks in lowland districts—common. Fl. July and Au .

3. C. soldanella *Linn.* SEASIDE BINDWEED.
Coast sandhills—very rare. Fl. July and Aug.
Down—Sandhills on the shore at Newcastle; *Templeton*, 1795. Benderg Bay near Ardglass; T.H.C. Sparingly on the sands at Ballykinler; S.A.S.
Antrim—Sands beside the salmon weir at Bushfoot; *Templeton.* (still found there). Shores at Cushendun, and Ballycastle; *Flor. Ulst.*

Holywood offers no *habitat* that would be accepted by this plant, and the station there assigned it by Flora of Ulster is doubtless an error.

The apparent absence of this bindweed from the wide expanse of sandhills on the coast of Derry is remarkable.

ORDER XXXIX. BORAGINACEÆ.

CYNOGLOSSUM *Tournefort.* HOUNDSTONGUE.

1. C. officinale *Linn.*

Sandy seashores, and dry sandy wastes—found only in Co. Down, where it is plentiful. Fl. mid. June till mid. Aug.

Down—Sandy shores at Kirkiston, and at Dundrum; *Templeton*, 1793. Sandy waste ground near the shore southeast of Portavo; *Miss Maffett.* Sandhills at Ballykinler, and at Ringsallin Point, also in abundance at Millquarter Bay, and Killard Point near Ardglass; T.H.C. Greencastle, and abundant on the sandy shores of the Ards from Ballywalter to Kearney's Point, also on roadsides, and sandy places by the coast from Newcastle northwards to the entrance to Strangford Lough; S.A.S.

ANCHUSA *Linn.* ALKANET.

1. A. sempervirens *Linn.*

Hedge banks, and waste ground—not rare; naturalised in some of its stations, but frequently only a garden escape. Fl. mid. May till mid. July.

Down—Hedge between Bangor and Portavo; *Templeton*, 1797. Ruins of Greyabbey; *Flor. Hib.* Roadsides at Stormount, Braniel, and Conlig; S.A.S.

Antrim—Near Portglenone; *Templeton.* Lisburn; *Flor. Belf.* Ballycastle; *Flor. Ulst.* Cushendun; *Rev. S. A. Brenan.* Duneane; *Cyb. Hib.* Roadside in Glendun; T.H.C. Ditch north of Larne, and in Rathlin Island; S.A.S.

Derry—Near Magherafelt; *Templeton.* Woods at Roepark, also near Ashpark in the parish of Banagher, and lane from Limavady to Roe Mill; D.M.

LYCOPSIS *Linn.*

1. L. arvensis *Linn.* BUGLOSS.

Sandy fields, and waste ground—common, especially abundant by the coast. Fl. May—Sept.

SYMPHYTUM *Tournefort.* COMFREY.

1. S. officinale *Linn.*

Fields, and waste ground in sandy or gravelly land—frequent. Fl. June and July.

Down—Near Dromore; *Harris's Down*, 1744. Strandtown, and Belmont; *Flor. Belf.* Warrenpoint, Ballygowan, Comber, Ballylesson, and Dundonald; S.A.S. Newtownards, Groomsport, Craigavad, and Giants' Ring; T.H.C. Holywood; R.Ll.P.
Antrim—Stranmillis, Cave Hill, and roadside going to Carnmoney; *Templeton*, 1809. Ballydrain, Whitewell, and Woodburn; *Flor. Belf.* Colin Glen; *Flor. Ulst.* Shane's Castle, and Glenarm; R.Ll.P.
Derry—Sides of the Roe, and frequent in the county; D.M. A white variety near Limavady.

ECHIUM *Tournefort.*

1. E. vulgare *Linn.* VIPER'S BUGLOSS.
Sandy, or gravelly fields and banks—rare. Fl. July and Aug. Native on sandy seashores, but elsewhere probably a casual.

Down—Sands at Newcastle, and Dundrum, and roadside near Holywood (casual here); *Flor. Ulst.* Roadside at Knock (casual), and sandy ground by shore at Ballykinler; T.H.C. Sandy fields by the shore southwest of Strangford, also south of Rathmullen Point, and occasionally to Dundrum; S.A.S. Downpatrick, R.Ll.P.
Antrim—Clover field near Belfast (casual); *Templeton*, 1802. Crumlin road, and York Street terminus (casuals); *Flor. Belf.* Gravel bank at the Curran of Larne; S.A.S. Cushendun; *Rev. S. A. Brenan.*
Derry—Moneymore; *Flor. Ulst.*

MERTENSIA *Roth.*

1. M. maritima (*Linn.*) *Don.*
Gravelly, and sandy seashores—very rare, apparently much more so than formerly. Fl. June—Aug. (*Cyb. Hib.*). First found in Ireland by Sherard.

Down—Shore half a mile from the bar of Dundrum, and in great profusion on the shore about five miles south of Newcastle, 1799, also at Rathmullen, 1810; *Templeton.* Gravelly shore between Greencastle and the lighthouse; *Ir. Flor.* Balloch; *Flor. Ulst.* A single plant on the shore at Rathmullen Point, T.H.C., 1878. Sparingly on the strand a short distance north of the lighthouse at Greencastle; S.A.S.
Antrim—Gravelly shore of Islandmagee at Brown's Bay; *Templeton*, 1794. Garron Head; *Cyb. Hib.*
Derry—Near Castlerock station; *Flor. Ulst.*

98   BORAGINACEÆ.

LITHOSPERMUM Tournefort. GROMWELL.

1. L. officinale *Linn.*
Gravelly fields, and banks—rare. Fl. June and July.
Antrim—Rocky places at the southwest corner of the Belfast Deerpark, and ditches in the Pound loaning (now built over); *Templeton,* 1797. Cave Hill quarries (Templeton's loc.); *Flor. Belf.* Whitehead; *Flor. Ulst.* Rocky slopes south of Whiterock; *C. Dickson.* Knockagh; S.A.S. Trench Falls Road; *Richard Hanna.*
Derry—Dry banks in Termonbacca; *Ord. Surv. Derry,* 1837. Side of the Foyle above Derry, and banks of the Bann near Coleraine; D.M. On the southern border of the county near Cookstown ; *W. MacMillan.*

2. L. arvense *Linn.* CORN GROMWELL.
Sandy fields—very rare. Fl. May—July *(Cyb. Hib.).*
Down—Railway bank about a mile from Belfast (Newtownards line), also near Holywood; *Flor. Ulst.* Fields between Bangor and Groomsport; S.A.S. Giants' Ring; *C. Dickson.*
Antrim—Among vetches at Cromac (now built over); *Templeton.* Whitehouse; *Flor. Belf.* Carrickfergus; *Flor. Ulst.*
Less common, and sometimes confounded with the preceding species.

MYOSOTIS *Linn.* SCORPION-GRASS.

1. M. palustris *With.* FORGET-ME-NOT.
Ponds, ditches, and sluggish streams—frequent. Fl. mid. June till mid. Sept.
Down—Ballygowan, Scrabo, Copeland Islands, and Victoria Park; T.H.C. By the Lagan canal, and Newry canal; S.A.S. Narrow-water, and slopes of Slieve Donard; R.Ll.P.
Antrim—About Lough Neagh; *Templeton,* 1797. Lagan canal, and Bog meadows; *Flor. Belf.* Lisburn, Antrim, and Portglenone; S.A.S. Shane's Castle, and Loughmourne; T.H.C.
Derry—By Enagh Lough, and frequent through the county; D.M.
In many other localities through the district, but not always distinguished from its allies.

2. M. repens *Don.*
Ditches, and marshes in hilly, or rocky districts—not common in Down, but increases to the northwest, becoming the most abundant species in the Sperrin Mountains.
Down—Marshy wood in Greyabbey demesne, 1864, and base of Slieve

Donard, 1869 ; *B.N.F.C.* Ditch on a bye road between Newcastle and Kilkeel, and marsh at Ballyalloly ; S.A.S. Rowantree bog near Ashfield Dromore ; *C. Dickson.* Carngaver, and Conlig hills ; R. Ll. P.

Antrim — In Co. Antrim ; D.M. (*Mackey—Nat. Hist. Rev.,* 1854). Carrickfergus Commons ; *Flor. Ulst.* Black Mountain, Templepatrick, Slemish, and Glendun ; S.A.S.

Derry—Plentiful on Slieve Gallion, and White Mountain, and the most common of the genus on the Sperrin range ; S.A.S.

First found in Ireland by Prof. Babington, in 1836.

3. M. cæspitosa *Schultz.*
Marshes, and ditches—common throughout the district. Fl. June—Aug.

4. M. arvensis *Hoffmann.*
Cultivated fields, and sandy waste ground—common. Fl. mid. May till end of Aug.

5. M. collina *Hoffmann.*
Dry sandy places—rare. Fl. April and May (*Cyb. Hib.*).
Down—Locally plentiful on sandhills at Newcastle ; *C. Dickson.*
Antrim—Abundant on the sandy warrens, resting on limestone, between Portrush and Dunluce ; *Flor. Hib. append.* Sands at Bushmills ; *Cyb. Hib.*
The Dundonald station of Flora of Ulster is improbable.

6. M. versicolor *Reich.*
Sandy wastes, coast sandhills, and dry gravelly heaths—common. Fl. mid. May till mid. Aug. Ranges from sea level to 1300 feet.

ORDER XL. **SOLANACEÆ.**

SOLANUM *Tournefort.* NIGHTSHADE.

1. S. nigrum *Linn.* BLACK NIGHTSHADE.
Very rare, and not now found in the undermentioned localities ; perhaps only a casual. Fl. July—Sept. (*Cyb. Hib.*).
Down—In Copeland Islands, and on the mainland opposite ; *Flor. Hib.*
Antrim —Sandy ground near Cushendun, from 1867 till 1871 ; *Rev. S. A. Brenan.*

2. S. dulcamara *Linn.* BITTERSWEET.

Damp shady places, and river banks—not common. Fl. mid. June till end of Aug.

Down—Ballygowan ; *Flor. Belf.* Castle Espie ; S.A.S. Gillhall near Dromore ; T.H.C. River bank at Tullycairne near Dromore ; *C. Dickson.*

Antrim—By the stream in the Bog-meadows ; *Templeton*, 1797. Banks of Lagan near Belfast, and in Islandmagee ; *Flor. Belf.* Malone, Shane's Castle, and shores of Lough Neagh ; *Flor. Ulst.* Shore of Larne Lough ; *Cyb. Hib.* Portmore, and Glynn ; S.A.S.

Derry—Walls of Londonderry ; *Ord. Surv. Derry.* Hedges in parish of Clondermot ; D.M.

HYOSCIAMUS *Tournefort.* HENBANE.

1. H. niger *Linn.*

Waste places in sandy ground—rare, and often uncertain in its appearance. Fl. mid. May till end of July.

Down—Shores about Greencastle near the tower ; *Ir. Flor.* Sparingly at Groomsport Bay ; *Flor. Belf.* Greyabbey, and west side of Bangor Bay (not there now) ; *Flor. Ulst.* About Mill-quarter Bay, and Killard Point ; T.H.C.

Antrim—Roadside near Ballycastle ; *G. C. Hyndman*, 1821. Shore near Carrickfergus ; *M'Skimmin's C.fergus.* Curran of Larne ; *Flor. Ulst.* Shore of Lough Neagh at south side of Toomebridge ; S.A.S.

Derry—Frequent in Magilligan ; *Templeton.* At Benone in Magilligan, and side of Lough Neagh above Ballyronan ; D.M.

Profusely abundant at the ruins of Arboe old church, in Tyrone, but close to the southern border of Derry.

ORDER XLI. **OROBANCHACEÆ**.

OROBANCHE *Linn.* BROOM-RAPE.

1. O. rubra *Smith.* RED BROOM-RAPE.

On basaltic rocks, parasitic on roots of wild thyme—frequent on the Trap hills of Antrim and Derry. Fl. July, and first half of Aug. Ranges from 200 to about 1000 feet.

Antrim—On the rocks of the Cave Hill in little parcels of earth on projecting parts of the rock, found before 1793; *Templeton.* Knockagh; *Flor. Belf.* Cliffs south of Glenarm, and on Rathlin (*Miss Gage*); *Flor. Ulst.* On white limestone at Ballintoy; *Prof. Tate.* Between Dunseverick and Bengore, and on basalt south of Carrick-a-rede; T.H.C. Cushendall, and Portbradden; S.A.S.

Derry—In considerable quantity on the basaltic range from Magilligan to Benbradagh; D.M. 1835.

First found in Britain by Mr. Templeton, and still plentiful in the original station on the Cave Hill.

LATHRÆA *Linn.*

1. L. squamaria *Linn.* TOOTHWORT.

Woods, and thickets—frequent; parasitic on roots of hazels, &c. Fl. mid. April till end of May.

Down—Belvoir Park; *Templeton,* 1798, and *Flor. Belf.,* 1863. Tollymore Park, Ballymenoch, and riverside at Annadale; *Flor. Ulst.* By the Lagan near Shaw's bridge; T.H.C. Woods at Newcastle; *J. J. Macaulay.* Glen at Dundonald; R.Ll.P.

Antrim—Riverside near Lambeg; *Templeton,* 1800. Glenarm Park; *Ir. Flor.* Sinclair's green (now Falls Park); *G. C. Hyndman,* 1828. Colin Glen, and Whitehouse; *Flor. Belf.* Jennymount, Macedon, and Cave Hill; *Flor. Ulst.* Glendivis; *Dr. H. Burden, M.R.I.A.* Shane's Castle; S.A.S.

Derry—Woods at Roepark, Springhill, and Garvagh; D.M.

ORDER XLII. **SCROPHULARIACEÆ.**

VERBASCUM *Linn.* MULLEIN.

1. V. thapsus *Linn.* HIGH TAPER.

Waste ground, and banks in sandy places—rare, and often only sporadic; Fl. July and Aug.

Down—In a gravel-hole near the "whinny hill," going to Holywood; *Templeton,* 1800. Knocknagoney; *Flor. Belf.* Dundrum, and railway bank near Bloomfield; *Flor. Ulst.* One plant on railway bank at Cultra, and sparingly on sandhills at Newcastle; S.A.S.

Antrim—Malone; *Flor. Belf.* Ballycastle; *Flor. Ulst.* Entrance to Sixmilewater at Antrim; *Smith's Lough Neagh.* Island-

magee ; T.H.C. Sparingly on the rocky shore at southwest of Rathlin ; S.A.S. Railway bank near Dunadry ; R.Ll.P.
Derry—By the side of Lough Neagh above Ballyronan, and sparingly by the Foyle above Londonderry; D.M. Near Downhill; *B.N.F.C.*, 1863. Salter's Castle ; *Flor. Ulst.* Magilligan ; S.A.S.

DIGITALIS *Tournefort.* FOXGLOVE.

1. D. purpurea *Linn.*

Borders of woods, heathy banks, rocks, and stone fences in hilly districts—common, especially on the basalt. Ascends to 1250 feet on Benbradagh (*Hart*). Fl. mid. June till mid. Aug. With white flowers occasionally, as at Dundrum.

LINARIA *Tournefort.* TOADFLAX.

1. L. cymbalaria *Mill.* IVY-LEAVED TOADFLAX.

Common on old walls throughout the district. Fl. mid. May till end of Nov.

Though not considered indigenous, this plant is so long established, and so firmly rooted, that it now ranks as an important, and ineradicable member of our flora.

2. L. minor *Desf.*

Gravelly waste ground—very rare. Fl. June—Aug. (*Cyb. Hib.*).

Down—On a steeply sloping waste on the ascent of Spalga Mountain above Killowen coastguard station south of Rostrevor ; S.A.S., 1886. The field was completely covered with stones and gravel, and showed signs of former unprofitable cultivation.

3. L. vulgaris *Mill.* YELLOW TOADFLAX.

Sandy banks, waste ground, and roadsides—rare. Fl. July and till mid. Sept.

Down—Ballyholme, and on railway banks near Newtownards, and Conlig ; *Flor. Belf.* Sands at Newcastle ; *G. C. Hyndman*, 1831. Crawfordsburn ; *Flor. Ulst.* Fields at base of Slievenagriddle; *B.N.F.C.*, 1867. Abundant at Warrenpoint, and about Downpatrick, and sparingly at Bryansford, also at Carngaver, and between Comber and Dundonald ; S.A.S. In fields on the shore near Strangford, and Killough ; T.H.C. Killowen near Rostrevor ; R.Ll.P.

Antrim—Sandy ground south of Lisburn, and on Lisburn road near Drumbridge, also a little north of Crumlin waterfoot ; *Templeton*, 1793. Roadside at Dunsilly ; *G. C. Hyndman*, 1829. Shane's

SCROPHULARIACEÆ.

Castle, Carnlough, and Portrush; *Flor. Ulst.* Sparingly by shore of Lough Neagh a half mile north of Lagan canal; S.A.S.

Derry—Roadsides at Magilligan, and Faughanvale, frequent; D.M. Banks at Magherafelt; *Flor. Ulst.*

SCROPHULARIA Tournefort. FIGWORT.

1. S. nodosa *Linn.* ROSENOBLE (*loc.*).
Woods, ditch banks, and by streams—common. Fl. mid. June till mid. Aug.

2. S. aquatica *Linn.*
Ditches, and margins of streams—very rare. Fl. mid. June till end of July.

Down—Tollymore Park; *Flor. Ulst.* Riverside between Edenderry and Purdysburn; *G. C. Hyndman (spec.).* Plentiful by Moygannon River east of Warrenpoint, about half a mile from seashore; S.A.S.

Antrim—Ditch by the roadside near Cidercourt below Crumlin, and margin of Doagh water near where it joins the Sixmilewater, rather plentiful in both localities; S.A.S. This species, and also *S. Ehrhartii* are stated by Flora of Ulster to grow in the Bog meadows, only sp. I. can be found there.

Derry—"Very common through the county"; D.M. At variance with other botanists, and probably an error.

MELAMPYRUM Tournefort. COW-WHEAT.

1. M. pratense *Linn.*
Shady rocks and heaths, and in bushy places, and sides of glens—frequent. Fl. mid. June till end of Aug.

Down—Ballygowan moss; *Templeton*, 1797. Tollymore Park; *Robt. Patterson, F.R.S.* Glens at Newtownards, and Drumbo; S.A.S. Slieve Donard; T.H.C. Slieve Bignian, and Slieve Bernagh, also on Slieve Croob; *C. Dickson.*

Antrim—Colin Glen, Knock, and Woodburn; *Flor. Belf.* Cave Hill, and Drumnasole; *Flor. Ulst.* Glenavy, and cliffs at Tor Head; T.H.C. Duneane, and Glens, and rocks between Larne and Glenarm, also in Glenariffe, Glenshesk, Fairhead, and by the river at Bushmills; S.A.S. Glenarm Park, Cushendall, and Cushendun; R.Ll.P. Near Kells; *D. Redmond.*

Derry—Frequent, especially in the mountains; D.M. Lignapeiste Glen; S.A.S.

SCROPHULARIACEÆ.

2. M. sylvaticum *Linn.*

Damp shady glens—rare. Fl. mid. June till end of July.

Down—Tollymore Park ; *Flor. Ulst.*

Antrim—By the Forth River near the fall, and in Glenarm Park, 1800 (*still there*), also in the glens at Magheramorne, and on shrubby ground by the Larne road a little south of Glenarm, 1808; *Templeton.* Cushendall, and Woodburn ; *Flor. Ulst.* Glenariffe ; *B.N.F.C.*, 1866.

Derry—South side of Erigal banks above Garvagh ; D.M.

PEDICULARIS *Tournefort.* LOUSEWORT.

1. P. palustris *Linn.*

Bogs, marshes, and wet heaths—common, especially on the hills. Fl. mid. June till mid. Aug. Ascends to 1500 feet.

2. P. sylvatica *Linn.*

Damp pastures, and heaths—very common at all elevations up to 2000 feet. Fl. May—Oct.

RHINANTHUS *Linn.*

1. R. crista-galli *Linn.* YELLOW-RATTLE.

Meadows, damp pastures, marshy places, and bogs—very common. Fl. June—Aug.

BARTSIA *Linn.*

1. B. odontites *Hudson.* RED BARTSIA.

Damp places by roadsides, and in wet, stony pasture land—common. Fl. mid. July till end of Sept.

EUPHRASIA *Linn.*

1. E. officinalis *Linn.* EYEBRIGHT.

Heaths, short pastures, and sandy wastes—very common. Fl. mid. June till end of Sept.

VERONICA *Tournefort.* SPEEDWELL.

1. V. scutellata *Linn.*

Bogs, and marshes—frequent, especially in the hills ; ranging from sea level to over 1000 feet. Fl. mid. June till mid. Aug.

Down—Kinnegar, and dam at Ballycroghan; *Flor. Belf.* Newcastle, and Greyabbey; *Flor. Ulst.* Kilkeel, Ballynahinch, Saintfield, Monlough, and bog at Moneyreagh; S.A.S. Slieve Croob, Rademon, and marshes in the Ballykinler sandhills; *C. Dickson.* Holywood Hill, and Cotton moss; R.Ll.P.
Antrim—Massereene Park; *G. C. Hyndman*, 1836; Bog meadows, King's moss, Carrickfergus commons, and Rathlin Island; *Flor. Ulst.* Bogs at Black Mountain, and Ballinderry; S.A.S. Laganside near Lisburn; T.H.C. Glenariffe, Cushendall, and Cushendun; R.Ll.P.
Derry—Common in the county; D.M. Moneymore; *Flor. Ulst.*

2. V. anagallis *Linn.* WATER SPEEDWELL.

Ditches, and bog drains—frequent, but not to be reckoned a very common plant. Fl. June—Aug.
Down—Tillysburn; *Flor. Belf.* Monlough, and Saintfield; S.A.S. Downpatrick, Ballykinler, Clandeboye, and Ballyholme; R.Ll.P.
Antrim—Ditches by Lagan canal, Bog meadows, and Whiteabbey; *Flor. Belf.* Bogs at Portmore, and Ballinderry, and ditches at Shawsbridge, and Oldpark; S.A.S. Ballycastle, and Ballycarry; R.Ll.P.
Derry—Benone in Magilligan, and near Toome, also at Kilrea, and by the side of the Bann; D.M.

3. V. Beccabunga *Linn.* BROOKLIME. WELL-INK *(loc.).*

Shallow streams, and drains where the water moves, and in wet places about springs—common. Fl. June—Aug. Sometimes flowers again late in autumn.

4. V. chamædrys *Linn.* GERMANDER.

Hedge banks, and grassy places in woods—very common. Fl. May—July.

5. V. montana *Linn.*

Woods, and bushy places—frequent. Fl. May—July, and occasionally, again, late in autumn.
Down—Newcastle, and by a stream near Holywood church; *Flor. Ulst.* Gillhall; T.H.C. Ballynahinch, Dundonald, and Stormount; S.A.S. Rademon; *C. Dickson.*
Antrim—Colin Glen, Springfield, and Woodburn; *Flor. Belf.* Wolfhill, Greencastle, Cave Hill, and roadside a half mile west of Dunmurry; *Flor. Ulst.* Shane's Castle; *Cyb. Hib.* Oldforge near Dunmurry; *Prof. Cunningham.* Black Mountain, Lisburn, Antrim, and Glenarm Park; S.A.S. Glenariffe; R.Ll.P.
Derry—By the Roe, above Limavady, and not unfrequent elsewhere; D.M. Benbradagh; *H. C. Hart.*

6. V. officinalis *Linn.*
Dry banks, heaths, and rocks—common. Fl. June—Aug. Altitudinal range—sea level to 1500 feet.

Var. β V. GLABRA *Bab.* 6th ed.—since abandoned, but more distinct than some plants, e.g. *Rubi* exalted to specific rank.

"Leaves smooth, shining, green, and rigid. On Sallagh Braes about the summer of 1793 "; *Templeton.* "Mourne Mountains near Warrenpoint, and Glens in Co. Antrim; *J. White*" *Templeton.* (*sub V. Allionii*). Still plenty on Sallagh Braes.

7. V. serpyllifolia *Linn.*
Dry banks, pastures, heaths, and walls—common. Fl. May—July.

8. V. arvensis *Linn.*
Cultivated ground—common. Fl. mid. April till mid. Aug.

9. V. agrestis *Linn.* GREEN FIELD SPEEDWELL.
Cultivated fields—common. Fl. April—Sept.

10. V. polita *Fries.* GRAY FIELD SPEEDWELL
Cultivated fields, and sandy waste ground—frequent, but much less common than the two preceding species. Fl. May—Sept.
Down—Bangor, and Crawfordsburn; *Flor. Belf.* Hillsborough, Dundela, Groomsport, and Millisle; S.A.S.
Antrim—Duncairn, now included in Belfast, and at Woodburn; *Flor. Belf.*

The distribution of this species is not sufficiently known. It is no doubt, more widely diffused, but we have no reliable note of its occurrence in Derry, or north Antrim.

11. V. Buxbaumii *Ten.*
Cultivated fields, roadsides, and waste places in sandy land—frequent. Fl. mid. April till end of Sept.

Said to have been originally imported with seed. It is now certainly naturalised, and spreading widely.
Down—Tillysburn, and Ballylesson; *Flor. Belf.* Comber; *Flor. Ulst.* Hillsborough, Belmont, and Scrabo; S.A.S. Holywood, and Dundonald; R.Ll.P.
Antrim—Cave Hill, and by the canal near Belfast; *Flor. Belf.* Newforge, Ballygomartin, Kilroot, and gravel banks at the Curran of Larne; S.A.S. Glenarm and Cairncastle; R.Ll.P.
Derry—Moneymore; *Flor. Ulst.* Coleraine; S.A.S.

12. V. hederæfolia *Linn.* IVY-LEAVED SPEEDWELL.
Hedge banks, and sandy waste ground—frequent. Fl. April—June.

LABIATÆ. 107

Down—Strandtown; *Flor. Belf.* Belmont, Ballylesson, Dundonald, Comber, Newtownards, and Groomsport; S.A.S.
Antrim—Malone; T.H.C. Dunmurry, Kilroot, Larne, and Glenarm; S.A.S.
Derry—Common in many parts of the county; D.M.

ORDER XLIII. **LABIATÆ.**

MENTHA *Tournefort.* MINT.

1. M. piperita *Hudson.* PEPPERMINT.
Wet places by streams—rare. Fl. mid. July till mid. Sept. If not indigenous certainly well naturalised.
Down—In a stream by the Moneyreagh road, beyond the corn mill at Comber; *Flor. Belf.*
Antrim—By the stream near Carnmoney church, and on the shore of Lough Neagh; at Massereene Park; *Flor. Ulst.*
Derry—Side of the Curlyburn in the parish of Balteagh, and on the roadside near Muff in Faughanvale; D.M.
Still plentiful on the rocky margin of the stream west of Carnmoney church, and near Antrim on the gravelly shore at the river foot.

2. M. aquatica *Linn.*
Marshes, and margins of streams—common. Fl. Aug. and Sept.

3. M. sativa *Linn.*
Marshes, and wet meadows—common. Fl. mid. July till end of Sept.

4. M. arvensis *Linn.* CORN MINT.
Borders of fields, and waste places in dry sandy land—frequent. Fl. mid. July till mid. Sept.

5. M. pulegium *Linn.* PENNYROYAL.
Marshy places—very rare. Fl. Aug. and Sept. (*Cyb. Hib.*).
Down—Abundant in wet pastures at the foot of Tullybranagan Mountain near Newcastle; *Ir. Flor.*
Antrim—Plentiful on the Creagh bog north of Toome; D.M.
Derry—Sparingly in wet meadows by the Bann near Toomebridge; D.M.

LYCOPUS. *Tournefort.*

108   *LABIATÆ.*

1. L. europæus *Linn.* GIPSYWORT.

Marshy margins of rivers and lakes—frequent, and in many places abundant. Fl. mid. July till end of Aug.

Down—Ballycroghan dam, and Ballyalloley Lake; *Flor. Belf.* Roadside south of Newcastle, and plentiful by the river at Downpatrick, also by lakes near Saintfield, and Ballynahinch; S.A.S.
Antrim—By the canal near Belfast; *Flor. Belf.* Shore of the lough near Shane's Castle, and near Toome; *Flor. Ulst.* Cushendun; T.H.C. By the Lagan from Lisburn upwards, and abundant by the shores of Lough Neagh from the Lagan canal to Toome, also common by the Bann; S.A.S. Glendun; *Rev. S. A. Brenan.*
Derry—Common through the county; D.M. Shores of Lough Beg; T.H.C. Abundant by Enagh Lough, and by the Bann; S.A.S.

ORIGANUM *Tournefort.* MARJORUM.

1. O. vulgare *Linn.*

Waste places—very rare. Fl. July and Aug. Perhaps introduced, or an escape, in many of the stations.

Down—Ruins of Portavo House; *Flor. Belf.* Top of orchard wall near Shawsbridge; T.H.C. Wall of Clandeboye garden, an escape; R.Ll.P.
Antrim—Roadside near Ballycarry; *Flor. Ulst.* Between Larne and Carrickfergus, and by the river in Shane's Castle grounds; *Cyb. Hib.*
Derry—By the Faughan near Fincairn Glen; D.M.

THYMUS *Linn.* THYME.

1. T. serpyllum *Linn.* WILD THYME.

Heaths, rocks, and dry banks—very common. Fl. June—Aug.

CALAMINTHA *Moench.* CALAMINT.

1. C. officinalis (*Linn.*) *Moench.*

River bank—very rare. Fl. July—Sept. (*Cyb. Hib.*).
Antrim—Left bank of Glendun River going from Knocknacarry to the sea; *Rev. S. A. Brenan.*

SCUTELLARIA *Linn.* SKULLCAP.

1. S. galericulata *Linn.*

Woods, and bushy places, and by lake shores—rare. Fl. July and Aug.

LABIATÆ.

Down—Belfast road a half mile north of Killyleagh ; *Templeton*, 1804. Sparingly at north end of Corbet Lake east of Banbridge, and on east shore of Aughnadarragh Lake near Saintfield; S.A.S.
Antrim—Derriaghy, and Portmore woods; *Templeton*, 1794. Shore of Lough Neagh at Shane's Castle ; *Flor. Hib.* (Still plentiful in Shane's Castle woods). Marsh near Whitehouse ; *Flor. Ulst.* Massereene Park, and shore of the lake two miles south of Antrim ; S.A.S. Two plants in a ditch by the Lagan canal near Belfast; *Jas. Nesbitt.* Glenariffe ; R.Ll.P.
Derry—Tamlaght; *Templeton.* Side of the Bann a little higher than the Cutts, and near Kilrea ; D.M. Ballygillan near Ballyronan ; *Flor. Ulst.*

BRUNELLA *Linn.*

1. B. vulgaris *Linn.* SELFHEAL.
Pastures, woods, hedge banks, and heaths—very common. Fl. July—Sept.

NEPETA *Linn.*

1. N. cataria *Linn.* CATMINT.
Dry waste ground—very rare, and doubtfully native. Fl. July and Aug. (*Cyb. Hib.*).
Down—Waste places by the shore north of Ardglass ; S.A.S.
Antrim—Ballycastle ; *Miss Hincks* (*Flor. Ulst.*).
Derry—Sandy hedge bank at Benone in the parish of Magilligan ; D.M. Sandhills at Magilligan (one plant) ; *B.N.F.C.*, 1863.

2. N. glechoma *Linn.* GROUND IVY.
Hedge banks, and in woods and bushy places—common. Fl. April—June.

LAMIUM *Linn.* DEADNETTLE.

1. L. amplexicaule *Linn.*
Roadsides, and fields in sandy ground—rare. Fl. May—Aug.
Down—Sandy fields near Bangor; *Flor. Belf.* Newcastle ; *Cyb. Hib* Cornfield at Giants' Ring ; S.A.S. At Kilkeel, and abundantly in fields at Tullycairne; *C. Dickson.* Groomsport; R.Ll.P.
Antrim—Carrickfergus ; *Flor. Belf.* Ballycastle; *Flor. Ulst.* Malone ; *Dr. J. S. Holden.* Roadside between Larne and Ballygally,

110   *LABIATÆ.*

and abundant on the Curran of Larne; S.A.S. Ballycarry, and Glynn; R.Ll.P.
Derry—By the side of Lough Foyle; D.M.

2. L. intermedium *Fries.*
Sandy waste ground, and fields—frequent. Fl. May—Oct.
Down—Common in sandy fields at Strandtown, Belmont, Holywood, Crawfordsburn, and Bangor; *Flor. Belf.* Sandpits at Newtownbreda, and Dundonald, also by the shore at Donaghadee, and sandy shores and fields in the Ards from Donaghadee to Kearney's Point; S.A.S. Newcastle, and Comber, frequent in Down; T.H.C. Hillsborough; R.Ll.P.
Antrim—Kilroot; *Flor. Belf.* Glenariffe, Bushmills, and abundant on Rathlin Island; S.A.S. Malone, Crumlin, Glynn, Curran of Larne, and Cushendun; T.H.C. Portrush, and Carnlough; R.Ll.P.
Derry—Coleraine; *Cyb. Hib.* On the coast from the Bann to the Foyle; S.A.S.

3. L. hybridum *Villars.* (*L.* INCISUM *Willd.*).
Sandy or gravelly fields, and wastes—rare, but sometimes locally abundant. Fl. May—Sept.
Down—Dry sandy soil near Comber nursery; *Templeton.* Killyleagh; *Flor. Ulst.* Groomsport; R.Ll.P.
Antrim—Ballinderry, 1810, and near Carrickfergus, 1812; *Templeton.* Ballycastle; *Flor. Ulst.* Sparingly at Glynn, and on gravelly shore at south side of Loughmourne; S.A.S. Abundant on gravels at the Curran of Larne; R.Ll.P.
Derry—Salterstown, Ballynascreen, and by the Foyle; D.M.

The recent records, only, are really reliable, as the earlier botanists did not distinguish this species from the preceding.

4. L. purpureum *Linn.* RED DEADNETTLE.
Waste ground, and cultivated fields—very common. Fl. April—Oct.

5. L. album *Linn.* WHITE DEADNETTLE.
Hedge banks, and waste ground—frequent. Fl. mid. May. till mid. Sept.
Down—On bye road from Newtownbreda to Castlereagh; *Flor. Belf.* Rostrevor and Newcastle; *Flor. Ulst.* Dundrum to Downpatrick, also near Ballyhalbert; S.A.S. Clough; R.Ll.P.
Antrim—Pound Loaning; *Templeton,* 1805 (now part of Belfast). Roadside between Belfast and Colin Glen, and on Rathlin Island; *Flor. Ulst.* Plentiful in the neighbourhood of Antrim; *Cyb. Hib.* Roadside near Doagh; T.H.C. Laganside opposite to the park, also at Ballypallady, Templepatrick, and Glenavy; S.A.S.

LABIATÆ.

Derry—Abundant about Coleraine, otherwise rare ; D.M.   Moneymore ;
    Flor. Ulst.   Coleraine ;   R.Ll.P.

GALEOPSIS *Linn.*   HEMPNETTLE.

1. G. tetrahit   *Linn.*
Cultivated fields, especially in light sandy soil—common.   Fl. July — Sept.

2. G. speciosa *Miller.*   (G. VERSICOLOR *Curt.*) BEE NETTLE.
Damp gravelly ground—rare.   Fl. July and Aug.
Antrim—In a field at Malone, and in corn west of Broughshane, also on the roadside one mile north of Glenarm ; *Templeton*, 1797.
Derry—Banks of the Faughan below Cumber ; *Templeton.* Frequent about Dungiven, and Kilrea ; D.M.   Very sparingly on gravel behind the railway station at Dungiven ; S.A.S., 1885.
A coarse form of the preceding, with larger cream coloured flowers, is frequently named *G. versicolor.* Such was the case with the Co. Down stations of Flora Belfastiensis, and probably such occurred as respects Flora of Ulster. It has been considered necessary to reject numerous notes of localities for this species, which were not accompanied by satisfactory specimens. The old notes, given above, are probably correct, yet some may be erroneous.

STACHYS *Linn.*   WOUNDWORT.

1. S. betonica (*Linn.*) *Benth.* BETONY.
Woods, and shady places—very rare.   Fl. July and Aug. (*Cyb. Hib.*).
Antrim—In the woods of Shane's Castle ; *Wade Rar.*
Derry—Sparingly by the Bann both above and below the bridge at Kilrea ; D.M.
The authority for Shane's Castle is not Templeton, but Wade. Not found there by any subsequent botanist, and may have been an error. Not to be found at Parkmount, and that record may be set aside. The bugle—locally misnamed betony, was no doubt, the plant noted in Flora of Ulster.

2. S. sylvatica *Linn.*   HEDGE WOUNDWORT.
Bushy, and shady places—common.   Fl. July and Aug.

3. S. palustris *Linn.*   MARSH WOUNDWORT.
Ditches, wet pastures, and by streams—common.   Fl. July and Aug.
    Var. β S. AMBIGUA *Smith.*

LABIATÆ.

Down—Near Ballynahinch ; *Miss E. Waddell.* Near the mill and bridge on the Seaforde road a little east of Ballynahinch ; *Templeton*, 1813. At Saul near Downpatrick ; *C. Dickson.* Plentiful by the Causeway Water ; S.A.S.
Antrim—Roadside near Colin Glen ; *Templeton.* Blackstaff Lane near Belfast ; *Flor. Ulst.* Islandmagee, and opposite shore of Larne Lough ; *Cyb. Hib.* Black Mountain, Antrim, Portglenone, and Rathlin ; S.A.S.
Derry—Near Muff in Faughanvale ; D.M. Moneymore ; *Flor. Ulst.* First found in Ireland by Miss Waddell.

4. S. arvensis *Linn.* CORN WOUNDWORT.
Cultivated fields, and sandy banks—rare. Fl. mid. July till mid. Sept.
Down—Roadsides, and fields below Kilkeel ; *Ir. Flor.* Sandy bank near the coastguard station at Greencastle, and in corn at Newcastle, also fields at Strangford, and one plant at Ballyholme ; S.A.S.
Antrim—Kilroot ; *Flor. Belf.* Railway bank at Whitehead ; *B.N.F.C.*, 1863. Abundant in Rathlin ; *Cyb. Hib.* (still plenty there).
Derry—Sandy cornfields by Lough Foyle, and near the Bann below Kilrea ; D.M.

BALLOTA *Linn.*

1. B. alba *Linn.* (FŒTIDA *Lamk.*) BLACK HOREHOUND.
Roadsides, and waste ground in dry places—rare, and not always native. Fl. mid. June till end of Aug.
Down—Shore near Ardglass ; *Harris's Down*, 1744. Near Ballyhalbert, and Portaferry, and near Comber ; *Templeton*, 1806. Bangor, and Castle Espie ; *Flor. Belf.* Roadsides near Greyabbey, and Ballylesson ; *Flor. Ulst.* Roadside between Comber and Newtownards ; T.H.C. Several places on the Down coast, as Mill Bay, and Killough Bay, roadsides near St. John's Point, and from Tyrella to Rathmullen, also at Saul, and between Crossgar and Lough Leagh ; S.A.S.
Antrim—Belfast road three-quarters of a mile from Lisburn ; *Templeton*, 1808. Rathlin Island ; *Miss Gage.* Ballinderry, and roadside near Aghalee ; *Cyb. Hib.* Cairncastle ; R.Ll.P.
Derry—Hedges about Coleraine, but only as a garden escape ; D.M.

TEUCRIUM *Linn.*

1. T. scorodonia *Linn.* WOODSAGE.
Rocky, and bushy places, and heaths—very common. Fl. from commencement of July till early in Sept. Ranges from sea level to 1200 feet.

LENTIBULARIACEÆ.

AJUGA *Linn.*

1. A. reptans *Linn.* BUGLE. Locally known as Betony. Woods, and damp shady places—common. Fl. May—July.

ORDER XLIV. **LENTIBULARIACEÆ.**

PINGUICULA *Tournefort.* BUTTERWORT.

1. P. vulgaris *Linn.*
Bogs, pastures, and wet heaths—common. Fl. June and July. Ranges from sea level to 1700 feet, (*Hart*).

2. P. lusitanica *Linn.*
Wet rocky, and heathy places—frequent. Fl. mid. June till mid. Aug. Ranges from sea level in Down, to over 1000 feet.
Down—Mourne Mountains; *Templeton*, 1794. Slieve Donard; *Flor. Ulst.* Heath above Newtownards; *W. Darragh.* Cove Mountain above Kilkeel, and by the shore about three miles south of Newcastle, also in various places by the coast near Ardglass; S.A.S. Rostrevor Mountain; *Rev. C. H. Waddell.* Slieve Bignian; R.Ll.P.
Antrim—Marshy places at the Headwood (head of Sixmilewater); *Templeton.* Glenarm deerpark; *Flor. Ulst.* Hill range between Larne and Glenarm, summit of Fairhead, and plentiful at the northwest end of Rathlin Island; S.A.S. Garron Point, R.Ll.P.
Derry—In Co. Derry; *Templeton.* Abundantly by the side of the Faughan above Clady, and in a glen in upper Camlin; D.M.

UTRICULARIA *Linn.* BLADDERWORT.

1. U. vulgaris *Linn.*
Ditches in peat mosses, and on marshy margins of lakes—not rare. Fl. mid. July till early in Sept.
Down—Ballygowan moss, and marsh at Ballyalloley; *Flor. Belf.* Lough Cowey near Portaferry; *Flor. Ulst.* By Annsboro' Lake east of Castlewellan, and by Monlough, also common about lakes near Ballynahinch, and Saintfield; S.A.S. Cotton moss; R.Ll.P.

Antrim—Kings' moss; *Flor. Belf.* By the Lagan canal above Lisburn; S.A.S.
Derry—In the racecourse bog near Derry, and bog holes near Garvagh; D.M.

2. U. intermedia *Hayne.*
Pools, and bog drains—very rare. Fl. Aug. (*Cyb. Hib.*).
Derry—Side of the Foyle, in the parish of Templemore above Londonderry; D.M.
There are several notes of the occurrence of this plant in the County of Down, but none by any recent observer. Old records of such a critical plant require confirmation and they are therefore left over for verification

3. U. minor *Linn.*
Bog drains, and stagnant ditches in moory places—not rare. Fl. July and Aug.
Down—In a bog north of Greyabbey, and in old peat holes east of Ballynahinch spa; *Templeton,* 1794. Bog holes at foot of Slieve Donard; *Wade Rar.* Mosses near Ballygowan, Donaghadee, and Holywood; *Flor. Belf.* Conlig; R.Ll.P.
Antrim—In old moss holes in a little bog about 200 yards north of the second lock of the Lagan canal; *Templeton* (bog disappeared long since). Fairhead; *F. Whitla.* Bog between Lisburn and Lough Neagh; *G. C. Hyndman,* 1830. Kings' moss, and Ballycastle; *Flor. Ulst.* Bog at the Lough rocks above Dunloy, and on the boggy summit of Fairhead; S.A.S.
Derry—Common throughout the county; D.M.

Order XLV. **PRIMULACEÆ.**

HOTTONIA *Linn.*

1. H. palustris *Linn.* WATER VIOLET.
Marshes, and stagnant ditches—very rare. Fl. mid. May till end of July.
Down—" Found by Mr. Richard Kennedy in the marshy ditches on the right side of the road at Everogue bridge near Downpatrick"; *Templeton.* On the lands of Inch near Downpatrick; *F. Whitla.* Still plentiful at Downpatrick; *Eds.*
Mr. Templeton planted the water-violet at Cranmore. More recently it was introduced to the Bog meadows, and has spread amazingly in the

drains there. Still later it has been brought to Holywood, and to Cushendun. The origin of this plant in Down was most probably through human agency, at no remote date.

PRIMULA *Linn.*

1. P. vulgaris *Hudson.* PRIMROSE.
Woods, hedge banks, and other shady places—very common. Fl. March till early in June. Ranges from sea level to 1200 feet.

2. P. veris *Linn.* COWSLIP.
Woods and banks—very rare in the wild state. Fl. April and May (*Cyb. Hib.*).
Down—Wood at Rostrevor; *Wm. Gray, M.R.I.A.*
Possibly introduced at Rostrevor, but this seems the only station in our district where it is probably native. The cowslip has been reported from Holywood, Cave Hill, Whitehouse, Castle Dobbs, Shane's Castle, and Crumlin, in addition to two stations in Derry—probably introduced by design, or accident in all of these localities.

LYSIMACHIA *Linn.*

1. L. vulgaris *Linn.* YELLOW LOOSETRIFE.
Marshes, and ditches by lakes and rivers—locally abundant. Fl. July and Aug.
Down—Watery places about Ballynahinch, and Lough Leagh; *Templeton*, 1793. Bow Lake near Saintfield; *Flor. Belf.*
Antrim—Sides of the lower Bann; *Templeton.* Massereene Park, and Shane's Castle, also on Ram's Island; *Flor. Ulst.* Islands in Lough Neagh south of Toome; T.H.C. River bank above Lisburn, sparingly, and about Portmore Lough, abundant, also margin of Lough Neagh near Toome; S.A.S.
Derry—In Myroe; *Samp's. Surv. Derry.* Confined to the south and east of the county, abundant by the side of Lough Neagh; D.M. Ballyronan; *Flor. Ulst.* Bannside from Coleraine to Toome; S.A.S.

2. L. nummularia *Linn.* MONEYWORT.
Wet grassy places by rivers and lakes—rare. Fl. July and Aug.
Down—Hedge side at Rathgael near Bangor; *Flor. Belf.* Stony banks of the Quoile near Downpatrick; *C. Dickson.* Marsh at Ballyalloley; R.Ll.P.
Antrim—Colin Glen, and Maryville; *Flor. Belf.* (escapes only). Dun-

silly near the town of Antrim; *G. C. Hyndman*, 1829. Ladyhill near Antrim, and run wild at Whitehouse; *Flor. Ulst.* Abundant on wet sandy shore a little west of Crumlin waterfoot; S.A.S.

Derry—On a bank near Ballinderry Rectory; *Flor. Ulst.* Abundant in marshes by the Bann above Coleraine; S.A.S.

Well naturalised by Lough Neagh, and the Quoile, but probably occurs in most other cases as merely a garden escape.

3. L. nemorum *Linn.* YELLOW PIMPERNEL.

Damp heathy or rocky places, and wet pastures—common. Fl. May—July. Ranges from sea level to over 1000 feet.

GLAUX *Tournefort.*

1. G. maritima *Linn.* BLACK SALTWORT.

On the entire coastline of the three counties, occurring at intervals—common at high water mark, and slightly above it, being especially abundant on gravelly margins, and flats which are occasionally washed by the sea. Fl. June and July.

ANAGALLIS *Tournefort.*

1. A. arvensis *Linn.* SCARLET PIMPERNEL. POOR-MAN'S WEATHER GLASS.

Sandy and gravelly waste ground, and fields—common. Fl. June—Sept.

Var. β. A. CŒRULEA *Schreb.*

Down—Narrow-water near Warrenpoint; *Templeton*, 1808.
Derry—Sandbanks at Portstewart; *Flor. Ulst.*

2. A. tenella *Linn.* BOG PIMPERNEL.

Wet sandy, or rocky places, and in bogs, and damp heaths—frequent. Fl. July and Aug.

Down—Holywood warren, and Newtownards Park; *Templeton*, 1800. Bangor; *Flor. Hib.* Scrabo; *Flor. Belf.* Castlewellan; *Flor. Ulst.* Greencastle, and wet places in sandhills at Newcastle and Ballykinler, also strand at Greyabbey, and Groomsport; S.A.S. Clandeboye; T.H.C. Bloody Bridge, Tollymore, and Ardglass; *C. Dickson.* Dundonald, Conlig, and Carngaver; R.Ll.P.

Antrim—South end of Islandmagee, also at Dervock, and on Rathlin; *Templeton.* Ballycastle, and Portrush; *Flor. Ulst.* Glen-

PRIMULACEÆ.

ariffe; *B.N.F.C.* Bogs above Carnlough, also Fairhead, Glendun, and Knocklayd; S.A.S.
Derry—In a bog southeast of Slieve Gallion, and at Magilligan; *Templeton.* Common throughout the county; D.M.

CENTUNCULUS *Linn.* BASTARD PIMPERNEL.

1. C. minimus *Linn.*

Damp gravelly places, lake shores, and sands on the coast—rare, but perhaps sometimes overlooked. Fl. July and Aug.

Down—Plentiful about the lake at Ballygowan bridge (Monlough); *Templeton,* 1807. Sandy shore three miles south of Newcastle, and on Dundrum sandhills, also shore of lake near Ballynahinch, shore of Carrickmannan Lake near Saintfield, margin of Magherascouse Lake, and gravel pit on the Newtownards road a half-mile from Comber; S.A.S. By the Quoile below Downpatrick; R.Ll.P.
Antrim—Sparingly on wet gravelly shore of Ushet Lake in Rathlin Island; S.A.S.
Derry—Abundant on sandy warrens between Downhill, and Portstewart; D.M. Benevenagh; *J. Ball.* Shore of Lough Beg; S.A.S.

SAMOLUS *Tournefort.*

1. S. valerandi *Linn.* BROOKWEED.

Marshy places by rivers and lakes, and on the coast—frequent. Fl. July and Aug.

Down—Sydenham, and Holywood; *Flor. Belf.* Groomsport, Copeland Islands, and Newcastle; *Flor. Ulst.* Ardglass; T.H.C. Strand at Quentin Castle, and Mountstewart, also Downpatrick, and Dundrum; S.A.S. Greypoint, and Comber; R.Ll.P. By the entrance of Lagan canal into Lough Neagh; *Rev. H. W. Lett.*
Antrim—Banks of the Lagan as far as the tide flows; *Templeton,* 1797. By Lough Neagh north of the Lagan canal, and at Giants' Causeway, and Rathlin Island; S.A.S.
Derry—Sides of the Foyle, and frequent on wet banks by the coast, especially near Londonderry; D.M. Castlerock; *Flor. Ulst.* Shore of Lough Beg, and by the Bann from Portstewart to Coleraine; S.A.S.

ORDER XLVI. **PLUMBAGINACEÆ.**

STATICE *Linn.* SEA LAVENDER.

1. S. bahusiensis *Fries.*
Muddy seashores—rather rare, but locally abundant. Fl. July and Aug. Down—Ardglass; *Harris's Down.* 1744. Shore of Newton Lough about a quarter mile below Newtownards; *Templeton.* Shore at Narrow-water; *Ir. Flor.* Strangford Lough near Comber; *Flor. Belf.* Holywood, Ballyholme, Castle Espie, salt marsh below Downpatrick, and at Dundrum Bay; *Flor. Ulst.* Formerly grew on shore at Sydenham; *W. Millen.* Muddy shore near the Black islands, Strangford; T.H.C. In profusion on the shore outside the railway from Warrenpoint to Newry; S.A.S. Gravelly shore on the west side of the Quoile; *C. Dickson.* Rockport; R.Ll.P. Ballymacormick Point; *Miss Reid.* Antrim—Larne Lough; *Cyb. Hib.* Muddy shore opposite the railway station at Larne; S.A.S.

ARMERIA *Willd.* THRIFT.

1. A. maritima (*Linn.*) *Willd.* SEA PINK.
Seashores, and maritime rocks—common around the coast. Fl. May till mid. Sept. Occurs at 1200 feet on Benevenagh; D.M.

ORDER XLVII. **PLANTAGINACEÆ.**

PLANTAGO *Linn.* PLANTAIN.

1. P. coronopus *Linn.* BUCKSHORN PLANTAIN.
Rocky and stony seashores—common. Fl. mid. May till mid. Aug. Ranges to 1000 feet on the basalt in Derry.

2. P. maritima *Linn.* SEA PLANTAIN.
Rocky and muddy seashores, and occasionally on mountains—very common. Fl. June—Aug. Occurs on the Carntogher Mountains at 1500 feet; D.M.

3. P. lanceolata *Linn.* COMMON PLANTAIN.

CHENOPODIACEÆ.

Roadsides, waste ground, meadows, and pastures—very common. Fl. May—July. Ranges from sea level to over 1000 feet.

4. P. major *Linn.* WAYBREAD. RATSTAIL (*loc.*).
Roadsides, pastures, and wastes—common. Fl. mid. June till mid. Aug.

LITTORELLA *Linn.* SHOREWEED.

1. L. lacustris *Linn.*
Sandy or gravelly margins of lakes—common. Fl. June and July. Ascends to 1500 feet in the Mourne Mountains (*Hart.*).

ORDER XLVIII. **CHENOPODIACEÆ.**

SUÆDA *Forsk.* SEA BLITE.

1. S. maritima (*Linn.*) *Dumort.*
Muddy seashores—common. Fl. mid. July till end of Aug.

SALSOLA *Linn.* SALTWORT.

1. S. kali *Linn.*
Sandy seashores—frequent. Fl. mid. July till end of Aug.
Down—Sands about Clough, and at Kirkiston in the Ards; *Templeton*, 1797. Along the gravelly shore from Greencastle to Kilkeel; *Ir. Flor.* Ballyholme, and Groomsport; *Flor. Belf.* Newcastle, and Dundrum Bay; *Flor. Ulst.* Ballykinler, and near Ardglass, and Millisle; S.A.S. Benderg Bay; T.H.C. Antrim—Ballycastle; *Flor. Ulst.* Cushendall; *Eds.* Cushendun; R.Ll.P.
Derry—Not common, but occurs occasionally on the shore from the river Roe to Portrush; D.M.

CHENOPODIUM *Tournefort.* GOOSEFOOT.

1. C. album *Linn.*
Cultivated fields, and waste places—common. Fl. mid. July till mid. Sept.
The variety β C. VIRENS *Linn.* seems the more common form in this district.

2. C. rubrum *Linn.*

Waste places, and fields near the sea—rare, and uncertain in its appearance; Fl. Aug. and Sept. (*Cyb. Hib.*).

Down—Side of the shore near Ballyhalbert; *Templeton*, 1799. Opposite the gasworks at Holywood; *Flor. Belf.*

Antrim--Waste ground near the new barracks, also near the dockyard at Belfast; *Templeton* (now built over).

We have other notes as to the occurrence of this species, but none that are altogether reliable. It is, however, a plant to be expected.

3. C. bonus-henricus *Linn.* ALLGOOD.

Hedge banks, and waste ground—not common—Fl. June and July.

Down—About Kirkiston, and Laganside at Lambeg; *Templeton*, 1793. Holywood, and near Newtownbreda; *Flor. Belf.* Downpatrick, and Portaferry; *Flor. Ulst.*

Antrim—Roadside to the west of Dunmurry; *Templeton.* Carrickfergus, and Ballycorr; *Flor. Ulst.* Donegore, 1865, and Glenoe, 1875; *B.N.F.C.* Roadside one mile south of Carnlough; T.H.C.

Derry—Near Muff, and near Dungiven; D.M. Roadside at Coleraine salmon leap, and at Bellarena, also near Dungiven old church; S.A.S.

BETA *Linn.* BEET.

1. B. maritima *Linn.*

Rocky, and gravelly seashores—rather rare. Fl. July—Sept.

Down—Gravelly shores between Greencastle and Kilkeel; *Ir. Flor.* Seashore near Newcastle; *Rev. W. M. Hind.* Sandy shores of Benderg, and Ballyhornan Bays; T.H.C. Killough Bay, and Kearney's Point in the Ards; S.A.S.

Antrim—Ballycastle, and Rathlin Island; *Flor. Ulst.* Shore at Magheramorne; T.H.C.

SALICORNIA *Linn.* GLASSWORT.

1. S. herbacea *Linn.*

At high-water mark on muddy seashores—common. Fl. Aug. and Sept.

ATRIPLEX *Tournefort.* ORACHE.

The Atriplices are not attractive plants, and they have secured but a small amount of attention from local botanists. The details of their distribution in this district have not been worked out, and the following notes are by no means complete.

1. A. littoralis *Linn.*
Stony or gravelly seashores—not common. Fl. mid. July till mid. Sept. The localities it will be seen are mostly about Belfast Bay.
Down—Seashore about one mile below Bangor; *Templeton*, 1797. Peoples' (now Victoria) Park; *Flor. Belf.* Dundrum Bay; *Flor. Ulst.* Seashore near Newcastle; *Rev. W. M. Hind.* Abundant on the Kinnegar of Holywood; S.A.S.
Antrim—Macedon Point, and Kilroot; S.A.S.

2. A. angustifolia *Smith.*
Fields, and waste ground—common. Fl. mid. July till mid. Sept.

3. A. erecta *Hudson.*
Marshy, and stony seashores—rare? Fl. July—Oct. (*Bab. Man.*).
Down—By the Quoile below Downpatrick; S.A.S.
Antrim—Shore at Macedon Point; T.H.C.
No doubt occurs elsewhere, but not distinguished from the other forms.

4. A. deltoidea *Bab.*
Seashores—perhaps not rare. Fl. Aug. and Sept.
  Var. β A. SALINA *Bab.*
Newtownards, and Rathlin Island; S.A.S. Probably common.

5. A. hastata *Linn.*
Seashores—common. Fl. July—Sept.

6. A. Babingtonii *Woods.*
Muddy seashores—common? Fl. July—Sept.

7. A. farinosa *Dumort.* (A. LACINIATA *Linn.* ?).
Sandy seashores—rare. Fl. mid. July till mid. Sept.
Down—Seashore at Ballyholme (still there), and sands north of Newcastle; *Templeton*, 1805. Holywood, and Groomsport; *Flor. Belf.* Sandy seashore opposite Gunn's Island south of Strangford, and sparingly on shore north of Greencastle; S.A.S.
Antrim—Shore of Islandmagee; *Templeton*, 1808.
This is a plant too well marked to be overlooked, and its rarity, or absence in Derry is real.

8 A. portulacoides *Linn.* (OBIONE *Gaert.*). SEA PURSLANE.

Muddy seashore— very rare. Fl. Aug. and Sept. (*Cyb. Hib.*).
Down—Plentiful by the canal a short distance below Narrow-water
Castle north of Warrenpoint; S.A.S., Sept., 1882. *Proc.
Belf. Nat. Field Club.*, 1882-83, p. 184.
Found subsequently, but independently, in same place by *H. C. Hart*,
in July, 1883.

ORDER XLIX. **POLYGONACEÆ.**

RUMEX *Linn.* DOCKEN.

1. R. conglomeratus *Murray.*
Roadsides, and waste ground—common. Fl. July and Aug.

2. R. sanguineus *Linn.*
Var. β R. VIRIDIS *Sibthorp.*
Ditch banks, and damp shady places—common. Fl. July and Aug.
The red-veined, typical form is quite rare.

3. R. obtusifolius *Linn.* COMMON DOCKEN.
Roadsides, waste places, and by streams—very common. Fl. July till
till mid. Sept. Ranges to 1000 feet on the hills.

4. R. crispus *Linn.* CURLED DOCKEN.
Roadsides, waste places, and banks of streams—very common. Fl.
June—Aug. Ranges from sea level to 1000 feet.

5. R. hydrolapathum *Hudson.* GREAT WATER DOCK.
Marshy places by rivers and ponds—rare. Fl. July and Aug.
Down—Ballygowan Lough; *Flor. Ulst.* Plentiful in a ditch near the
windmill at Bangor; T.H.C. Abundant on old mill dam
behind Ballyholme; R.Ll.P. Wooded banks of the Quoile
below Inch Abbey; S.A.S.
Antrim—In marshes near Lough Neagh; *Templeton*, 1793. By the
Bann; *Cyb. Hib.*
Derry—By Clady River west of Portglenone; *Wade Rar.* In a small
lake at Ballymacpeak; *Flor. Ulst.*

6. R. acetosa *Linn.* SORREL.
Damp shady places in pastures, and woods, and among rocks—common.
Fl. June—Aug.

7. R. acetosella *Linn.* SHEEPS' SORREL.
Heaths, bogs, and bare stony pastures—very common. Fl. June—Aug.

POLYGONUM *Linn.*

1. P. amphibium *Linn.*
Lakes, ponds, ditches, slow streams, and marshes—common. Fl. mid. June till end of Aug.

2. P. lapathifolium *Linn.*
Peaty, and damp sandy fields—frequent in the south of the district. Fl. July and Aug.
Down—Mountpottinger, Ballygowan, and Cotton moss; *Flor. Belf.* (*sub P. nodosum*). Belmont, Dundonald, Newtownards, Monlough, and bogs about Ballynahinch; S.A.S.
Antrim—Near Belfast; *Flor. Hib.* Shore road; *Flor. Belf.* Abundant by Lagan canal above Moira station; S.A.S. Whitewell; R.Ll.P.
Derry—Ballyronan, and Moneymore; *Flor. Ulst.*

3. P. persicaria *Linn.* REDSHANK (loc.).
Cultivated fields, and waste places—very common. Fl. mid. June till mid. Sept.

4. P. hydropiper *Linn.* WATER PEPPER.
In marshes, and by streams—common. Fl. Aug. and Sept.

5. P. minus *Hudson.*
Sandy and gravelly lake, and river margins—rare. Fl. mid. July till mid. Sept.
Down—Ballygowan Lake; *Templeton*, 1807. Margin of Lough Henny near Saintfield; *Flor. Belf.* Abundant by the margin of Monlough; *Cyb. Hib.* Downpatrick, and by Newry canal occasionally; S.A.S Ballynahinch and by the lake at Clandeboye; R.Ll.P.
Antrim—Dam above Whitehouse; *Flor. Ulst.*
Derry—Side of the Bann below Coleraine; *Flor. Hib.* Sparingly by the Bann above the Cutts; D.M.

6. P. aviculare *Linn.* KNOTGRASS.
Waysides, and waste ground—very common. Fl. June—Aug.
Var. β P. LITTORALE *Link*.
This form, which is often taken for the following species, occurs on

sandy, and gravelly shores at Groomsport, Ballyholme, Cultra, an Portstewart.

7. P. Raii *Babington.*
Sandy seashores—rather rare. Fl. Aug. and Sept.
Down—Shore at Holywood, and among drifting sand on the seashore about Kirkiston, and other places; *Templeton*,1800. Sandy shores at Ballyholme, and Groomsport; *Flor. Belf.* Newcastle, and Dundrum; *Flor. Ulst.* Sandy shores at Greencastle, and north of Ardglass; S.A.S.
Antrim—Sands near Ballintoy; *Cyb. Hib.*
Derry—Shores of Lough Foyle; *Cyb. Hib.*

8. P. convolvulus *Linn.* BLACK BINDWEED.
Cultivated fields, and waste places, especially in sandy land—common. Fl. July and Aug.

ORDER L. **EMPETRACEÆ.**

EMPETRUM *Linn.*

1. E. nigrum *Linn.* CROWBERRY.
Heaths—frequent, especially on the hills. Fl. April and May. Ranges from less than 100, to over 2700 feet.
Down—Top of Slieve Donard; *Templeton*, 1793. Cotton moss; T.H.C. Frequent on the Mourne Mountains, and on Slieve Croob; S.A.S.
Antrim—Summit of Divis, and among heath on hills from Portrush to Glenarm; S A.S. Fairhead, and Slievenanee; T.H.C.
Derry—At 2236 feet on the top of Sawel, and frequent on mountains; D.M. Abundant on the Sperrin Mountains; S.A.S. Mullaghmore; *H. C. Hart, Proc. R.I.A.*

ORDER LI. **EUPHORBIACEÆ.**

EUPHORBIA *Linn.* SPURGE.

1. E. helioscopia *Linn.* SUN SPURGE. DEVILS' CHURNSTAFF (*loc.*).

EUPHORBIACEÆ.

Borders of fields, mainly in cultivated land—common. Fl. mid. June till mid. Sept.

2. E. paralias *Linn.* SEA SPURGE.

Sand dunes, and loose drifting sands by the coast—not uncommon in suitable places. Fl. mid. July till mid. Sept.

Down—Shore between Greencastle and Kilkeel; *Ir. Flor.* Shore at Newcastle; *Flor. Ulst.* Sandhills at Ballykinler; T.H.C. Abundant on shore north of Greencastle, and at Ballykinler; S.A.S.

Antrim—Sandhills at Bushfoot; *B.N.F.C.*, 1868.

3. E. portlandica *Linn.*

Loose sands on the coast—frequent where suitable *habitat* offers. Fl. June—Aug.

Down—Greencastle; *Ir. Flor.* Dundrum; *Flor. Ulst.* Ringsallin Point; T.H.C. Sparingly on the shore north of Greencastle, also at Ballykinler, and occasionally to Rathmullen; S A.S.

Antrim—Sandy warren at Bushmills; *Cyb. Hib.* Plentiful at the Bushfoot, and at Portrush; S.A.S.

4. E. peplus *Linn.* PETTY SPURGE.

Borders of fields, and on waste ground near cultivation—common. Fl. June—Aug.

5. E. exigua *Linn.*

Cultivated fields, and sandy banks—rare, but locally abundant. Fl. mid. July till mid. Sept.

Down—Cherryvalley near Comber, and at Groomsport; *Flor. Belf.* Pastures about Holywood and Bangor; *Flor. Ulst.* Ballyholme, and Ballycroghan; T.H.C. Abundant in cornfields about Newcastle, Bryansford, and Dundrum; *C. Dickson.* Dundonald; R.Ll.P.

Antrim—Gravel pit at Lambeg, and field at Derriaghy; *Templeton*, 1804. Near Randalstown, Carrickfergus, and Glenarm; *Flor. Ulst.* Railway banks from Antrim to Ballinderry, roadside near Ballymena, and fields at north end of Islandmagee; S.A.S. Sandy fields by Crumlin River; T.H.C. Donegore; R.Ll.P.

Derry—Common in cornfields, and in sandy or gravelly land; D.M.

MERCURIALIS *Tournefort.* MERCURY.

1. M. perennis *Linn.*

Bushy, and shady places—very rare. Fl. April and May.

Down—Marino, near Holywood; (*Grainger*) *Flor. Belf.* (not seen recently). Glen near Loughbrickland; *Rev. H. W. Lett.*
Antrim—Altaferna Glen near Ballycastle; *Flor. Ulst.* By Lough Neagh at Langford Lodge; *Miss Maffett.* Among bushes by the rocky river bank at Glynn near Larne; *Rev. Geo. Robinson.*

Order LII. **CERATOPHYLLACEÆ**.

CERATOPHYLLUM *Linn.* Hornwort.

1. C. demersum *Linn.*
Lakes, and drains—rare. Fl. June and July (*Cyb. Hib.*).
Down—In Lough Leagh near Killyleagh, 1804, and Lake near Ballynahinch; *Templeton*, 1808, also *Hyndman*, 1832 (spec.). Ditches by the Quoile below Downpatrick, and in Lough Aghery, which is four miles west of Ballynahinch; S.A.S.
Antrim—In the large lake at the southern end of Rathlin Island; *Templeton*, 1795. Still to be found in the lake referred to, which is Rinalvin or Ushet Lake; S.A.S.
Derry—By the Foyle above Derry; *Flor. Ulst.* In a ditch by the Bann below Coleraine; *Cyb. Hib.*

The Lough Neagh record was an error. Mr. Templeton's note is Lough Leagh.

Order LIII. **CALLITRICHACEÆ**.

CALLITRICHE *Linn.* Water Starwort.

1. C. verna *Linn.*
Ditches, pools, and slow streams—common. Fl. May—Sept. In mild seasons flowers may be found in March.

2. C. stagnalis *Scop.* (C. PLATYCARPA *Kuetz.*).
Marshes, and stagnant ditches—frequent. Fl. May—Sept.

3. C. hamulata *Kuetz.*
Lakes, streams, and ditches—common. Fl. June—Sept. Sometimes very luxuriant, and then usually barren,

Var. β C. PEDUNCULATA D.C.
Down—Marshes by the Quoile below Downpatrick, very small—perhaps not rare.

4. C. autumnalis *Linn.*
Lakes, and slow streams—rare. Fl. mid. June till end of Aug.
Down—Canal at Newry; *W. Thompson.* Plentiful in Lough Aghery near Ballynahinch, and sparingly in Carrickmannan Lake near Saintfield; S.A.S. Dunbeg Lake southwest of Ballynahinch; R.Ll.P.
Derry—Abundant by the Bann between Lough Beg and Coleraine; D.M. In the Bann at the salmon leap above Coleraine; S.A.S.

Templeton noted this plant as occurring in ditches by Lough Neagh, but there is reason for thinking that he was mistaken. Similar stations were assigned by Whitla, and Moore, but it is probable that *C. hamulata*, which they do not mention, was the plant really seen in the Portmore ditches, and not *C. autumnalis*, which is a denizen of clear water.

ORDER LIV. **URTICACEÆ.**

PARIETARIA *Tournefort.* PELLITORY.

1. P. officinalis *Linn.* WALL PELLITORY.
Old walls—frequent, and occasionally on stony banks. Fl. June—Sept.
Down—Walls of the Long Bridge; *Templeton,* 1797. Among stones in the Peoples (Victoria) Park; *Flor. Belf.* Newtownards; *Flor. Ulst.* Walls of Greyabbey, and at Warrenpoint, and Rostrevor; S.A.S. Conlig, and shore at Craigdarragh; R.Ll.P. Bridge at Tullycairne, and walls of old castles of Dromore, and Dundrum; T.H.C.
Antrim—Common on old walls; *Templeton.* Walls at Carrickfergus, and Lisburn; *Flor. Belf.* Larne, and Antrim; *Flor. Ulst.* In great abundance on walls of bridge, and other old walls by the river at Antrim, and on old Shane's Castle; S.A.S.
Derry—Abundant on the old walls of Londonderry, and of Salterstown Castle; D M.

URTICA *Tournefort.* NETTLE.

1. U. urens *Linn.*

Waste ground, and about houses and gardens—common, but prefers a light sandy soil, and is absent from stiff clay lands. Fl. June—Aug.

2. U. dioica *Linn.*
Waste ground, and about houses and roadsides—very common. Fl. June—Aug. Ranges from sea level to over 1000 feet on the mountains.

ORDER LV. **ULMACEÆ**.

ULMUS *Linn.* ELM.

1. U. montana *With.* WYCH ELM.
Woods, glens, and river banks—frequent. Fl. March and April. "Has all appearance of being native"; *Templeton*. It is difficult to distinguish the native from the introduced localities occupied by this tree, but it seems certainly wild in many glens and uncultivated places.

ORDER LVI. **AMENTIFERÆ**.

SALIX *Tournefort.* WILLOW, SALLOW.

This was a pet genus with the last generation of botanists, and has been overdone. A large proportion of the varieties noted are unrecognisable, and at present it would be useless to attempt to indicate, in many cases, what plant was really observed. There is also much difficulty is distinguishing the native from the cultivated forms. More careful work is needed, and notes to be of any value, should be fortified by satisfactory specimens.

1. S. pentandra *Linn.* BAY WILLOW.
Glens, and bushy places—frequent, especially in hilly and rocky districts. Fl. mid. May till end of June. Ranges from sea level to 800 feet.
Down—Ballyalloley, and Ballycroghan; *Flor. Belf.* Groomsport, and roadside near Moneyreagh; S.A.S. By the Kilbroney River, and at Dundonald, and Craigauntlet; R.Ll.P. Between Kilkeel and Newcastle; *C. Dickson.*

AMENTIFERÆ.

Antrim—In a bog west of Ballycastle, and amongst rocks at Fairhead; *Templeton*, 1793. Derriaghy *Flor. Belf.* Roadside between Glenarm and Cushendall; *Flor. Ulst.* Glenarm; *R. Tate.* By the river in Glenshesk; T.H.C. Frequent by the roadside from Ligoniel to Crumlin, also in Crow Glen, and several places about Belfast; S.A.S.
Derry—Very general through the county; D.M.

2. S. alba *Linn.* WHITE WILLOW.
River banks, and in glens—frequent. Fl. mid. April till end of May.
Often planted, but often spontaneous, and if not native is certainly completely naturalised.

3. S. triandra *Linn.* FRENCH WILLOW.
Hedges and thickets—rare. Fl. mid. April till end of May.
Down—Ballyalloley; *Flor. Belf.*
Antrim—Portmore, and fields between Malone and the Falls; *F. Whitla.* In the Bog meadows near Balmoral, and (*var. Hoffmani*) amongst natural wood by the Sixmilewater at Antrim; S.A.S.
The stations assigned to this willow, at the Bann, and Fairhead, by Flora Hibernica, were errors, which were copied into the Flora of Ulster. Mr. Templeton quoted those localities for sp. 1, and seems only to have known the present plant as cultivated in his garden.

4. S. purpurea *Linn.* ROSE WILLOW.
Roadsides, ditches, and other wet places—rare. Fl. April and May.
Often planted, but seems wild in some of the stations.
Antrim—Wet places above Carrickfergus; *Flor. Ulst.* Shores of Lough Neagh at Shane's Castle, and near Ballycastle (*S. helix*), and also by the river at Dervock (*var. Lambertiana*); *Cyb. Hib.* By the Lagan below the first lock, and by the Forth River, also at Whiterock, base of Knocklayd, and south of Giants' Causeway; S.A.S. Ballycastle; R.Ll.P.
Derry—River side in Glendermot, and by the Faughan (*S. helix*); D.M.

5. S. viminalis *Linn.* OSIER.
By streams, and roadside ditches—not rare. Naturalised in many marshes, possibly native in the mountain districts. Cultivated in osier grounds. Fl. April and May.

6. S. Smithiana *Willd.*
Hedge banks, thickets, and glens—not rare, especially in hilly districts. Fl. April and May. Has all appearance of being a native.
Down—Ballycroghan near Bangor, also in marshes at Knock, and Castle-

reagh; *Flor. Belf.* Near Newtownbreda; T.H.C. Cregagh Glen; S.A.S.
Antrim—Bog meadows; T.H.C. By the Forth River, also Killead, Carnmoney, Kilroot, and southern slopes of Cave Hill; S.A.S.
Derry—By the side of the Curlyburn near Drumachose old church, and not unfrequent in the county; D.M. Toome; S.A.S.
[S. ACUMINATA *Smith.*
By the Lagan; *Templeton.* The notes respecting this plant in the Flora of Ulster, and in Flora Belfastiensis are of no authority, and if Mr. Templeton's plant was correctly named it is now lost. There is, however, some reason to think it was really the preceding species.]

7. S. cinerea *Linn.* COMMON SALLOW.
In hedges and thickets, and by streams—very common. Fl. March and April.
Var. β S. AQUATICA *Smith.*
Common throughout the district.

8. S. aurita *Linn.*
Ditch banks, and waste ground in sandy, rocky, and heathy places—very common. Fl. April and May. Ranges from sea level to 1500 feet.

9. S. caprea *Linn.*
Rocks, and banks—frequent. Fl. April and May. Rises to 1000 feet on the hills. There seem to be few notes of definite localities for this willow, but it grows mostly in mountain districts, as Knock Glen; T.H.C. Colin Glen; *Templeton.* Hannahstown, Crow Glen, Oldpark, Woodburn, and Sallagh Braes; S.A.S. Cushendall; R.Ll.P.

10. S. nigricans *Smith.*
Glens, and rocky places—rare. Fl. April—June (*Cyb. Hib.*).
Antrim—Rocks at Skerries, in the Braid; *Templeton,* 1809 (var. *rupestris*). Kilwaughter, and Glenoe; *Flor. Ulst.* (var. *hirta*). Between Larne and Ballymena, also in Glendun, and Glenariffe; *Cyb. Hib.*
Derry—Bennedy Glen near Dungiven; D.M. (var. *Forsterianum*).

11. S. phylicifolia *Linn.*
Hedges and thickets—rare. Fl. April and May (*Cyb. Hib.*).
Antrim—Glenballyeamon; *Cyb. Hib.*
Derry—Shore of Enagh Lough; D.M.
Var. S. LAURINA *Smith.*
Antrim—Shore of Lough Neagh near Massereene Park; *Cyb. Hib.*

Derry—In a moist bushy place on the roadside between Castledawson and Bellaghy ; D.M.

12. S. repens *Linn.*

Bare sandy, or heathy ground on the coast, and by lakes, and on rocky mountain pastures—common. Fl. April and May. There are several puzzling varieties of this species, and we have not sufficiently distinguished them. The following have been noted.

Var. S. PROSTRATA *Smith.*
Common in rushy ground ; *Templeton.* Cavehill ; *Flor. Hib.* Scrabo, and Rathlin ; *Flor. Ulst.*

Var. S. INCUBACEA *Linn.*
Near the lint mill at Errigal banks near Garvagh, Co. Derry ; D.M.

Var. S. ARGENTEA *Smith.*
In the bog at Kirkiston, Co. Down, and on blown sands at same place ; *Templeton*, 1810.

13. S. herbacea *Linn.* DWARF MOUNTAIN WILLOW.

Crevices of rocks on the higher mountains—rare. Fl. June and early half of July. Altitudinal range 1000 feet on Clontygearagh, Co. Derry ; D.M., to 2700 on Slieve Donard.

Down.—Scallaree, and Cairn rocks west of Bencrom, and rocks on Slievenamady ; *Wade Rar.* Slieve Donard, and the Diamond Mountain ; *Templeton*, 1808. Summit of Slieve Bignian ; S.A.S. Slieve Bernagh ; *C. Dickson.* Slieve Commedagh ; *H. C. Hart, Proc. R.I.A.*

Antrim—Slievenanee ; *Flor. Hib.* North side of Slievenamon ; *Cyb. Hib.* (? Slievenanee).

Derry—Abundant on Dart Mountain, also occurs on Clontygeragh in Ballynascreen ; D.M. Benevenagh ; *J. Ball.* Mullaghmore ; *H. C. Hart, op. cit.*

POPULUS *Tournefort.* POPLAR.

1. P. tremula *Linn.* ASPEN.

Rocks, and glens in mountain districts—frequent. Fl. from late in March till early in May.

Antrim—Rocks on east side of Agnew's Hill, and rocks at Redbay ; *Templeton,* 1794. Cave Hill, and cliffs at Glenarm ; *Flor. Ulst.* On cliffs at Sallagh Braes, and at north end of Rathlin (stunted) ; S.A.S. Carrick-a-rede ; T.H.C. Rocks by the Inver River, and at Knockagh, Fairhead, and Glenariffe ; R.Ll.P.

Derry—At the waterfall in the Ness Glen ; *Templeton.* Altahara Glen Draperstown, also Errigal banks. and Umbra rocks ; D.M.

AMENTIFERÆ.

MYRICA *Linn.* BOG MYRTLE.

1. M. gale *Linn.* SWEET GALE.
Peat bogs—common. Fl. May and June.

BETULA *Tournefort.* BIRCH.

1. B. glutinosa *Fries.*
Rocks, and glens, and by streams in hilly districts—frequent. Fl. May and June. Ascends to 1200 feet in Derry; D.M.
Down—Tollymore Park; *Templeton*, 1794. By the Connswater; S.A.S.
Antrim—Shores of Lough Neagh, and in the park at Glenarm, and on the mountain road above Carnlough; *Templeton*, 1810. Colin Glen; *Flor. Belf.* Rocks near Carrick-a-rede; T.H.C. Whitewell, Glendun, and on the rocks and undercliff at Fairhead; S.A.S.
Derry—Common by mountain streams, and in glens; D.M. Rathmelton; *Flor. Ulst.*

Var. β B. PUBESCENS *Ehr.*
On the basaltic cliffs of Sallagh Braes; S.A.S.

We have not seen any birches that could be referred to B. VERRUCOSA *Ehr.* (B. ALBA *Koch*). The segregate forms were not distinguished by the older botanists, and we have placed their records to that species which we know does exist here. The var. *pubescens* no doubt occurs in other localities than that given above. Mr. Templeton says of the birch, "There appear evidently two kinds, one with smooth leaves, toothed to the peduncle, and smooth twigs. The other, the most common, having downy leaves, entire at the base, and downy twigs. There is also a variety, along the river at Tollymore, which I think like the var. *pendula*."

ALNUS *Tournefort.* ALDER.

1. A. glutinosa (*Linn.*) *Gaert.*
In damp thickets, and by streams and lake shores—very common. Fl. mid. Feb. till mid. April.

QUERCUS *Tournefort.* OAK.

1. Q. robur *Linn.*
Woods, thickets, and mountain glens—common, an ancient native, though often planted in places where now found. Fl. mid. April till late in May. The former abundance of the oak is shown by its frequent

remains in bogs, and by ancient local names. The var. *sessiliflora* is said to occur, and Templeton states that the old wood of Glenarm deerpark had been of this variety, but long since cut down.

CORYLUS *Linn.* HAZEL.

1. C. avellana *Linn.*
Shady glens, by streams, and on the lower slopes of the mountains— common. Fl. mid. Feb. till mid. April.

ORDER LVII. **CONIFERÆ.**

TAXUS *Tournefort.* YEW.

1. T. baccata *Linn.*
Rocks and cliffs—extremely rare, not seen recently. Fl. March (*Cyb. Hib.*).
Down—We find no record of the occurrence of the yew in Down, but the town of Newry (formerly the Newries) derives its name from this tree, and thus we know that at an ancient date, it flourished in south Down.
Antrim—"Among the rocks at Glenariffe on the north side about half way up the glen. According to tradition it was plentiful about this place, being used to timber cottages. It is now (1795) reduced to a few stunted plants growing out of the crevices of the rocks, only accessible to a man lowered from above on a rope to gather branches, in order to decorate the chapels and churches at Christmas"; *Templeton.* Rocks overhanging the sea in the deerpark at Glenarm; *Ir. Flor.* As far as we can ascertain was last seen about the commencement of the century, and is most probably now extinct in the county (*Eds.*).
Derry—At 1000 to 1200 feet on Benevenagh, where it spreads close to the face of the rock. Formerly very abundant, trunks and roots of large size being still dug up in the sandy flat of Magilligan; D.M., 1835.

JUNIPERUS *Linn.* JUNIPER.

1. J. nana *Willd.* DWARF MOUNTAIN JUNIPER.
Heaths, and rocks—not common, though in some places abundant.

Fl. May and June. Ranges from 300 to over 2000 feet. We refer all previous notes to the above, typical *J. communis* not being found.

Down—" In great plenty ;" *Harris's Down*, 1744. In several places of the Mourne Mountains ; *Templeton*, 1793. Slieve Donard; *B.N.F.C.* Cove Mountain ; S.A.S. Eagle cliffs Slieve Bernagh, and on rocks of Commedagh; T.H.C.

Antrim—* Shane's Hill, Knocklayd, and rocks about Glenariffe ; *Templeton*. About Cairncastle, and at Ballylig above Ballycastle ; *Ir. Flor.* Glen of Altmore ; *Flor. Ulst.* Black Mountain near Belfast ; *Flor. Belf.* (limited here, apparently, to three or four plants above the "windy gap." *Eds.*). Cliffs south of Carrick-a-rede ; T.H.C. Glendun ; R.Ll.P. Garron Head; *W. H. Phillips*.

Derry—Ballygaylagh on the coast; *Samp's. Surv.* Only observed on Benevenagh ; D.M.

* Not Shane's Castle as stated in the Flora of Ulster.

PINUS *Linn.* PINE.

[P. SYLVESTRIS *Linn.* SCOTCH FIR.

Abundant in prehistoric times, as shown by the profusion in which it is found, below the peat, in our bogs. There is, however, very little evidence to prove that any trees of the Scotch fir, now growing in Ireland, have descended from the indigenous stock. As we learn from Joyce, the Irish term for this tree—*Giumhas (guse)* has seldom been used in naming localities ; and he further points out, that, in the few cases where the name does occur, the reference may have been to the abundance of fir-wood in the bogs. The philological argument, which, when applied to the yew, proves its former abundance, would, as respects the Scotch fir, indicate its comparative rarity, during the times when this country was occupied by the races who imposed our local Celtic names.]

ORDER LVIII. **HYDROCHARIDACEÆ.**

HYDROCHARIS *Linn.*

1. H. morsus-ranæ *Linn.* FROGBIT.

Stagnant ditches —very rare. We have not met with any flowers, though frequently sought for in July and August.

Down—In a bog hole near Portaferry ; *Templeton*.

Antrim—In the drains at Portmore Park ; *Templeton*, 1793. Templeton's locality covers all subsequent records. Not in *all* the drains at Portmore, as stated in Cybele Hibernica, but abundant in *some* of the drains at the south and west of the little lake.

ELODEA *Michx.*

[E. CANADENSIS *Mich.* (Anacharis alsinastrum *Bab.*). WATER THYME. Rivers, lakes, dams, and canals—abundant. Fl. July and Aug.

This American immigrant was, no doubt, introduced by means of square timber imported into Belfast from Canada, but the exact date of its appearance in our waters cannot be fixed with certainty. In Hooker's Students' Flora it is stated to have been first observed in a pond at Waringstown, Co. Down, in 1836. Dr. Dickie is cited as authority, his note on the subject having appeared in the Phylologist, vol. 5, 1854. This statement seems improbable, and in Dickie's Flora of Ulster, published in 1864, no notice is taken of it. The *Anacharis* is there referred to as having appeared in the canal near Lisburn about the year 1844.

Belfast was probably the original Irish centre of this plant, but it must have been imported to England independently. If brought to Ireland with American wood, as suggested above, it is easy to understand its spread from north to south. The logs are launched out of the ship's hold, and an incoming tide carries a sprig, or sprigs, of the plant above the bridge, to pass thence into the freshwater of the canal; or logs towed up the river could carry water plants, or seeds, adhering to them. Once in the canal the way was clear to Lough Neagh, and thence transport by birds or other agencies was easy.

The *Anacharis* has not yet made its way to Rathlin Island, but stated in Allins' Flora to have arrived at Cork about 1851. The following localities are put on record, the better to enable botanists, at a future time, to decide whether this plant holds the ground which it has gained.

Lagan canal, Lough Neagh, Portmore Lake, the Bann, Springfield, and other mill dams about Belfast. Ballynahinch Lake, Lough Aghery, the Quoile, Newry canal, and many other waters.]

ORDER LIX.   **ORCHIDACEÆ.**

ORCHIS *Linn.*

1. O. mascula *Linn.* EARLY PURPLE ORCHIS.
Damp pastures, and heaths—common. Fl. mid. April till early in June. Altitudinal range, from sea level to 1000 feet.

2. O. maculata *Linn.* SPOTTED ORCHIS.
Meadows, wet pastures, and heaths—common. Fl. June and July. Ranges from sea level to over 2000 feet.

3. O. incarnata *Linn.* MARSH ORCHIS.
Wet meadows, and marshy pastures—frequent. Fl. mid. June till end of July. Ranges up to 1000 feet.

The records here quoted from various authorities were originally attached to *O. latifolia.* Satisfactory specimens of that form are, however, not forthcoming, and we do not hesitate to transfer the notes to the present species, which was, no doubt, the plant observed.

Down—Holywood ; *Flor. Ulst.* Hillsborough, Conlig, Groomsport, and Donaghadee ; S.A.S.
Antrim—Fields near Cushendall ; *G. C. Hyndman.* Loughmourne, and marshes on the Belfast hill range ; *Flor. Belf.* Colin Glen, and Kings' Moss ; *Flor. Ulst.* Templepatrick, Knockagh, Carrick commons, and frequent by Lough Neagh, and in marshy ground all along the coast ; S.A.S.
Derry—Magilligan, and common throughout the county ; D.M. Toome ; T.H.C. Slieve Gallion ; S.A.S.

Var. β *O.* ANGUSTIFOLIA *Bab.*
Down—Sparingly on Carngaver Hill ; S.A.S.

GYMNADENIA *R. Brown.*

1. G. conopsea (*Linn.*) *Brown.*
Grassy heaths, and damp pastures in moory districts—rare. Fl. mid. June till end of July. Ranges from sea level to nearly 1000 feet.

Down—Moist places in deerpark at Newtownards ; *Templeton*, 1803. Conlig Hill ; *Flor. Belf.* Carngaver Hill ; R.Ll.P. West side of Conlig Hill (white flowered) ; S.A.S.
Antrim—Seaside at Cushendall ; *Templeton,* 1809. Shore of Lough Neagh at Massereene Park ; *G. C. Hyndman.* Islandmagee ; *Flor. Ulst.* Damp field above the fall at Glenoe, also at Garron Head, and Redbay, T.H.C. Northern slopes of Glenariffe, and pastures in the deerpark at Glenarm ; S.A.S. Carnlough, Altmore, Lurigethan, and Murlough ; R.Ll.P.
Derry—Abundant at the base of Umbra rocks ; D.M.

2. G. albida (*Swartz*) *Rich.*
Mountain pastures—rare. Fl. mid. June till end of July. Ranges from 200 to 800 feet.

Down—Sparingly on Carngaver Hill above Holywood ; S.A.S., 1883.
Antrim—Southern side of Black Mountain, and at the Headwood near the

source of the Sixmilewater, and on Slemish; *Templeton*, 1793, and 1803. Cave Hill; *Flor. Belf.* Squire's Hill, Carrick commons, and pastures one mile east of Ballynure; *Flor. Ulst.* One plant seen on Cave Hill below the cliffs close to Thronemount; *C. Dickson*, 1883. Scawt Hill north of Larne; R.Ll.P.

Derry—Side of a glen in Ballynascreen mountains; *Templeton*, 1800. Rare, but plentiful in the townland of Carus, parish of Dungiven; D.M.

HABENARIA *Swartz.*

1. H. viridis (*Linn.*) *Brown.*

Short heathy pastures—locally frequent. It is rare on the silicious rocks of Down, and frequent, but not abundant on the basaltic range of Antrim and Derry. Fl. July and first half of Aug.

Down—Carngaver Hill; R.Ll.P.

Antrim—Face of the Black Mountain, and other like places; *Templeton*, 1797. Belfast Hills; *Flor. Belf.* Kings' Moss, Woodburn, and Straid; *Flor. Ulst.* Knockagh; *W. Millen.* Glenariffe; *B.N.F.C.*, 1875. Shane's Castle, Broughshane, and Glenarm Park; S.A.S. Scawt Hill; R.Ll.P.

Derry—Most abundant (?) all over the basaltic district; D.M. Sparingly on Benbradagh Mountain, and Keady Hill; S.A.S.

2. H. bifolia *R. Brown.*

Marshy pastures, and heaths, mainly on the mountains—rather rare. Fl. mid. June till end of July. Ranges from near sea level to over 1000 feet.

Down—Carngaver Hill; S.A.S. Ballymenoch near Holywood; R.Ll.P.

Antrim—On the hills above Hannahstown; *Flor. Belf.* Black Mountain, Carmavy, marshy places in Glendun, and hills above Ballycastle; S.A.S. Cushendall; R.Ll.P.

Derry—In the Bennedy Glen near Dungiven; S.A.S. Doubtless will be found in many other parts of the county.

3. H. chlorantha *Babington.*

Wet pastures in moory districts—frequent. Fl. July and early part of Aug. Ranges from 100 to over 1000 feet.

Down—Aughnadarragh Lake near Saintfield, Lough Aghery, Hillsborough Park, Clandeboye, and hills above Dundonald, and Holywood; S.A.S.

Antrim—Belfast hills; *Flor. Belf.* Carrick commons; *B.N.F.C.*, 1863. Shane's Castle, Drumadarragh, Carmavy, Sallagh Braes, Glenarm, and Glendun; S.A.S.

138   *ORCHIDACEÆ.*

Derry—Slieve Gallion, and Benbradagh; S.A.S.

The notes respecting this and the preceding species are insufficient, mainly because the old records were not available. Only in the Flora Belfastiensis and the Cybele Hibernica were the two plants clearly distinguished. The notes in the Flora of Ulster are misleading, and the names require to be transposed.

LISTERA *R. Brown.*

1. L. ovata *(Linn.) Brown.* TWAYBLADE.

Woods, glens, and shady rocks—common. Fl. mid. June till mid. Aug. Ranges from sea level to over 1000 feet.

2. L. cordata *(Linn.) Brown.* LESSER TWAYBLADE.

Amongst moss on wet heaths at elevations up to 1700 feet in the Sperrin range, and 2000 in Mourne Mountains—rather rare. Fl. June and July.

Down—On the heath of Newtownards Park; *Templeton.* Slieve Donard, and among heath on hills above Dundonald; *Flor. Ulst.* Luxuriant, and sporting in a bog (Thorn Island) 2½ miles southwest of Donaghadee; S.A.S.
Antrim—Among heath on the Clough side of Slievenanee, and on heathy grounds below the rocks of Agnew's Hill; *Templeton,* 1804. Summit of Black Mountain; *Flor. Ulst.* Cushendall; *Cyb. Hib.* Black Mountain above "windy gap"; S.A.S. Refound on Slievenanee; R.Ll.P., 1886.
Derry—Northeast side of Slieve Gallion near the Cairn; D.M. Near Moneymore; *Flor. Hib.* On the Sperrin Mountains, and plentiful on White Mountain, Brown Hill, and Mullaghmore to the southwest of Dungiven; S.A.S.

First found in Ireland by Dr. Whitley Stokes.

NEOTTIA *Linn.*

1. N. nidus-avis *(Linn.) Rich.*

In woods, parasitic on the roots of various trees—rare. Fl. June and July.

Down—Tollymore Park, and one plant seen at Belvoir Park near Belfast; *C. Dickson.*
Antrim—At the root of a Scotch pine on the southern side of the glen at Sinclaire's green (now Falls Park); *Miss Ellen Templeton,* 1807. Refound at same place, by Mr. Templeton, and Dr. Taylor, 1814. Colin Glen; *Flor. Hib.* (still there). Near Crumlin; *F. Whitla.* Glenarm; *Robt. Patterson,* and subsequently (1866); *B.N.F.C.* Glen of Altmore near Cushen-

ORCHIDACEÆ. 139

dall, and at Drumnasole; *W. Thompson.* Shane's Castle; *B.N.F.C.*, 1865. Massereene Park; *A. G. More.* Near Randalstown; T.H.C. On beech at Muckamore; S.A.S.
Derry—Errigal banks, Castledawson woods, and at Garvagh; D.M.

EPIPACTIS *Rich.* HELLEBORINE.

1. E. latifolia (*Linn.*) *Swartz.*
Woods, thickets, and shady roadside banks—frequent. Fl. mid. July till mid. Sept.
Down—Plantations at Belvoir; *Templeton,* 1793. Marino; *Flor. Belf.* Holywood; *Flor. Ulst.* Hillsborough; *Cyb. Hib.* Clandeboye, and Dundonald, also at Gillhall near Dromore; T.H.C. Cultra; R.Ll.P. Rademon; *C. Dickson.*
Antrim—Malone, and Colin Glen; *Flor. Belf.* Greencastle, Whitehead, Redhall, and abundant in Glenarm Park; *Flor. Ulst.* At Muckamore, and abundant from Crumlin to Langford Lodge; S.A.S. Ligoniel, and Shane's Castle; T.H.C.
Derry—Abundant throughout the county; D.M. Woods about Coleraine, and in Bond's glen; *Cyb. Hib.*

Var. β E. MEDIA *Fries.*
Damp woods—rare.
Sparingly in Glenarm Park; S.A.S.

2. E. palustris *Swartz.*
Wet boggy places—very rare, cannot now be found, and is probably extinct here.
Down—"In a marshy field on the shore of Belfast Lough half a mile above Knockmagunny (Knocknagoney) Hill, coming into flower on July 5th, 1820"; *Templeton.*
Antrim—In a low meadow south of Ballintrae; D.M. in *Cyb. Hib.*

CEPHALANTHERA *Rich.*

1. C. ensifolia *Rich.*
Wet places in old woods—extremely rare. Not found recently, though diligently sought for by several botanists. Fl. June and July.
Down—By the margin of Lough Derry about two miles north of Ballynahinch; *Sherard,* 1694 (now extinct). "*Helleborine foliis prælongis angustis acutis.* Found on a rotten bog, by a lough side, near the Dairy House in Creveteneau, Ballynahinch, Ireland"; *Raii Synopsis, ed.* II., 1696.

The above quoted Latin descriptive name is the synonym used by Ray for *C. ensifolia,* and for this, in addition to weightier reasons, we cannot concur with the authors of the Cybele in transferring Sherard's station to *Epipactis palustris*. First, to correct the local terms cited by Ray. The original and proper name of the Lough is Derry Lough, and the house, Derry House. The townland name is Creevytenent, and not the French-looking "Creveteneau," as quoted in the Synopsis. Derry (Daire) we need scarcely say is the Irish term for oak. Creevagh or Creevy signifies bushy or branchy land (Joyce); the termination—tenent is probably a corruption of teinte—tires. Perhaps a place of resort for fuel, though the name for firewood is different. The rotten bog meant not a peat moss, but a black mud, made up of rotten leaves and branches; just such a place as affords a proper *habitat* for *C. ensifolia*, and similar to the ground on which it now flourishes at Killarney. *Epipactis palustris*, as the name indicates, is a plant of ordinary marshes. In addition to the foregoing considerations, the diagnostic terms *foliis prælongis angustis acutis* point to *C. ensifolia*, rather than *E. palustris*.

Antrim—*At Duneane near Antrim ; *Flor. Ulst.* Found by Mr. Whitla in the County of Antrim in July, 1835 ; *Flor. Hib.* (*sub. E. grandiflora*) In woods at Muckamore, and at Glenavy (*Whitla*), also in Shane's Castle woods on the shore of Lough Neagh (D.M.); *Cyb. Hib.*

* A very vague locality, Duneane is not near Antrim, but is a parish which touches the County of Derry at Toome, and stretches thence east and southeast for several miles.

MALAXIS *Swartz.*

1. M. paludosa (*Linn.*) *Swartz.* BOG ORCHIS.

Amongst Sphagnum, and other mosses in wet peaty bogs—very rare. Fl. Aug. and early part of Sept.

Antrim—Found on marshy ground ascending Slievenanee ; *Templeton*, 1809. Rathlin Island ; (*Miss Gage*) *Flor. Ulst.* On an elevated bog above Dunloy ; *Cyb. Hib.* Sparingly on Fairhead between Loughnacranagh and the cliffs, also a little west of Lough Naroon, which lies on the margin of an extensive bog, and is about two miles southwest of Dunloy ; S.A.S.

ORDER LX. **IRIDACEÆ.**

IRIS *Linn.* FLAG.

1. I. pseudacorus *Linn.* SAGGON (*loc.*).
Marshy places—very common. Fl. June and July.

ALISMACEÆ.

Order LXI. ALISMACEÆ.

ALISMA *Linn.*

1. A. plantago *Linn.* WATER PLANTAIN.
Ditches, and ponds—very common. Fl. mid. June till mid. Aug.

2. A. ranunculoides *Linn.*
Marshes, ponds, and lake margins—common, especially abundant in sandy districts. Fl. June and July.

SAGITTARIA *Linn.* ARROWHEAD.

1. S. sagittifolia *Linn.*
Rivers, lakes, and ditches—rare. Fl. July and Aug.
Down—Pond at Belmont, and many places in the Newry canal, also frequent in the Lagan canal from Lough Neagh to near Belfast; S.A.S. The Lagan stations belong equally to Antrim.
Antrim—In the large drain at Portmore, and other places about Lough Neagh, also in the Bann at Portglenone; *Templeton*, 1794 (still found about these places). In Lough Neagh near Antrim; *Flor. Ulst.* In Lough Neagh north of Lurgan; S.A.S.
Derry—In a small lake above Kilrea, and abundant by the Bann from Toome to near Coleraine; D.M. In the Bann at Toome; S.A.S.

BUTOMUS *Linn.* FLOWERING RUSH.

1. B. umbellatus *Linn.*
Ditches, rivers, and lakes—rare. Fl. mid. June till mid. Aug.
Down—South end of Lough Leagh; *Templeton*, 1804. Lagan near Belfast; *Flor. Ulst.* Abundant in Lagan canal near Moira and elsewhere, also in the lake at Ballynahinch; S.A.S. Marshes in the neglected Victoria Park; R.Rl.P.
Antrim—Lagan canal, as above.
Not known to Mr. Templeton as occurring in the canal, and doubtless introduced there as stated in Flora of Ulster.

TRIGLOCHIN *Linn.* ARROWGRASS.

1. T. maritimum *Linn.*
Muddy seashores—common. Fl. latter end of May till mid. Aug.

LILLIACEÆ.

2. T. palustre *Linn.*
Marshes—common everywhere. Fl. June and July.

ORDER LXII. **LILLIACEÆ.**

SCILLA *Linn.* SQUILL.

1. S. verna *Hudson.* VERNAL SQUILL.
Rocky and sandy seashores—frequent. Fl. mid. April till mid. June.
Down—Rocks along the shore from Donaghadee to Bangor, and, with pale red flowers, on the west side of Ardglass harbour ; *Templeton*, 1794. Rocky shore at Holywood, and the Copelands ; *F. Whitla.* Seacoast between Cultra and Bangor ; *Flor. Belf.* Groomsport ; *Flor. Ulst.* Black Island in Strangford Lough ; T.H.C. Rocky shore at Quentin Castle in the Ards ; S.A.S.
Antrim—Rocky headlands about the Giants' Causeway ; *Templeton,* and *Wade Rar.* Rathlin ; *Miss Gage.* Shore near Carrick-a-rede ; *G. C. Hyndman.* Portrush ; *Flor. Ulst.* At the Bushfoot, and at Whitepark Bay ; S.A.S.
Derry—Plentiful on rocks by the coast from Downhill to Portrush ; D.M.

ALLIUM *Linn.* GARLIC.

1. A. vineale *Linn.* CROW GARLIC.
Damp shady places—very rare. Fl. June and early July.
Down—At Rockport below Holywood ; *Flor. Belf.*
Antrim—Shore of Lough Neagh near Shane's Castle ; *Flor. Ulst.* Shane's Castle Park ; T.H.C. In some plenty by the Sixmilewater above Antrim ; *D. Redmond.*

2. A. ursinum *Linn.* RAMSONS, WILD GARLIC.
Woods, and shady places in glens—frequent. Fl. May and June.
Down—Crawfordsburn ; *Flor. Ulst.* Rostrevor, Portavo, Conlig, Drumbo, Newtownards, and Newtownbreda ; S.A.S.
Antrim—Woods at Parkmount and Macedon ; *Templeton.* Knockagh ; *M'Skimmin's C.fergus.* Murlough, and glens above Ballycastle ; *Ir. Flor.* Colin Glen ; *Flor. Belf.* Cave Hill ; *Flor. Ulst.* Castle Dobbs ; S.A.S. Shane's Castle ; T.H.C. Glenarm, and Carnlough ; R.Ll.P.
Derry—Abounds in Ballymaghir ; *Samp's. Surv.* In the woods at Springhill, and Castledawson, but not common ; D.M.

ENDYMION *Dumort.*

1. E. nutans (*Linn.*) *Dum.* BLUEBELL. WILD HYACINTH.
Woods, and shady banks—very common. Fl. mid. April till mid. June.

ORDER LXIII. **MELANTHACEÆ**.

NARTHECIUM BOG ASPHODEL.

1. N. ossifragum *Hudson.*
Peat bogs, and wet heaths—common. Fl. July and Aug. Ranges from sea level to 1500 feet in Down and Antrim (2000 in Derry; D.M.).

ORDER LXIV. **JUNCACEÆ**.

JUNCUS *Linn.* RUSH.

1. J. maritimus *Smith.* SEA-RUSH.
Salt marshes—frequent. Fl. mid. July till end of Aug.
Down—Shores of the loughs of Belfast and Strangford; *Templeton*, 1800. Holywood, Dundrum, and the Quoile; *Flor. Ulst.* Common, and abundant around the coast; S.A.S.
Derry—Very abundant at the mouth of the Bann, below Coleraine, and only there; D.M. Abundant by the Bann at Portstewart; S.A.S.
It is singular that we can find no record, or note of this plant in Co. Antrim.

2. J. effusus *Linn.*
Damp pastures, and undrained fields—very common. Fl. July and Aug. Ranges from sea level to over 1000 feet.

3. J. conglomeratus *Linn.*
Damp untilled fields and pastures—very common. Fl. July and Aug. Ranges to 1500 feet on the mountains.

4. J. glaucus (*Huds.*) *Ehr.*

Sandy or calcareous lands, at low elevations—rare. Fl. July and Aug.

Down—By the Quoile at Downpatrick ; *Flor. Ulst.* Marsh by the shore a little south of Killough ; S.A.S.

Antrim—In several places about Larne ; *Templeton*, 1805. By the Lagan (near Bot. Gardens) ; *Flor. Ulst.* Shore of the lough between Crumlin waterfoot and Glenavy River ; S.A.S. Cairncastle ; R.Ll.P.

Derry—Near Coleraine ; D.M.

5. J. obtusiflorus *Ehr.*

Marshy places—very rare. Fl. mid. July till mid. Sept.

Down—Sparingly on the muddy shore at the upper end of the inlet of Dundrum Bay near Clough ; S.A.S., 1865.

6. J. acutiflorus *Ehr.*

Marshy land, and sides of streams—very common. Fl. July and Aug.

7. J. lamprocarpus *Ehr.*

Meadows, and wet pastures—very common. Fl. mid. June till mid. Aug.

8. J. supinus *Moench.*

Ditches, and pools—not common. Fl. July—Sept. Ascends to 1200 feet.

Down—Cotton Moss ; *Flor. Belf.* In a small stream at foot of Slievenabrock above Tollymore Park, and on wet rocks above Bryansford ; S.A.S.

Antrim—Woodburn, and frequent at the Kenramer end of Rathlin ; S.A.S.

Derry—Very common ; D.M. (? *Eds.*).

No doubt more frequent than appears from the above account, but reliable notes, and authentic specimens are rare. A mountain form of the preceding species is sometimes mistaken for this plant. Our usual form seems to be the J. SUBVERTICILLATUS *Wulf.*

9. J. squarrosus *Linn.*

Moory and heathy places—common. Fl. mid. June till mid. Aug. Ranges from a few hundred feet to the tops of our highest mountains.

10. J. Gerardi *Loisel.*

Marshy shores—common on the coast, especially at the upper ends of bays and loughs. Fl. July and Aug.

11. J. bufonius *Linn.* TOADRUSH.

Wet places by roadsides, and marshy, sandy, waste ground—very common. Fl. July and Aug.

LUZULA *D. C.* WOODRUSH.

1. L. maxima *D.C.* (L. SYLVATICA *Bicheno*).
Wet places in woods and glens—common. Fl. mid. April till mid. June. Ranges from sea level to 2700 feet.

2. L. vernalis *D.C.* (L. PILOSA *Willd.*).
Woods, and damp shady places—common. Fl. April and May. From near sea level to 1000 feet.

3. L. campestris *Willd.*
Pastures, and heaths—common, especially on the hills. Fl. April and May. Ascends to 2350 feet (*Hart*).

4. L. erecta *Desv.* (L. MULTIFLORA *Lej.*).
Heaths, and peat bogs—common. Fl. May and June. Ranges from 100 to 1500 feet.

ORDER LXV. **TYPHACEÆ**.

TYPHA *Linn.* REEDMACE.

1. T. latifolia *Linn.* BLACKHEADS (loc.).
Marshes, ditches, ponds, and lakes—frequent. Fl. July and Aug.

Down—Belmont, Ballyalloley, and Ballygowan; *Flor. Belf.* Abundant in marshes at Downpatrick, and in ditches on roads thence to Strangford and Newcastle; *Flor. Ulst.* Clandeboye Lake, and ditches inland from Cloughey; S.A.S. Marsh three miles north of Downpatrick; T.H.C. Dams at Dundonald, Carngaver, and Conlig; R.Ll.P. Drumskee bog near Tullycairne, and in the Lagan below the bridge at Gillhall; *C. Dickson.*

Antrim—In the Bog meadows, and dams at Whitehouse and Whiteabbey; *W. Millen.* In the Main at Shane's Castle, and in ditches a half mile west of Carrickfergus; *Flor. Ulst.* Ballyfinaghey, and by the Lagan above the first lock, also at the mouth of the Sixmilewater, and by the same river below Templepatrick; S.A.S.

Derry—Lough Beg, lakes above Kilrea, and marsh west of Enagh Lough; D.M.

2. T. angustifolia *Linn.*

Ditches, and lake margins—very rare, and now lost in several of the old stations. Fl. July.

Down—Lough Henney near Saintfield (*Campbell*) ; *Flor. Hib.* On the west side of Long Lough three miles north of Ballynahinch ; S.A.S., 1884.

Antrim—In a ditch on the right hand side of Carrickfergus road about 200 yards above the Milewater; *Templeton*, 1818 (loc. now obsolete). Abundant by the shores of the lough at Portmore, and by Lough Neagh at Selshan; *Cyb. Hib.* Not seen recently in the latter locality, but re-discovered at Portmore, by *Rev. H. W. Lett*, in 1882.

SPARGANIUM *Linn.* BUR REED.

1. S. ramosum *Hudson.*

Ditches, and margins of sluggish streams—common. Fl. mid. June till end of July.

2. S. simplex *Hudson.*

Ditches, slow streams, and lake-margins—frequent. Fl. July and till mid. Aug.

Down—Watery places in the Ards; *Templeton*, 1806. Lakes near Ballynahinch; *G. C. Hyndman.* Ballyalloley; *Flor. Belf.* Derry Lough north of Ballynahinch, Monlough, and Saintfield lakes; S.A.S. Ballykinler, Crossgar, and Dundonald; R.Ll.P.

Antrim—Bog meadows; *Templeton.* Near Belfast; *Flor. Hib.* Laganside; *Flor. Ulst.* Several places by Lough Neagh, as Portmore, Crumlin waterfoot, Ballinderry, and north of Toome; S.A.S.

Derry—By the Bann, and at Ballyarnot near Derry; D.M. Lough Beg near Toome; S.A.S.

3. S. natans *Linn.*

Rivers, and lakes—rare. Fl. from early in July till late in Aug.

Down—Castlewellan Lake; *Templeton*, 1808. Annsboro' Lake northeast of Castlewellan; S.A.S.

Antrim—Drains at Portmore; *Dr. Scott* (*fide Templeton*). Cranmore; *Templeton*, 1803. Lakes near Fairhead; *Cyb. Hib.* Plentiful in the Sixmilewater where it empties into Lough Neagh; S.A.S.

Derry—In Lough Beg, and plentiful in the Bann above Kilrea; D.M. Ballyarnot Lake; *Flor. Ulst.*

The long floating leaves so abundant in the Lagan, and some other streams, may belong to this species, but in the absence of flowers, which seem never to form, the identification is uncertain.

4. S. minimum *Fries.*
Lakes and other still-water—very rare. Fl. mid. June till mid. Aug.
Down—Conlig Hill; R.Ll.P.
Antrim—In some plenty in lakelets at south end of Rathlin Island; S.A.S. Loughnacranagh on Fairhead; R.Ll.P.

ORDER LXVI. **ARACEÆ.**

ARUM *Linn.* WAKE ROBIN.

1. A. maculatum *Linn.* LORDS AND LADIES.
Woods, and damp shady hedge banks—common. Fl. May and early June.

ORDER LXVII. **LEMNACEÆ.**

LEMNA *Linn.* DUCKWEED.

1. L. trisulca *Linn.*
In ditches, lakes, and stagnant or very slowly moving water—common, but much less general than the next species. Fl. June (*Cyb. Hib.*).

2. L. minor *Linn.*
Marshes, ditches, and still places in sluggish streams—very common. Rarely flowers, but found in flower by Mr. Templeton, Aug. 6th, 1802.

3. L. gibba *Linn.*
Stagnant ditches—rare, apparently, though profusely abundant in some places. Fl. June and July (*Cyb. Hib.*).
Down—Newry canal, occasionally, and by the Quoile below Downpatrick, also in brackish ditches between Queen's Quay and Connswater; S.A.S. Ballycroghan near Bangor; R.Ll.P.
Antrim—By the Lagan canal near Belfast; *Flor. Ulst.* By the Lagan

canal near Lough Neagh; *Cyb. Hib.* Bog meadows; T.H.C. Mr. Templeton found this species growing in M'Clean's fields, where Bedford Street now stands.

ORDER LXVIII. **POTAMOGETONACEÆ.**

POTAMOGETON *Linn.* PONDWEED.

1. P. natans *Linn.*
Still water of ditches, lakes, and reservoirs—very common. Fl. mid. June till mid. Aug.
  Var. α P. OVALIFORMIS *Fieber.*
Drains by Portmore Lake, Co. Antrim; S.A.S. The foliage of this form somewhat resembles that of *P. plantagineus.*

2. P. polygonifolius *Pourr.*
Shallow pools, and ditches—common, especially on bogs and heaths. Fl. mid. June till mid. Aug. Rises to over 1200 feet on the mountains.

3. P. rufescens *Schrad.*
Slow running streams, rarely in water entirely still—frequent. Fl. mid. June till end of July.
Down—Newry canal, and drains by Magherascouse Lake; S.A.S.
Antrim—Malone; *Templeton*, 1809. Lagan canal near Belfast; *Flor. Ulst.* Ballymena; *Cyb. Hib.* Loughmourne; S.A.S. In the Sixmilewater at Templepatrick; T.H.C. Glendun; *Rev. S. A. Brenan.*
Derry—West side of Enagh Lough; D.M. In the river Roe; *Cyb. Hib.*
 More frequent than appears from the above, but there is a lack of notes referring to definite localities.

4. P. heterophyllus *Schreb.*
Rivers, and lakes—rare. Fl. mid. June till mid. Aug.
Down—In the river near Ballynahinch; *Templeton*, 1813.
Antrim—Lagan canal above Lisburn, Sixmilewater from Ballyclare to Dunadry, Lough Neagh north of Toome, and in Loughmourne above Carrickfergus; S.A.S.
 Now lost at Ballyclare through fouling of the water, and in Loughmourne by conversion of the lake into a reservoir.
Derry—Abundant in the Bann near Kilrea; D.M.

Var. *a* P. PSEUDO-NITENS *Bennett.*
Very rare. Found in the large lake at Ushet in the Island of Rathlin;
S.A.S.

5. P. nitens *Weber.*
In lakes—very rare. Fl. June and Juiy.
Down—North side of Lough Aghery, plentiful in one spot where the bottom is peaty, but fruit stems scarce; S.A.S., July, 1887.
Antrim—"There is a specimen of this species among the plants collected by D.M. many years ago in Antrim. It was probably gathered in Lough Neagh"; *Cyb. Hibernica.*
It is only since going to press that we have any information as to this species beyond the doubtful note above quoted.

6. P. lucens *Linn.*
In rather deep water of lakes, streams, and reservoirs—frequent. Fl. mid. June till mid. Aug.
Down—Newry canal in many places; S.A.S.
Antrim—Lagan canal above the second lock; *Templeton,* 1801 (still found at various points in the Lagan). Loughmourne (now lost), and plentiful in Lough Neagh; S.A.S.
Derry—Lough Beg, and common in many parts of the county; D.M. In the Bann at Toomebridge; *W. Thompson.* Streams on west side of Lough Neagh; *Flor. Ulst.* In the lough from Toome to Ballyronan; S.A.S.

Var. P. BOREALIS *Tiselius.*
In Loughmorne above Carrickfergus, but now lost in consequence of operations by the Water Commissioners.

7. P. Zizii *Roth.*
Lakes, and slow streams—rare. Fl. July and Aug.
Antrim—Loughmourne; S.A.S. East side of Lough Beg near Toome; T.H.C.
Derry—West side of Lough Beg; T.H.C. In the Bann above Coleraine; S.A.S.
First found in Ireland by S.A.S. in Fermanagh.

8. P. prælongus *Wulfen.*
Lakes, and canals—rare. Fl. June and July.
Antrim—Loughmourne; *Flor. Belf.* Ditches by the Lagan between first and second lock of the canal; *Flor. Ulst.* In Lough Neagh; *Cyb. Hib.* In the canal at several points between Belfast and Lisburn, and plentiful in the large lake (Ushet)

on Rathlin Island; S.A.S. Was found in Loughmourne until 1880.
Derry—Rare, only observed in one of the small lakes above Kilrea; D.M.

9. P. perfoliatus *Linn.*
Lakes, streams, and ditches—common. Fl. July and first half of Aug.

10. P. crispus *Linn.*
Lakes, ditches, and quiet streams—common. Fl. mid. June till mid. Aug.

Var. β P. SERRATUS *Hudson.*

In the Bann near Portglenone, Antrim and Derry; perhaps not rare; S.A.S. Under the name *P. compressum* Mr. Templeton notices a pondweed occurring in the Bann. His note says "leaves finely serrate," and hence we conclude that the plant was incorrectly determined, and was also Hudson's *P. serratus.*

11. P. obtusifolius *Mert. & Koch.*
Lakes, and drains—rare. Fl. July and early part of Aug.
Down—Ballyalloley near Comber; *Flor. Belf.* (sub *P. compressus*). Aughnadarragh Lake near Saintfield, Derry Lake, and Magherascouse Lake; S.A.S. Munce's dam near Dundonald; R.Ll.P.
Antrim—Drains at Portmore; *Templeton,* 1795. Abundant at Portmore; *Cyb. Hib.*
Derry—Enagh Lough, and Lough Lug near Moneymore; D.M.

The original authority for the reported occurrence of this species in the Lagan was by no means a critical botanist, and as the plant cannot be found there that record may be expunged. There is also a note of its occurrence in Islandmagee, but for a somewhat similar reason this station is likewise omitted.

12. P. pusillus *Linn.*
Lakes, ponds, ditches, and streams—common. Fl. July and Aug. Occurs at 1150 feet in the Mourne Mountains; *Hart.* The var. *P. tenuissima* is much the more common form.

13. P. pectinatus *Linn.*
Lakes and rivers—frequent. Fl. July and Aug.
Down—Pond at Castle Espie, Saintfield lakes, Lough Aghery near Ballynahinch, Annsboro' Lake near Castlewellan, and abundant in the Quoile at Downpatrick; S.A.S. Marshes by the Strand Lake near Killough; T.H.C.
Antrim—Crumlin River near the waterfoot, and in drains between Portmore and Lough Neagh; *Templeton,* 1808. Lagan canal;

Flor. Belf. Ditches by Lough Neagh in Massereene Park; *Flor. Ulst.* At various points in the course of the Lagan canal to Lough Neagh; S.A.S. Pollanburn near Glenarm; R.Ll.P.
Derry—Abundant between Coleraine and the sea; D.M.

RUPPIA *Linn.* TASSEL PONDWEED.

1. R. maritima *Linn.*
Salt marshes, and brackish ditches—rare. Fl. mid. July till mid. Sept.
Down—Brackish ponds at Castle Espie, salt marsh at Newtownards, also abundant in brackish marsh by the railway bridge at Connswater, and on the Kinnegar at Holywood; S.A.S.
Antrim—Among *Zostera* on the Milewater shore of Belfast Lough; *Templeton*, 1794. Abundant in salt marshes at Glynn, and Ballycarry; R.Ll.P.

2. R. rostellata *Koch.*
Salt marshes—rare. Fl. July and Aug. (*Cyb. Hib.*).
Derry—Plentiful in muddy salt marsh above the Bannfoot near Portstewart; S.A.S.
The two forms of *Ruppia* have been so much confounded that great part of the former records are useless, specimens, also, are of no value unless they bear mature fruit. The stations accredited, in Flora Belfastiensis, to the present species belong really to the preceding, and there is every reason to distrust the notes on these plants as recorded in Flora of Ulster.

ZANNICHELLIA *Linn.* HORNED PONDWEED.

1. Z. palustris *Linn.*
Marshes, and ditches, mostly by the sea—rare. Fl. July and Aug.
Down—Holywood; *Flor. Ulst.* In the Quoile below the rath at Downpatrick; S.A.S.
Antrim—Ditches by the Lagan; *Flor. Ulst.*, and subsequently; T.H.C.
Derry—Common in many places; D.M.

2. Z. polycarpa *Nolte.*
Brackish ditches—very rare. Fl. mid. June till end of Aug.
Down—Abundant in brackish ditches in the long-neglected "Victoria Park," and in the reclaimed land between Queen's Quay and the Park; S.A.S., 1872.

First found in Britain by S.A.S. in the station last cited. *Vide* Bot. Exchange Club Rep., 1876, and Jour. Botany, 1878.

Possibly imported with Baltic timber, but it grows on the opposite side of the harbour from the timber ponds.

ORDER LXIX. **NAIDACEÆ.**

ZOSTERA *Linn.* GRASSWRACK.

1. Z. marina *Linn.*

In salt water near low water mark on muddy shores, and inlets of the sea—common. Fl. July and Aug. (*Cyb. Hib.*). Dredged up from a depth of several fathoms off Glenarm; S.A.S.

ORDER LXX. **CYPERACEÆ.**

SCHŒNUS *Linn.*

1. S. nigricans *Linn.* BOGRUSH.

Bogs, and wet heaths—frequent. Fl. June and July. Ranges from sea level to 1500 feet.

Down—Tollymore Park, bog at Kirkiston, and marshy field on the shore near Knocknagoney; *Templeton*, 1793. Salt marsh beyond Bangor; *Flor. Belf.* Slieve Donard; *Flor. Ulst.* Boggy places by the coast from Strangford to Newcastle, and abundant on the Mourne Mountains; S.A.S. Heath at Killough Bay, and on mountains above Bryansford; T.H.C.

Antrim—Shores of Lough Neagh; *Templeton.* Islandmagee, and Portrush; *Flor. Ulst.* Shores of the lough at Selshan, and Langford Lodge, also top of Fairhead, and heath above Giants' Causeway; S.A.S.

Derry—In Glenchain, and abundant by the Bann below Coleraine; D.M. Slieve Gallion; *Flor. Ulst.* Portstewart; S.A.S.

CLADIUM *P. Brown.*

1. C. mariscus (*Linn.*) *R. Brown.* TWIGRUSH.

CYPERACEÆ. 153

Margins of lakes—very rare, not found recently. Fl. July (*Cyb. Hib.*).
Down—"In a lake about half a mile along the road going from Castlewellan to Rathfryland"; *Templeton*. Another note in Mr. Templeton's hand says, In the second lake on the road from Castlewellan to Rathfriland.
Antrim—Lough Neagh at Selshan; *Cyb. Hib.*

There is no lake at half a mile on the road from Castlewellan to Rathfriland, but possibly a formerly existing small lake may have been since drained. If Mr. Templeton's second note be the correct one, it indicates Lough Islandreavy, which is two miles from Castlewellan. This lake has been utilised as a reservoir, and there are now very few aquatic plants to be found about its margins. It is, at all events, nearly certain that the plant is now extinct in Antrim and Down. The notes in Flora of Ulster are erroneous.

RHYNCHOSPORA *Vahl.* BEAKRUSH.

1. R. alba (*Linn.*) *Vahl.*
Peat bogs at low elevations—frequent. Fl. mid. July till end of Aug.
Down—Ballygowan bog, and Cotton moss; *Flor. Belf.* Bogs near Newtownards; *Flor. Ulst.* Heathy places in Ballynahinch bog; S.A.S.
Antrim—Bogs by the Bann at Portglenone, and above Dunloy; S.A.S.
Derry—Common on peat bogs which have not been disturbed; D.M. Bogs by the Bann in several places; S.A.S.

R. sordida, a very slight variety, occurs in a portion of the Cotton moss southwest of Donaghadee.

ELEOCHARIS *R. Brown.* SPIKERUSH.

1. E. palustris (*Linn.*) *Brown.*
Ditches, ponds, and marshy margins of streams—very common. Fl. June and July. Occurs at 1450 feet in Down; *Hart.*

2. E. uniglumis *Link.*
Sandy marshy waste ground—very rare. Fl. July and till mid. Aug.
Down—Very sparingly in a marshy field by the shore a half mile east of Bangor; S.A.S., 1867.

3. E. multicaulis *Smith.*
Marshes in sandy and moory ground—not common. Fl. June and July. Ranges from sea level to 1200 feet on Mourne Mountains.

Down—Ballygowan bog ; *Templeton*, 1807. Ballyholme, and common on mountains above Newcastle, and Rostrevor ; S.A.S.
Antrim—Marshes by the canal at Stranmillis ; *Flor. Belf.* Fairhead, and by Lough Neagh at Antrim, and occasionally to Crumlin, S.A.S.
Derry—Rare, only observed sparingly at the mouth of the Roe ; D.M. By the Bann below Coleraine ; S.A.S.
First found in Ireland by Mr. Templeton at the station quoted at top.

4. E. acicularis (*Linn.*) *Smith.*
Wet sandy ground by lakes and rivers—not common. Fl. July and Aug.
Antrim—Plentiful at Portmore, and by the Bann below Portglenone ; *Templeton*, 1796. By the Lagan, and on shores of Lough Neagh ; *Cyb. Hib.* Banks of Lagan canal above Lisburn, shore of Lough Neagh north of Lurgan, and marshy places close to Portmore old church ; S.A.S.
Derry—Not general, but plentiful by the Bann in several places ; D.M. Lough Foy ; *Cyb. Hib.* Shore of Lough Beg ; S.A.S.

SCIRPUS *Linn.* CLUBRUSH.

1. S. maritimus *Linn.*
Brackish ditches and marshes—frequent. Fl. mid. June till mid. Aug.
Down—Marshy places about the sea ; *Templeton*, 1797. Tillysburn ; *Flor. Belf.* Copeland Islands ; *Flor. Ulst.* By the Quoile at Downpatrick, and abundant on shores of Strangford Lough ; S.A.S. Groomsport ; T.H.C.
Antrim—In the Lagan at Ormeau, and in ditches by the shore road ; *Flor. Belf.* Lagan side below the first lock ; *Flor. Ulst.* Larne Lough, and Giants' Causeway ; S.A.S. By Lough Neagh at Shane's Castle ; T.H.C.
Derry—Very abundant by the Foyle, and at the Bannfoot ; D.M. Abundant at Coleraine, and Limavady ; S.A.S.

2. S. sylvaticus *Linn.*
Damp woods and thickets, and bushy banks of rivers and lakes—frequent. Fl. June and July.
Down—Banks of the Bann near Corbet Lake east of Banbridge ; *G. C. Hyndman.* By the Lagan near Shaw's Bridge ; S.A.S.
Antrim—Derriaghy ; *Flor. Belf.* Kilroot, and shores of Lough Neagh at Massereene Park and Shane's Castle ; *Flor. Ulst.* By the river in Glenariffe ; *Cyb. Hib.* Glenarm deerpark ; *R. Tate.* Banks of the Bush at Bushfoot ; *B.N.F.C.*, 1868. Bog

CYPERACEÆ. 155

meadows, Glenavy, Crumlin, and banks of Sixmilewater near Templepatrick; S.A.S. Larne; R.Ll.P.

Derry—Confined to the southern baronies where it is abundant by all the rivers; D.M. By Lough Neagh at Ballyronan; *Flor. Ulst.* Woods at Castledawson; S.A.S.

3. S. lacustris *Linn.* BULRUSH.

In shallow water of lakes and slow flowing streams—frequent. Fl. mid. June till mid. Aug.

Down—Lakes at Ballyalloley and Magherascouse, and bog drains at Ballygowan; *Flor. Belf.* Lough Cowey near Portaferry; *Flor. Ulst.* Clandeboye Lake, Magheralagan, and Loughinisland, and lakes near Saintfield; S.A.S. Lake near Killough; T.H.C. Ballynahinch; R.Ll.P.

Antrim—Massereene Park; *Flor. Ulst.* Portmore Lake, Sixmilewater at Templepatrick, lake at summit of Fairhead, and river at Bushfoot; S.A.S.

Derry—Common in Lough Beg, and by the sides of the Bann, and the Foyle; D.M.

4. S. Tabernæmontani *Gmelin.*

Muddy ditches by the seacoast—rare, very rare inland, and only where the sea has reached in late Quaternary times. Fl. July and Aug.

Down—By the Quoile at Downpatrick; S.A.S. Salt marsh by the Donaghadee line at south side of Comber; R.Ll.P.

Antrim—"It grows on the shore of the Lagan a little below Stranmillis limekiln, where the tide constantly flows about it"; *Templeton,* July, 1810 (still there). Larne Lough; *Flor. Ulst.* Shane's Castle; T.H C.

Derry—By the Bann above Portstewart; S.A.S.

First found in Ireland by Mr. Templeton at the Belfast station above noted.

5. S. cæspitosus *Linn.*

Peat bogs, and wet heaths—common. Fl. May—July. Ranges from sea level to 2500 feet.

6. S. pauciflorus *Lightfoot.*

Wet places, and marshes in sandy land—rare. Fl. July and Aug.

Down—Boggy fields on the shore of Belfast Lough about 200 yards above Knocknagoney, and shore four or five miles south of Newcastle; *Templeton,* 1797.

Antrim—Around the shores of Belfast Lough; *Templeton.* Rams Island, and Portrush; *Flor. Ulst.* Sandy shore of Lough Neagh opposite Portmore; S.A.S.

Derry—Very common in the county; D.M. (? *Eds.*).

CYPERACEÆ.

7. S. fluitans *Linn.*

Ponds, ditches, and slow streams—not common. Fl. June and July. From sea level to 1400 feet in the mountains.

Down—In a marshy hollow on Holywood warren; *Templeton*, 1797, and *Flor. Belf.*, 1863 (but perhaps now lost). Ballyalloley marsh, Newcastle warren, and stream at lower slopes of Slievenabrock above Bryansford; S.A.S. Conlig, and Dundonald; R.Ll.P.
Antrim—Bog meadows; *Flor. Ulst.* (now lost there). Portmore Lake, and in profusion on Rathlin Island, S.A.S. Garron Head; R.Ll.P.
Derry—Common, particularly abundant in a marsh on the west side of Enagh Lough; D.M.

8. S. setaceus *Linn.*

Wet plashy places, especially where gravelly—not common. Fl. July —Sept.

Down—Knock, and in the Kinnegar at Holywood; *Flor. Belf.* Copeland Islands, and coast west of Bangor; *Flor. Ulst.* Mealough Hill, Monlough, Groomsport, Greencastle, and in marshy spots all round the coast; S.A.S. Comber; R.Ll.P.
Antrim—Newforge, and very luxuriant in Colin Glen, also Ballinderry, Knocklayd, and Rathlin; S.A.S.
Derry—Black rocks near Portstewart, not very common in this county; D.M. About Toome; S.A.S.

9. S. Savii *Seb. & Maur.*

Marshy places, principally by the coast—not common. Fl. July and Aug.

Down—Crawfordsburn, Bangor, and Groomsport; *Flor. Belf.* Kinnegar at Holywood, and abundant about Newcastle; S.A.S.
Antrim—Giants' Causeway, Ballintoy, and frequent on boggy ground in the Island of Rathlin; S.A.S.
Derry—Common near the seacoast; D.M. Portstewart; S.A.S.

BLYSMUS *Panz.*

1. B. rufus (*Huds.*) *Link.*

Marshy seashores—rather rare. Fl. June—Aug.

Down—Shore about half a mile south of Donaghadee, and by the shore some 300 yards above Knocknagoney; *Templeton.* Salt marshes beyond Bangor; *Flor. Belf.* Shore at Newtownards, and Holywood; *Flor. Ulst.* Warrenpoint; *Rev. G. Robinson.* Groomsport, and North of Ardglass; S.A.S.
Antrim—In a marshy place by the shore between Larne and Ballygally; *Templeton*, 1808. Larne Lough, and Cushendall; *Cyb. Hib.*

CYPERACEÆ. 157

Derry--Rare, but observed plentifully by the side of the Foyle at Brookhall; D.M. By the Bann a short distance above Portstewart; S.A.S.

ERIOPHORUM *Linn.* COTTON GRASS.

1. E. vaginatum *Linn.*
Bogs, and wet peaty heaths—common, especially on the mountains. Fl. April and May. Ranges from near sea level to 2000 feet.

2. E. polystachion *Linn.* (E. ANGUSTIFOLIA *Roth*).
Bogs, and wet heaths—very common. Fl. mid. April till end of June. At all levels up to 2200 feet.

3. E. latifolium *Hoppe.*
Peat bogs—very rare, if not extinct. Fl. April and May (*Cyb. Hib.*).
Antrim—On a large flow bog near Rasharkin; *Cyb. Hib.*
Reported also as abundant in a bog between Ballygowan and Comber, but cannot be found, and the authority not considered sufficient.

CAREX *Linn.* SEDGE.

1. C. dioica *Linn.*
Marshes, and wet peat bogs—frequent. Fl. mid. May till end of June. Ranges from sea level to 1500 feet (*Cyb. Hib.*).
Down—In the bogs of the county; *Harris's Down,* 1744. (No doubt this sp. but recorded as "*Gramen cyperoides minus ranunculi capitulo longiore.*"). Kirkiston bog, and frequent in boggy situations; *Templeton,* 1797 (*sub C. Davalliana*). Slieve Donard; *Flor. Ulst.* Slievemartin, and frequent at about 1000 feet on the Mourne Mountains; *Hart, Proc. R.I.A.,* 1884.
Antrim—Marsh between Divis and Black Mountain, back of the Knockagh, and abundant on the commons of Carrickfergus, also on Tardree Hill, Rathlin Island, and marshy margin of river at Bushmills; S.A.S.
Derry—Glenchain, and in Altahan Glen near Draperstown, not rare; D.M.

2. C. pulicaris *Linn.*
Heaths, and short pastures—common. Fl. June and July. From sea level to 1500 feet; *Hart, op. cit.*

3. C. disticha *Hudson.*

Wet gravelly or stony places—frequent. Fl. mid. May till end of June.

Down—Plentiful in a meadow halfway between Dundrum and Bryansford; *Templeton*, 1799. Cultra; *Flor. Ulst.* By the Quoile near Inch Abbey, and by the margin of Loughinisland; S.A.S.

Antrim—Bog meadows, Stranmillis, meadow below Shawsbridge, and by the Lagan above Lisburn; *Templeton*, 1809. Meadows on borders of Loughmourne; *Flor. Ulst.* Shores of Lough Neagh at Shane's Castle, abundant; T.H.C. Shore below Whiteabbey, Cave Hill west of the quarries, and at Crumlin waterfoot; S.A.S.

Derry—Wet meadows by the Bann, above Coleraine; D.M. Near Toome; T.H.C.

4. C. arenaria *Linn.*

Sandy seashores—frequent. Fl. mid. May till mid. July.

Down—Blown sand on the shore below Bangor; *Templeton*, 1794 (still there). Sandy shores east of Greencastle; *Ir. Flor.* Sandy coast between Holywood and Bangor; *Flor. Belf.* Dry, loose sea sands at Sydenham, Dundrum, Greencastle, and Millisle, also on marshy and gravelly shore at Quentin Castle (*forma coarctata*); S.A.S. Shore at Kilclief Bay; T.H.C.

Antrim—Among loose sand on the shore at Ballycastle; *Templeton.* Redbay; S.A.S.

Derry—Abundant along the coast from the Foyle to the Bann; D.M.

5. C. vulpina *Linn.*

Marshes, and drains, especially by the coast—frequent. Fl. June and July.

Down—Copeland Islands, Crawfordsburn, and shore east of Bangor; *Flor. Ulst.* Sydenham, shores from Comber to Mountstewart, and in Ballygowan bog; S.A.S.

Antrim—Banks of the Lagan near Belfast; *Templeton*, 1796. Springfield, Monkstown, Rathlin, and Giants' Causeway; S.A.S. Drains near Whitehouse; T.H.C.

Derry—Abundant along the shores of Lough Foyle; D.M. In profusion close to Limavady Junction; S.A.S. Magilligan; *H. C. Hart, op. cit.*

6. C. muricata *Linn.*

Wet pastures—very rare. Fl. June and July.

Antrim—Plentiful in a damp pasture field by the shore a little northeast of the coastguard station at Macedon; S.A.S., 1868. One stem gathered at the end of July, 1885, measured four feet in length.

This sedge grows in a field which was tenanted by two other of our rarest plants, *Hypericum hirsutum*, and *Thrincia hirta*. The former of these is now almost lost, but both still survive, along with the present plant. They are, however, reduced in quantity, by reason of the breaking up and cultivation of a portion of the field. Another step in this direction, and they will disappear.

7. C. teretiuscula *Good.*

Boggy places—very rare, possibly extinct. Fl. June (*Cyb. Hib.*).

[Down—Reported as occurring in this county, but on insufficient authority. There is no bog or marsh at or near Giants' Ring, as mentioned in Cybele Hibernica.]

Antrim—In old moss holes at Cranmore; *Templeton*. Marshy borders of a small lake at Killymurry; *Cyb. Hib.*

Mr. Templeton's original *habitat*, moss holes, Cranmore, has been transcribed marl hole, and then transformed from marl hole, Cranmore, to Marble Hall, Carnmoney, a locality which we, and others, in vain sought to find.

8. C. paniculata *Linn.*

Marshes, and ditches—frequent. Fl. May and June.

Down—Ballyalloley Lake, and by lakes at Magheralagan, and Loughinisland; S.A.S. Downpatrick; *C. Dickson*.

Antrim—Near second lock of Lagan canal; *Flor. Ulst*. By the Lagan opposite Edenderry, and thence downwards, Massareene and Shane's Castle parks, Cave Hill near head of the Tramway, Loughmourne, and boggy ditch at Carmavy; S.A.S.

Derry—Common in many places, especially near Moneymore; D.M. By the Bann near Coleraine; S.A.S.

9. C. remota *Linn.*

Woods, and shady banks—common. Fl. June and July.

10. C. echinata *Murr.* (C. STELLULATA *Good.*).

Wet heaths, peat bogs, and damp mountain pastures—very common. Fl. June and July. Ranges from sea level to over 2000 feet.

11. C. elongata *Linn.*

Damp wood, very rare, or most likely extinct. Fl. June (*Cyb. Hyb.*).

Antrim—Banks of Lough Neagh near Gally's Gate, *D. Moore*; *Nat. Hist. Rev.*, vol. VI., p. 538. In a wet boggy wood near Derrymore House at Selshan in the parish of Aghagallon, D.M., 1838; *Cyb. Hib.*

The two notes, above quoted, refer to one station where, however, the plant is not now found, nor is there, at present, any wet boggy wood at the place mentioned.

12. C. canescens *Linn.* (C. CURTA *Good.*).

Turf bogs, and wet peaty heaths—frequent. Fl. June and July. Ranges from sea level to 1500 feet.

Down—Birky moss south of Castlereagh Hill; *Flor. Ulst.* In bogs at Moneyreagh, Brown's Island near Newtownards, Thornisland near Donaghadee, Monlough, and Cotton moss; S.A.S. Ballygowan bog; T.H.C. Rowantree bog at Ashfield near Dromore; *C. Dickson.*

Antrim—Carnmoney, and Kings' moss; *Flor. Ulst.* Plentiful in marshy meadows, and bog holes about Ballinderry, and near Ballymena; *Cyb. Hib.* At Portmore, and in bog northwest of the cairn on the summit of Divis; S.A.S.

Derry—Rare, only observed on the northern slopes of Slieve Gallion, near the base; D.M. In a bog west of Toome; S.A.S.

13. C. leporina *Linn.* (C. OVALIS *Good.*).

Meadows and pastures—frequent. Fl. June and July. Occurs at 1500 feet on the Mourne range; *Hart.*

Down—Greyabbey; *Ir. Flor.* Magheralagan, Loughinisland, Saintfield, Hillsborough Park, Cherryvalley, and a very tall slender form in a marshy hollow of the Newcastle sandhills; S.A.S. Holywood; R.Ll.P. Carn Mountain; *H. C. Hart, op. cit.*

Antrim—Wet meadows by Lagan canal, and by the Hannahstown road; *Flor. Belf.* Cave Hill, Colin Glen, and Massereene Park; *Flor. Ulst.* Templepatrick, wet meadows at Portmore, by the canal above Lisburn, Castlerobin, Carnmoney Hill, Loughmourne, and Portrush; S.A.S. Cushendall, and Glendun; R.Ll.P.

Derry—Very common; D.M. Toome, and Draperstown; S.A.S.

14. C. stricta *Gooden.*

Marshes—very rare. Fl. May and June (*Cyb. Hib.*).

The localities specified below may be right, but there are no specimens by which the correctness of the determination may be tested.

Antrim—Side of the Lagan at Lambeg, and by Lough Neagh at Portmore, and banks of the Tunny drain at same place; *Templeton.*

15. C. acuta *Linn.*

Marshy margins of rivers and lakes—rather rare. Fl. June and first half of July.

Down—By the side of the Lagan near Dromore; *Templeton,* 1803. Left side of Newtownards road near Belfast; *Flor. Ulst.* Shore

of Lough Neagh north of Lurgan, and abundant by the Lagan at the west end of Gillhall demesne near Dromore; S.A.S.
Antrim—Loonburn near Doagh; *Templeton*, 1795. At Sinclair's green near Belfast (present Falls Park), and lake shore at Massereene Park; *Flor. Ulst.* Langford Lodge; *B.N.F.C.*, 1864. Ram's Island, and abundant in marshes at the south side of Portmore Lake; S.A.S. Banks of the river at Randalstown; *Cyb. Hib.*
Derry—Marshy meadows west of Toomebridge; D.M. Moneymore; *Flor. Ulst.*

16. C. rigida *Gooden.*

Mountains—very rare. Fl. June and July (*Cyb. Hib.*).
[Down—Summit of Slieve Donard (Dickie); *Flor. Ulst.* Not found by any other botanist on this oft searched summit, and probably erroneously determined.]
Derry—Lagnagapagh (Robt. Brown); *Templeton*. From 1730 feet to summit of Sawel; *Hart, Proc. R.I.A.*, 1884.

17. C. Goodenovii *Gay* (C. VULGARIS *Fries*).

Meadows, wet pastures, and marshy margins of streams, and drains— very common. Fl. mid. May till mid. July. A very variable plant, specimens have been named as var. *gracilis*, and others as var. *brachystachya*.

18. C. Buxbaumii *Wahlenberg.*

Wet gravelly lake shores—extremely rare. Fl. June and early part of July.
Antrim—In one of the small islands in Lough Neagh near Toome; D.M. Found by Mr. D. Moore in July, 1835, on a small island in Lough Neagh near Toomebridge; *Flor. Hib.*
Collected again by D.M. in 1836, and by Joseph Woods in 1855 (*Cyb. Hib.*). On 1st June, 1867, two isolated specimens were gathered by S.A.S. On 14th June, 1878, one large tuft, with many flower stems, was found by T.H.C. Finally on 27th June, 1886, it was met with again by S.A.S. One little patch of about two feet square was seen on which there were a number of stems, some immature, others trampled down, or eaten by cattle, and a few perfect.

The above notes refer to one locality, the spot where this sedge was originally discovered. It is not known to occur elsewhere in Britain, and it is to be feared that its existence as a British plant is near its close. The dense, almost impenetrable thicket of shrubby wood which, a few years since, afforded excellent cover for many aquatic birds, and also sheltered *C. Buxbaumii* has recently been cut down. The little island is now a bare exposed pasturage where, in the struggle for existence, only the more hardy and aggressive are likely to maintain their hold.

19. C. pallescens *Linn.*
Wet pastures, and margins of lakes and streams—rather rare. Fl. mid. May till mid. July.
Down—Ballyalloley Lake, and People's Park; *Flor. Belf.*
Antrim—In a meadow about 100 yards below the second lock of the Lagan canal, and in other meadows; *Templeton*, 1797. Marshes at Stranmillis; *Flor. Belf.* Woodburn Glen; *Flor. Ulst.* Crumlin, and island in Lough Neagh; *Cyb. Hib.* Glenarm Park; *R. Tate.* Meadows near Glenavy River; T.H.C. Muckamore, Massereene Park, and in bog on Black Mountain; S.A.S.
Derry—Frequent at Magilligan, and in meadows by the Bann; D.M. Shores of Lough Neagh at Toome; *Flor. Ulst.*

20. C. panicea *Linn.*
Wet pastures, and heaths—very common. Fl. mid. May till mid. July Ranges from sea level to about 1400 feet.

21. C. limosa *Linn.*
Wet peat bogs—very rare. Fl. June (*Cyb. Hib.*).
Down—Ballygowan bog near the middle, and in Donaghadee bog; *Templeton*, 1804. Bog near Saintfield; *Flor. Ulst.*
Antrim—Between Fairhead and Murlough, and on the large bog at Rasharkin; *Cyb. Hib.* The Portmore record of *Cyb. Hib.* (*Stewart*) is most likely an error. The original specimen was very far advanced, and owing to the loss of the seed cannot now be referred with certainty to any species. It has not been again found at Portmore.

22. C. strigosa *Hudson.*
Woods, and bushy places—rare. Fl. mid. May till mid. July.
Down—Crawfordsburn; *Flor. Ulst.* Belvoir Park; *B.N.F.C.*, 1870.
Antrim—At a ditch near the banks of the Forth River about 100 or 150 yards above Clowney bridge on the Falls road; *Templeton*, 1797 (now lost). Shane's Castle, Colin Glen, Knockagh, Woodburn, and Glenarm Park; *Flor. Ulst.* By the Lagan canal at the lower side of Drumbridge; S.A.S.

23. C. pendula *Hudson.*
Damp woods and glens—rare. Fl. mid. May till mid. July.
Down—In the grounds of Montalto at Ballynahinch; S.A.S.
Antrim—By a little rivulet running down to the Bog meadows; *Templeton* 1806. Colin Glen; *Flor. Belf.* Near Glenarm; *Cyb. Hib.* Abundant in Glenarm Park; S.A.S.

CYPERACEÆ.

24. C. præcox *Jacq.*

Dry grassy heaths—frequent. Fl. May and June. Ranges to 1500 feet in the Mourne Mountains.

Down—By the Quoile at Downpatrick, also on Spalga Mountain near Rostrevor, and frequent on the Mourne range; S.A.S.
Antrim—Belfast hills; *Flor. Belf.* Colin Mountain, Cave Hill, and Knockagh; *Flor. Ulst.* Derriaghy, slopes of Black Mountain, Kilroot, and Sallagh Braes; S.A.S. Tannaghmore near Kells, and lake shore at Antrim; *D. Redmond.*
Derry—Abundant throughout the county, on dry pastures and heaths; D.M.

25. C. pilulifera *Linn.*

Dry stony heaths—frequent. Fl. mid. May till mid. July. Ranges from sea level to 2700 feet.

Down—Top of Slieve Donard; *Flor. Ulst.* Sandhills at Newcastle; T.H.C. Frequent in Mourne Mountains; S.A.S.
Antrim—Top of Black Mountain; *Templeton*, 1798. Boughal Hill, Drumadarragh, Cave Hill, Gobbins, Sallagh Braes, Slemish, summit of Fairhead, and scarce and stunted on Rathlin; S.A.S. A fine tufted form, with stems two feet high, by a little rivulet under trees at upper end of Colin Glen; S.A.S. Cushendall; R.Ll.P. Tannaghmore wood near Kells; *D. Redmond.*
Derry—Benevenagh, and frequent on moory ground; D.M. Slieve Gallion; *Flor. Ulst.* Benbradagh; S.A.S. Dart Mountain; *H. C. Hart, Proc. R.I.A.*

26. C. glauca

Meadows, wet pastures, heaths, and by streams—very common. Fl. May—June. Ranges from sea level to 1600 feet.

27. C. flava *Linn.*

Wet pastures, and marshy heaths—common, especially on the hills. Fl. mid. May till mid. July. Ascends to 2000 feet, but rarely flowers at the upper limits.

Var. β C. LEPIDOCARPA *Tausch.*

Perhaps frequent, but except the following there do not seem to be specimens verified authoritatively.

Antrim—Marshes at Portmore, and wet margin of Lough Aghery; S.A.S. Specimen determined by Dr. Christ, of Basle.

28. C. Œderi *Ehr.*

Wet sandy or gravelly places on lake shores, and on stony or gravelly heaths—rare. Fl. June and July.

Antrim—Belfast hills; *Templeton*, 1808. Stony shores of Loughmourne, and sparingly on gravelly heath at Drumadarragh Hill; S.A.S.
Derry—Frequent on bogs, and moist heaths; D.M. Occasionally on sandy shores of Lough Neagh from Toome to Ballyronan; S.A.S.

29. C. extensa *Gooden.*

Salt marshes—frequent. Fl. June and July.
Down—About Donaghadee, and in the salt marsh at Knocknagoney; *Templeton*, 1808. Kinnegar, Bangor, and Comber; *Flor. Belf.* Copeland Islands; *Flor. Ulst.* Greencastle, Newcastle, and frequent by the coast from Strangford to Dundrum, also at Castle Espie, and Groomsport; S.A.S.
Antrim—Larne Lough, and Cushendall; *Cyb. Hib.*
Derry—Salt marsh at the mouth of the Roe, and by the Foyle near Culmore; D.M.

30. C. Hornschuchiana *Hoppe.*

Grassy heaths—frequent. Fl. June and July. Ranges from sea level to 1500 feet.
Down—Mourne Mountains; *Flor. Hib.* Heath on Conlig Hill, mountains above Tollymore, and by the river at Rostrevor; S.A.S.
Antrim—In a marshy field south of Stranmillis dam; *Templeton*, 1808. Slemish, and commons of Carrickfergus; *Flor. Ulst.* Black Mountain, lower slopes of Divis, and frequent on the moors in Rathlin; S.A.S.
Derry—Slievegallion, and north side Muinard Mountain; D.M. Side of the Bann above Coleraine, and heathy hills above Dungiven; S.A.S.

31. C. distans *Linn.*

Wet sandy or gravelly seashores—frequent. Fl. June and July.
Down—In the salt marsh at Knocknagoney; *Templeton*, 1808. Between Bangor and Crawfordsburn; *Flor. Ulst.* Shores of the bay at Sydenham, and shores of Strangford Lough at Newtownards, and between Comber and Castle Espie, also at Donaghadee, and frequent from Strangford to Dundrum; S.A.S. Shore between Newcastle and Annalong; *C. Dickson.*
Antrim—Shore at the south end of Rathlin, but rare; S.A.S.
Derry—Rare, but occurs in a salt marsh at the mouth of the Roe; D.M.

32. C. binervis *Smith.*

Heaths, and rocks, and occasionally in woods, and on river banks—frequent. Fl. June and July. Ranges from sea level to 2000 feet.
Down—Kinnegar at Holywood, and on Conlig Hill; *Flor. Ulst.* Woods

CYPERACEÆ. 165

at Greyabbey, rocks at Groomsport, shore from Comber to Newtownards, Carngaver, and Holywood hills, also on Bignian, and frequent on the Mourne Mountains; S.A.S. Scrabo; T.H.C.

Antrim—Wolfhill, and drier parts of Divis; *Flor. Belf.* Colin Mountain, and Knockagh; *Flor. Ulst.* Banks of the Lagan near Belfast, Black Mountain, Colinward, Knocklayd, hills above Cushendall, and sparingly on Rathlin; S.A.S.

Derry—Common through the county; D.M. Benbradagh, and frequent on the Sperrin range; S.A.S.

33. C. lævigata *Smith.*

Damp woods, and river banks—rare. Fl. June and July.

Down—In the wood at Belvoir growing beside the rivulet which runs down to the garden; *Templeton*, 1814. Hillsborough Park; *Cyb. Hib.* By the Quoile at Downpatrick, also roadside hedge halfway between Ballynahinch and Dromore, and very luxuriant by the Spinkwee at upper end of Tollymore Park; S.A.S. Wet banks by Moneycara River near Newcastle ; *C. Dickson.*

Antrim—Banks of the Lagan near Belfast; *Flor. Hib.* Colin Glen, and at the head of Woodburn Glen; *Flor. Ulst.* Boggy wood near Ballycastle; *Cyb. Hib.* In a little marsh among trees at Stranmillis; S.A.S.

Derry—Very rare, only observed in boggy thickets in Bond's Glen parish of Upper Cumber; D.M.

34. C. sylvatica *Hudson.*

Woods, and shady banks—common. Fl. mid. May till mid. July.

35. C. hirta *Linn.*

Damp sandy or gravelly pastures and wastes—frequent. Fl. June and July.

Down—Sydenham ; *Flor. Belf.* Hillsborough, Newtownards, Crawfordsburn, and various places on the coast to Newcastle ; S.A.S.

Antrim—By the Lagan canal, and at the town reservoirs; *Flor. Belf.* By Lough Neagh at Selshan and Langford Lodge, Ballinderry, Lisburn, Colin Glen, Belfast hills, Monkstown, Loughmourne, Islandmagee, Knocklayd, and in small quantity on Rathlin ; S.A.S.

Derry—Side of Lough Neagh, and not unfrequent in wet pastures ; D.M. By Lough Neagh at Toome; *Flor. Ulst.* Marshy ground at Toome ; S.A.S.

36. C. pseudo-cyperus *Linn.*

Marshes—very rare. Fl. June (*Cyb. Hib.*).

Antrim—Sides of drains in the Bog meadows; *Templeton*, 1797. A very
few plants in a drain in the Bog meadows near Balmoral;
R.Ll.P., 1886.
Until rediscovered by Mr. Praeger it was thought that this fine sedge
had become extinct in our district.

37. C. rostrata *Stokes*. (C. AMPULLACEA *Good.*).
Marshes, bog drains, and margins of lakes and rivers—common. Fl.
mid. May till mid. July. Ranges from sea level to 1000 feet.

38. C. vesicaria *Linn*.
Marshes and ditches, also lake and river margins at low elevations—
frequent. Fl. mid. May till mid. July.
Down—Sides of Ballynahinch River; *Templeton*, 1800. Ballyalloley
Lake, and at Knock; *Flor. Belf.* Near Moira; *Cyb. Hib.*
Lough Aghery, Loughinisland, Ballynahinch, Monlough,
Hillsborough, and Lambeg; S.A.S. Orlock; R.Ll.P. North
of Newcastle; *H. C. Hart, op. cit.*
Antrim—Malone, and Belfast hills; *Flor. Belf.* Glenshesk; *Cyb. Hib.*
Portmore, and islands south of Toome, also by the Lagan
above and below Lisburn, and on the Knockagh; S.A.S.
Crumlin waterfoot; T.H.C.
Derry—Frequent in bogs and marshes; D.M. Shores of Lough Neagh
near Toome; *Flor. Ulst.* Magilligan; *G. C. Hyndman*.

39. C. paludosa *Gooden*.
Marshy places, and lake margins—extremely rare. Fl. May (*Cyb. Hib.*).
Down—In an old moss hole near Kilmore, and in a little lake two miles
southeast of Kilmore; *Templeton*, 1810. This is believed to be
the townland of Kilmore which touches the shores of Lough
Neagh, and the station is therefore near to the boundaries of
the three counties of Down, Antrim, and Armagh.
The authority for the Ballynockan station of Flora of Ulster is not
considered sufficient.

40. C. riparia *Curtis*.
River banks, and lake shores—rather rare. Fl. mid. May till. mid.
July.
Down—On the left side of the Newtownards road near Belfast, 1806, and
along with the preceeding sp. in a little lake two miles southeast of Kilmore, 1810; *Templeton*. Banks of the Lagan canal
where it enters Lough Neagh; *G. C. Hyndman*, 1838. Gillhall; T.H.C.
Antrim—Shores of Lough Neagh, and by the Carrickfergus road near

GRAMINEÆ.

Whitehouse; *Templeton*, 1804. Whitehouse dams; *Flor. Belf.* Portmore Lake; *Cyb. Hib.* Banks of the Sixmilewater above and below Templepatrick, and Bannside near Portglenone; S.A.S.

ORDER LXXI. **GRAMINEÆ.**

PHALARIS *Linn.*

1. P. arundinacea *Linn.* REED GRASS.
Ditches, and sides of rivers—common. Fl. mid. June till end of July.

ANTHOXANTHUM *Linn.* VERNAL GRASS.

1. A. odoratum *Linn.*
Meadows, pastures, and heaths—very common. Fl. May and June. Ranges from sea level to nearly 2000 feet.

PHLEUM *Linn.* CATSTAIL GRASS.

1. P. arenarium *Linn.*
Coast sandhills—rare, though locally abundant. Fl. June and early part of July.
Down—On the sands at Newcastle; *Templeton*, 1797. Dundrum sandhills; S.A.S.
Antrim—Abundant on sandhills at Portrush; S.A.S.
Derry—Magilligan sands; *Templeton*. Abundant on sandy warrens along the coast from Lough Foyle to Portstewart; D.M.

2. P. pratense *Linn.* TIMOTHY.
Meadows, pastures, and banks—very common. Fl. mid. May till mid. July.

ALOPECURUS *Linn.* FOXTAIL GRASS.

1. A. pratensis *Linn.*
Banks, and pastures—frequent. Fl. June and July.
Down—Pastures at Dromore, and Clandeboye Demesne; S.A.S.
Antrim—Portmore, Newforge, Belfast waterworks, fields beyond Colin

Glen, by the Sixmilewater, pastures at Glynn, and sparingly on Rathlin; S.A.S. Balmoral; T.H.C.

Derry—Very common near Londonderry; *Ord. Surv. Derry.* Not very common, except near Londonderry, and Salterstown; D.M.

2. A. geniculatus *Linn.*

Marshy pastures, and wet sandy places—common, especially near the coast. Fl. June and July.

NARDUS *Linn.* MAT GRASS.

1. N. stricta *Linn.*

Rocky, or stony heaths—common. Fl. mid. June till mid. Aug. Ranges from sea level to near 2000 feet.

MILIUM *Linn.* MILLET.

1. M. effusum *Linn.*

Woods, and shady glens—rare. Fl. mid. June till end of July.

Down—Upper end of Killeen Glen, and under trees in Stormount Glen; S.A.S. Tollymore Park; *H. C. Hart, op. cit.*

Antrim—In Carr's Glen; *Templeton,* 1794 (not there now). Old Lodge Road; *Flor Ulst.* (loc. obsolete).

PHRAGMITES *Trin.* REED.

1. P. communis *Trin.* (ARUNDO PHRAGMITES *Linn.*)

Ditches, lake margins, and by rivers—frequent. Fl. July and early part of Aug.

Down—Connswater; *Flor. Ulst.* Castlewellan, Ballynahinch, Saintfield, Cherryvalley near Comber, and at Tillysburn; S.A.S.

Antrim—By the Lagan canal; *Flor. Belf.* Portmore, Whiteabbey, Blackhead, Ballycastle, and plentiful on Rathlin; S.A.S.

Derry—Common in wet ground at Magilligan; *Samp's. Surv. Derry.* Common throughout the county; D.M. By the Bann at Portstewart; S.A.S.

PSAMMA *Beauv.* MARRAM.

1. P. arenaria (*Linn.*) *R. & S.* BENT (loc.).

Sand dunes, and loose shifting sands—common on the seacoast. Fl. July and early part of Aug. Extensively used for the manufacture of a common sort of door mats.

GRAMINEÆ.

CALAMAGROSTIS *Adanson*. SMALLREED.

1. C. epigejos (*Linn.*) *Roth.*
Damp shady places—extremely rare. Fl. July (*Cyb. Hib.*).
Derry—Shady, moist, rocky places on Formoyle Hill midway between Coleraine and Limavady; D.M. Between Maghera and Garvagh; *Cyb. Hib.*
First found in Ireland by Dr. Moore, and not known to occur in any other Irish localities.

2. C. Hookeri *Syme.* (*C. stricta* var. *auct.*).
Gravelly lake shores—very rare. Fl. June and first half of July.
Antrim—On a small island in Lough Neagh, and in several localities along the shores of the Lough near Antrim; *Ord. Surv. Londonderry.* (*as C. lapponica*). Island in Lough Neagh near Toome; *Flor. Ulst.* (*sub C. stricta*). On moist flat land on the shore of Lough Neagh west of the old castle in Shane's Castle Park; *Prof. Babington.* Refound by R.Ll.P. near the shore at Shane's Castle, 1886.
Derry—By the shores of Lough Beg near Coney Island northwest of Toome; S.A.S.
First found by Dr. Moore in 1836. This grass is confined, in Britain, to the immediate neighbourhood of Lough Neagh. It is not now found on the islet near Toome where it was originally discovered, and is exceedingly rare on the Antrim, and Derry shores. Rev. George Robinson found it on Scawdy Island in Co. Tyrone, and pointed it out to the Belfast Naturalists' Field Club in 1870, and to T.H.C. in 1878. Seen also, in plenty, on Scawdy by Rev. S. A. Brenan in July, 1887. In addition Mr. Robinson has observed it near Washing Bay, also in Tyrone.
The necessity for the removal of this species to *Deyeuxia* is not apparent.

AGROSTIS *Linn.* BENT GRASS.

1. A. canina *Linn.*
Moors, and heathy bogs—not common, though abundant in some places. Fl. July. Ascends to 1500 feet, or perhaps higher in the Mourne range.
Down—On rocky ground by the sea below Holywood; *Templeton.* Bog at Ballynahinch, and heaths on mountains above Newcastle; S.A.S.
Antrim—Cranmore, and near the second lock of the Lagan canal; *Templeton*, 1808. Cave Hill; S.A.S.
Derry—Heaths, and moory places, not common; D.M.
Mr. Templeton doubted if this could be separated from *A. alba*, and as no one seems to have found *A. canina* in the stations he records, it is not improbable that his plant was incorrectly named.

GRAMINEÆ.

2. A. vulgaris *Withering.*
Meadows, pastures, and banks—very common. Fl. mid. June till mid. Aug.
 Var. β A. PUMILA *Lightfoot.*
Down—Sandy, hilly ground north of Belvoir; *Templeton.* Sandhills at Newcastle; S.A.S.

3. A. alba *Linn.*
Seashores, and wet rocky heaths—common. Fl. July.
 Var. β A. STOLONIFERA *Linn.*
Muddy seashores—frequent.
Occurs on the marshy margins of Strangford, Belfast, and Larne Loughs.

HOLCUS *Linn.* SOFT GRASS.

1. H. lanatus *Linn.*
Pastures, meadows, hedge banks, and waste ground—very common. Fl. June and July.

2. H. mollis *Linn.*
Bogs, and wet moory pastures—frequent, but much less common than the preceding species. Fl. June—Aug.
Down—Hillsborough Park, Ballygowan bog, Cotton moss, and bogs at Ballynahinch, and Moneyrea; S.A.S.
Antrim—Woodburn Glen, and heath on Brean Mountain above Ballycastle; S.A.S.
Derry—Common; D.M.

More frequent than would appear from the above notes, but some collectors confound these two species; others do not sufficiently distinguish between them, and few take pains to note separate localities for each.

AIRA *Linn.* HAIR GRASS.

1. A. cæspitosa *Linn.*
Woods, glens, and damp shady places—common. Fl. July and Aug.

2. A. flexuosa *Linn.*
Mountain heaths—frequent. Fl. July and Aug. Ascends to 2000 feet on the mountains, and increases in abundance to the northwest.

GRAMINEÆ. 171

Down—Slieve Croob ; *Dubord. Surv. Down.*
Antrim—Mountain glens and woods near Belfast ; *Flor. Ulst.* Divis Mountain sparingly, Cave Hill, Sallagh Braes, heath on Fairhead, Knocklayd, mountains above Ballycastle, and heaths on west end of Rathlin ; S.A.S.
Derry—Sawel Mountain, and common on heaths ; D.M. Sluggada and Straw mountains, and frequent in the Sperrin range ; S.A.S.

3. A. caryophyllea *Linn.*
Sandy fields, dry banks, and walls—common. Fl. June and July. Ascends to 1800 feet (*Hart*).

4. A. præcox *Linn.*
Bogs, heaths, and bare pastures—common. Fl. mid. May till end of June.
From sea level to 1200 feet on Spalga.

TRISETUM *Persoon.*

1. T. flavescens (*Linn.*) *Beauv.*
Banks, and pastures—rather rare. Fl. July and Aug.
Down—Close to Sydenham station, and by the canal near Moira ; S.A.S.
Antrim—Meadow about 100 yards below the second lock of Lagan canal ; *Templeton,* 1797 (still lingers there). Cave Hill above the Limestone Quarries ; *B.N.F.C.,* 1871. Seaview ; *W. Darragh.* Bog meadows, Templepatrick, and abundant in pastures at Shane's Castle ; S.A.S. Southern slopes of Black Mountain ; *Richd. Hanna.*

AVENA *Linn.* OAT.

1. A. pubescens *Linn.*
Shady banks, and short dry rocky pastures—not common. Fl. June and July.
Down—On rough banks south of Newcastle ; *Templeton,* 1799.
Antrim—North side of Slemish ; *Templeton.* Rocks by the waterfall at head of Colin Glen, also Knockagh, Woodburn, Ballycastle, and cliffs south of Glenarm ; *Flor. Ulst.* By the Black cave south of the Gobbins, and on heights between Larne and Ballygalley, also at Sallagh Braes, Glenarm deerpark, and Rathlin ; S.A.S.
Derry—On barren lands in Magilligan ; *Samp's. Surv.* Not common, but plentiful about Garvagh ; D.M. Castlerock ; R.Ll.P.

172   *GRAMINEÆ.*

ARRHENATHERUM *Pal. de Beauv.* OAT GRASS.

1. A. elatius (*Linn.*) *M. & K.*
Meadows, pastures, and hedge banks—very common. Fl. June and July.

The var. *bulbosum* is the common form, and on account of its knobbed stems is known to the country people as "pearls." It is a well known cattle remedy.

TRIODA *R. Brown.* HEATH GRASS.

1. T. decumbens *Beauv.*
Heaths, and stony or sandy places—frequent. Fl. mid. June till end of July. Ranges from sea level to 1400 feet.
Down—Kinnegar at Holywood; *Flor. Belf.* Slieve Donard, and in stony and heathy places on the coast from Strangford to Kilkeel; S.A.S. Ballyholme, and Groomsport; T.H.C. Slievemartin; *H. C. Hart, Proc. R.I.A.*
Antrim—Sandy pastures in the Plains; *Templeton*, 1797 (now built over). Belfast hills; *Flor. Belf.* Black Mountain, and Cave Hill; *Flor. Ulst.* Fairhead, Knocklayd, and Rathlin; S.A.S.
Derry—Abundant on moory pastures near the mountains; D.M. Very luxuriant on the shore of Lough Beg near Toome; S.A.S. Benbradagh; *H. C. Hart, op. cit.*

KOELERIA *Persoon.*

1. K. cristata (*Linn.*) *Pers.* CRESTED HAIR GRASS.
Sandy seaside pastures—rather rare. Fl. mid. June till end of July.
Down—High grassy banks above the rocks on the shore about one mile south of Newcastle; *Templeton*, 1799. Shore at Strangford, and Castleward; *Flor. Ulst.* Shore between Bangor and Ballyholme; S.A.S.
Antrim—Islandmagee, and pastures on the shore near Giants' Causeway; *Templeton.* Sandy seashore at Garronpoint, and specially abundant on Rathlin; S.A.S.
Derry—Common on sandy shores, and in rocky places along the coast from Magilligan to Portrush; D.M. Castlerock; R.Ll.P.
Cannot be found on Cave Hill as stated in Flora of Ulster—some error.

MELICA *Linn.* MELIC.

1. M. uniflora *Retz.*
Shady banks in glens and woods—frequent but seldom abundant. Fl.

GRAMINEÆ. 173

mid. May till mid. July.

Down—Dundonald; *Flor. Belf.* Warrenpoint, Tollymore, and Crawfordsburn; *Flor. Ulst.* Abundant in woods at Rostrevor, and less so in glens at Newtownards, Drumbo, Holywood, and Cregagh; S.A.S. Cultra; R.Ll.P. Benson's glen near Newry; *C. Dickson.*

Antrim—Colin Glen, and Maryburn; *Templeton*, 1793. Springfield, and Woodburn Glen; *Flor. Belf.* By the Lagan canal above Shaw's Bridge, Carr's Glen, and glen at Magheramorne; S.A.S. Glendun; R.Ll.P.

Derry—Common in shady dry woods; D.M. Moneymore; *Flor. Ulst.* Lignapeiste Glen; *B.N.F.C.*, 1876.

MOLINIA Schrank.

1. M. cœrulea (*Linn.*) *Moench.*

Bogs, and wet heaths—frequent. Fl. July and Aug.

Down—Cotton moss, Ballycroghan dam, and Ballygowan bog; *Flor. Belf.* Bog at Ballynahinch, and in profusion on Mourne Mountains; S.A.S.

Antrim—Colin Mountain near Belfast; *Flor. Ulst.* Banks of Lough Neagh at Massereene Park, and Shane's Castle; S.A.S.

Derry—Creevagh Hill near Londonderry; *Ord. Surv. Londonderry.* Very abundant in wet heathy and moory places; D.M. Margins of Lough Beg near Toome; S.A.S.

Plants are frequent which seem referrible to the var. *depauperata*, but not authoritatively determined.

POA Linn. MEADOW GRASS.

1. P. annua *Linn.*

Waste ground, banks, and pastures—very common. Fl. Spring to Autumn.

2. P. nemoralis *Linn.*

Mountainous woods—very rare, not seen by us. Fl. June and July (*Cyb. Hib.*).

Down—In a wood at the side of Knockchree Mountain (three miles northwest of Kilkeel); *Wade Rar.* Donard Dodge; *Flor. Ulst.* Woods about Rostrevor; *Cyb. Hib.*

Antrim—Found on the rocks of the northern branch of Glenarve River below the waterfall at the bridge on the Cushendall road (about five miles southwest of Cushendall and one and a half east of Slievenanee); *Templeton*, 1809. Wooded mountains about

Knappen near Garron Head ; *Cyb. Hib.*
Derry—Woods, and thickets near Limavady, but not common ; D.M.

3. P. trivialis *Linn.*
Meadows, and pasture fields—very common. Fl. June and July. Ranges from sea level to 1500 feet (*Cyb. Hib.*).

4. P. pratensis *Linn.*
Hedge banks, roadsides, and pastures—very common. Fl. June—Aug.

5. P. compressa *Linn.*
Walls, and dry banks—extremely rare. Fl. June and July (*Cyb. Hib.*).
Antrim—Roadside between Ballycastle and Ballintoy ; *Cyb. Hib.*
Derry—Rare, only observed on the old walls of Londonderry ; D.M. Plentiful along the south wall of Derry ; *Flor. Ulst.* In several places on the old walls of Londonderry ; *Cyb. Hib.*

The note in Flora of Ulster " walls at Larne, F. Whitla " is not quoted above, as it seems to be incorrect. Mr. Whitla's note really attributes the discovery to Templeton " walls of a tower at Larne, T." (Templeton)
There does not however appear to be any such record in Templeton's MS.

GLYCERIA *R. Brown.*

1. G. aquatica (*Linn.*) *Smith.*
Deep ditches, and marshes—very rare. Fl. July and Aug.
Down—Plentiful in a ditch, and adjoining marsh by the Quoile between Downpatrick, and the bridge ; S.A.S., 1880.

2. G. fluitans (*Linn.*) *Br.*
Ditches, and margins of streams—common. Fl. June till mid. Aug.

SCLEROCHLOA *Pal. de Beauv.*

1. S. maritima (*Huds.*) *Lindley.*
Salt marshes, and muddy, or wet sandy seashores—common all around the coast. Fl. mid. June till end of July.

2. S. procumbens (*Curtis*) *Beauv.*
Salt marshes—very rare. Fl. mid. June till end of July.
Antrim—Near Larne ; *Flor. Ulst.* In a little drain in an unpaved portion of Albert Quay, Belfast ; S.A.S., 1870.

This grass was luxuriant in 1871, and plentiful for a few years until the spot was paved. Since that time the plant has been lost, nor can it now be found at Larne—possibly a casual in both places.
The reported occurrence at Comber (*Flor. Belf.*) was an error.

3. S. distans (*Linn.*) *Bab.*
Salt marshes—frequent. Fl. mid. June till end of July.
Down—Kinnegar, and Victoria Park; *Flor. Belf.* Shore at Portaferry; *Flor. Ulst.*
Antrim—Point fields (loc. obsolete), and plentiful on the shores; *Templeton*, 1794. Brackish ditches at Greencastle, and abundant on muddy shores of Belfast Bay; S.A.S.
Reported in Samp's. Surv. Londonderry as common. The *habitat* stated, however, is dry shady places, and this note may be assumed to be one of the very numerous errors which render the lists of plants, in that work, of little value.

4. S. rigida (*Linn.*) *Link.*
Dry banks—very rare. Fl. mid. June till end of July.
Down—Plentiful in one spot, among whins, at foot of a stone dyke close to the sea about two miles northeast of St. John's Point; S.A.S. Dry sandy bank in sandhills at Newcastle; T.H.C.

5. S. loliacea (*Huds.*) *Woods.*
Dry rocks, walls, and gravelly seashores—rather rare. Fl. mid. June till end of July.
Down—On rocks at Portaferry; *Templeton.* Graypoint, pier at Donaghadee, and in rocky and gravelly places by the shore one mile south of Strangford; S.A.S. Rathmullen point, and grassy heaths at Ardglass; T.H.C.

BRIZA *Linn.* QUAKING GRASS.

1. B. media *Linn.*
Meadows, and damp pastures—frequent. Fl. June and July.
Down—In the lawn on the north front of Belvoir; *Templeton*, 1797. Clandeboye demesne, and Drumbo glen; S.A.S. Holywood; R.Ll.P.
Antrim—By the Lagan canal at several points near Belfast; *Templeton.* Tildarg. Ballycorr, by the Sixmilewater at Ballyclare, shore below Whiteabbey, Knockagh, and meadows near Whitehead; *Flor. Ulst.* Glenarm Park; *B.N.F.C.*, 1866. Banks of Crumlin River, and near the old castle at Shane's Castle; T.H.C. Crumlin waterfoot, Whiterock, Carnmoney Hill, and at Gleno; S.A.S.

Derry—In pastures, rare; *Samp's. Surv.* Only observed in the neighbourhood of Coleraine, where it is abundant; D.M. By the Bann below the Cutts; S A.S.

CATABROSA *Pal. de Beauv.*

1. C. aquatica *Beauv.*
Ditches, and marshes—frequent. Fl. June and July.
Down—"Victoria Park," Drumbo, Ballygowan bog, and marsh near Carngaver; S.A.S.
Antrim—By the side of a road from Cromac to Malone; *Flor. Ulst.* Marsh by the canal two miles above Lisburn, ditch by the railway beyond Jennymount, ditch by the old road from Hyde Park to the mountain, Cave Hill, Black Mountain, ditch on Carnmoney Hill, Gobbins, marsh between Larne and the Curran, foot of Slemish, and at Redbay; S.A.S.
Derry—Common in ditches, and boggy drains through the county; D.M. On Slieve Gallion, and beyond Banagher old Church; S.A.S.

CYNOSURUS *Linn.* DOGSTAIL GRASS.

1. C. cristatus *Linn.*
Meadows, pastures, and waste ground—very common. Fl. July and Aug. Ranges to 1500 feet on the Sperrin Mountains.

DACTYLUS *Linn.*

1. D. glomerata *Linn.*
Meadows, pastures, roadsides, and wastes—very common. Fl. June and July.

FESTUCA *Linn.* FESCUE GRASS.

1. F. sciuroides *Roth.*
Walls, and sandy waste places—frequent. Fl. June and July.
Down—Marino, and Holywood Hill; *Flor. Belf.* Newcastle, and Bryansford; *Flor. Ulst.* Sandy fields at Dundela, and roadsides about Comber, also near Groomsport, and Saintfield; S.A.S.
Antrim—Malone, and Point fields; *Templeton*, 1797. By the Forth River bridge; *Flor. Belf.* Knockagh, and Ballycastle; *Flor. Ulst.* (sub. *F. myurus*). Glenariffe, Rathlin, Templepatrick, and frequent in sandy waste places, and walls, around the shores of Lough Neagh, and on the coast; S.A.S.

GRAMINEÆ. 177

Derry—Near Downhill, and abundant through the county; D.M.

2. F. ovina *Linn.* SHEEP'S FESCUE.
Sandy, maritime and lacustrine shores, and on heaths—common. Fl. June and July. Ascends to 2700 feet on the mountains.

The viviparous form, F. VIVIPARA *Sm.*, is plentiful on the Mourne range in Down, and on Slievenanee, Fairhead, and Rathlin heights in Antrim, also Dart, and Clontygearagh in Derry.

3. F. rubra *Linn.*
Roadsides, dry banks, and sandy wastes and shores—common. Fl. June and July. A form which is found on coast sandhills seems different.

4. F. sylvatica *Villars.*
Woods, and glens—very rare. Fl. July (*Cyb. Hib.*).
Down—Woods at Rostrevor; *Wade Rar.*
Antrim—Colin Glen; *Templeton,* 1802. Glenarm, and Glendun; *Cyb. Hib.*
Derry—In the Ness Glen, side of the Faughan about two miles below Cumber, and in a little glen near Londonderry; *Templeton,* 1800. By the Bann at Mountsandel wood above Coleraine; D.M.
Whiterock station of Flor. Ulst. certainly wrong.

5. F. gigantea *Villars.*
Woods, and shady glens—frequent. Fl. July and first half of Aug. The var. F. TRIFLORA *Sm.* occurs occasionally.
Down—Cregagh Glen, Stormount, and Crawfordsburn; S.A.S.
Antrim—Woody places, and hedges; *Templeton,* 1797. Abundant on the limestone rocks near Garron Head, and at demesne of Drumnasole (var. *triflora*); *Ord. Surv. Londonderry.* Colin Glen, and Woodburn (*triflora*); *Flor. Belf.* Blackstaff lane; *Flor. Ulst.* Springfield, Cave Hill, and Glenarm Park; S.A.S.
Derry—Neighbourhood of Garvagh, and common in some districts; D.M. Demesne at Moneymore; *Flor. Ulst.*

6. F. arundinacea *Schreb.* (F. ELATIOR *Linn.*).
Marshes, and ditches in woody, or shady places—rare. Fl. mid. June till end of July.
Down—Newcastle; *Rev. W. M. Hind.*
Antrim—Ditch at the Plains; *Templeton,* 1797. Shores of Lough Neagh at Massereene Park, and near Toome; *Flor. Ulst.* Shane's Castle woods; *B.N.F.C.,* 1866. Marsh at Stranmillis, ditch by the railway between Belfast and Greencastle, stream north

M

of Kilroot station, and frequent among shady rocks in Rathlin; S.A.S.

Derry—Meadows of Aghanloo, but seldom met with; *Samp's. Surv.* Rather rare, but plentiful in meadows about Garvagh, and by the sides of the Agivey River; D.M.

7. F. pratensis *Hudson.*

Marshes, and wet meadows, and by margins of lakes, and rivers—frequent through the district. Fl. June and July.

The var. F. LOLIACEA *Hudson* was found by Templeton at Cranmore, and by the Lagan canal.

BROMUS *Linn.* BROME GRASS.

1. B. asper *Murr.*
Woods, and bushy glens—common. Fl. July and early part of Aug.

2. B. sterilis *Linn.*
Sandy fields and banks—rare. Fl. June and July.
Down—Lambeg bridge, and southwest side of Lisburn; *Templeton,* 1794. Ballyhornan Bay near Ardglass; T.H.C. Roadside by the shore north of Ardglass, also sandy ground near Knockinelder in the Ards, and occasionally thence to Cloughey; S.A.S. Railway bank at Marino; R.Ll P.
Antrim—Sandy places, and ditches at the Curran of Larne; *Templeton.* Roadside near Ballygomartin schoolhouse; *Flor. Ulst.*
Derry—Waste ground by the roadside between Limavady and Artikelly bridge, but not common; D.M.

3. B. racemosus *Linn.*
Wet meadows, and by streams and ditches—not uncommon. Fl. June and July.

In consequence of not being sufficiently distinguished from the succeeding species but few localities can be given. In Antrim, however, it occurs plentifully in the Bog meadows, and is abundant by the Sixmilewater above Templepatrick (S.A.S.). In Derry it is found in plenty in meadows by the Bann at Mountsandel above Coleraine (D.M.).

4. B. mollis *Linn.* LOP GRASS.
Roadsides, waste ground, and borders of fields—common. Fl. June and July.

BRACHYPODIUM *Pal. de Beauv.*

1. B. sylvaticum (*Huds.*) *R. & S.*
Woods, and shady banks—common. Fl. July and Aug.

GRAMINEÆ.

TRITICUM *Linn.* WHEAT GRASS.

1. T. caninum *Linn.*
Waste ground and banks—seems rare. Fl. July (*Cyb. Hib.*).
This grass has been so much confounded with the following species that its prevalence cannot be accurately defined. Templeton recorded it as occuring about the river at Montalto, Co. Down, and among shrubs in Grogan's Glen, also in Glenarm deerpark. By S.A.S. it has been doubtfully noted as at Sydenham, Colin Glen, and Ballycarry.

2. T. repens *Linn.* COUCH GRASS (*locally corrupted to squitch*).
Fields, waste ground, and banks, also on seashores—common. Fl. June and July.

3. T. junceum *Linn.*
Above high water mark on sandy strands—frequent. Fl. mid. June till end of July.
Down—In blowing or loose sands on the seashore; *Templeton.* Holywood, Graypoint, and Ballyholme; *Flor. Belf.* Portaferry, and shore north of Newcastle; *Flor. Ulst.* Sandy shores at Donaghadee, Ballykinler, and opposite Gunn's Island; S A.S.
Antrim—Eden near Carrickfergus; *Flor. Ulst.* Sands at Redbay, and Ballintoy; S.A.S.
Derry—Plentiful along the sandy coast from the Foyle to the Bann; D.M. Castlerock; R.Ll.P.

HORDEUM *Linn.* BARLEY.

1. H. pratense *Hudson.*
Dry banks—very rare. Fl. mid. June till end of July.
Down—In the meadows going to Connswater; *Templeton,* 1806.
Antrim.--On the banks of the canal, and on the roadside going to Whitehouse; *Templeton,* 1799. Ditch bank near Whitehouse; *W. MacMillan,* 1869. Ditch bank on the north side of Springfield Road east of the Forth River bridge; S.A.S., 1866 to 1886. The plant was in small quantity at this last station, and can scarcely survive the alterations proceeding in that locality.

LEPTURUS *R. Brown.*

1. L. filiforme *Trin.*
Wet sandy ground by the sea—rare. Fl. latter part of July till end of Aug.

GRAMINEÆ.

Down—Kinnegar, and Tillysburn; *Flor. Belf.* Shore of the Lough near Sydenham station, and north-east corner of Ballyholme Bay; *Flor. Ulst.* Plentiful at south corner of the uncompleted Victoria Park; S.A.S.

Antrim—In the Point fields (now made into docks); *Templeton*, 1794.

Derry—Plentiful on the shores of Lough Foyle, and at the mouth of the Roe; D.M.

LOLIUM. *Linn.* RYE GRASS.

1. L. perenne *Linn.*

Meadows, pastures, and banks—very common. Fl. June and July.

2. L. temulentum *Linn.* DARNEL, STURDY (loc.).

Corn fields in sandy land—rare, and uncertain in its appearance. Fl. June—Aug. (*Cyb. Hib.*).

Mostly planted with the seed, and nowhere established. Recorded as found in the following localities:—Newcastle, and Holywood in Down, also Ballyronan, and frequent in Derry.

Var. β L. ARVENSE *Withering.*

Saintfield; *Flor. Ulst.* Malone; *Templeton*, 1812. By the Roe between Dungiven and Limavady; D.M.

CRYPTOGAMIC ORDERS.

SERIES II. ORDER I. **EQUISETACEÆ.**

EQUISETUM *Linn.* HORSETAIL.

1. E. arvense *Linn.*
Ditch banks, and damp places—very common. Fl. April and May. Ranges from sea level to 1000 feet.

2. E. pratense *Ehr.* (E. UMBROSUM *Willd.*).
Damp heaths, and rocky places—rather rare. Fl. April and May.
Antrim—South side of Wolfhill Glen (*T. Drummond*); *Phytologist, O.S. vol. IV.* Woodburn Glen; *Flor. Belf.* Base of Divis, and at Loonburn near Donegore; *Flor. Ulst.* At the base of Slemish above Buckna, also head of Glenariffe, and of Glenballyeamon; *Cyb. Hib.*

3. E. maximum *Lamk.*
Wet borders of fields, and by streams, and ditches—common, especially in clay land. Fl. April till mid. May.

4. E. sylvaticum *Linn.*
Wet pastures, thickets, and heaths—frequent. Fl. mid. April till end of May.
Down—Woods, and moist ground; *Templeton,* 1797. Between Comber and Castle Espie; *B.N.F.C.,* 1863. Knock Glen, and railway banks at same place; T.H.C. Carrickmannan, Moneyreagh, Purdysburn, and Ballynahinch; S.A.S. Clandeboye, Holywood hills, and Trassyburn above Bryansford; R.Ll.P.
Antrim—Springfield, Colin Glen, and Belfast waterworks; *Flor. Belf.* By the Lagan canal, Crow Glen, Woodburn, and on heaths in Rathlin; S.A.S. Cairncastle, and Carnlough; R.Ll.P.
Derry—Frequent in moist woods and hedge banks throughout the county; D.M. Desertmartin; S.A.S. Benbradagh; *H. C. Hart.*

5. E. limosum *Linn.*
Ditches, marshes, and lake margins—common. Fl. May and June.
Var. β E. FLUVIATILE *Linn.*
Common in similar places as occupied by this type.

6. E. palustre *Linn.*
Marshes, and banks of streams—common from sea level to 1000 feet. Fl. June and July.

7. E. hyemale *Linn.* DUTCH RUSH.
Wet rocky places, and marshy thickets—rare. Fl. June and July. Ascends from sea level to nearly 1000 feet.

Down—Marshy, and shady spots among the sandhills at Dundrum; *Flor. Ulst.* (must be very rare at this place, as no other botanist has succeeded in finding it there).
Antrim—By the side of Lough Neagh near Shane's Castle; *Templeton.* Banks of Forth River at Springfield; *Flor. Belf.* (not found there now). Glenarm, and many glens in Antrim; *Cyb. Hib.* Wet rocky places both above and below the waterfall in Crow Glen, and by a small stream which flows down from above Hannahstown to Colin Glen; S.A.S.
Derry—Woods at Mountsandel above Coleraine, and in woods by the Agivey River; D.M.

8. E. trachyodon *A. Braun.*
Wet rocks—very rare. Fl. mid. June till end of July.

Antrim—Moist rocks near the head of Colin Glen; *Templeton*, 1793. (as *E. hyemale*). Above Colin Glen by the stream that flows down from Divis, and also in Glenarm Glen; *F. Whitla.* Glens near Glenarm; *Cyb. Hib.* Still found on the wet rocks by the waterfall at upper end of Colin Glen; *Eds.*
Derry—In Ballyharrigan Glen near Dungiven; *Cyb. Hib.*

First found in Ireland by Mr. Templeton, but first recognised as distinct by Mr. F. Whitla, before 1830. This plant was pointed out, to the author of Flora Hibernica, at the Colin Glen station, and also at Glenarm, by Mr. Whitla in 1833.

9. E. variegatum *Schleich.*
Sandy seashores—extremely rare. Fl. July and Aug. (Bab. Manual).
Derry—Sandy shore at Benone in Magilligan; D.M.

FILICES.

SERIES II. ORDER II. **FILICES.**

CRYPTOGAMME *R. Brown.* ROCK BRAKE.

1. C. crispa *R. Br.* PARSLEY FERN.
Fissures and ledges of rock—very rare.
Down—Slieve Bignian Mountain above Kilkeel, *Templeton, Stokes, and Mackey, Aug.* 18*th* & 19*th*, 1808; *Herb. Belf. Mus.* Recently refound on Bignian by *Rev. H. W. Lett*, and R.Ll.P. Cliffs between Cove and Commedagh mountains, and cliffs southeast of Blue Lough; R.Ll.P. Shanslieve, Slievenabrock, and north side of Donard; (*Lett*) *Proc. B.N.F.C.*, 1885-6, *appendix.* Very scarce in all these stations.
Antrim—Crevices of rocks on south brow of Slievenance, at 1500 feet; *Flor. Ulst.* Just below the cairn on Slievenance; T.H.C., 1878. Found also on the same mountain by *Rev. S. A. Brenan*, 1884. A few specimens on Carrickfergus Commons; D.M. North side of Knocklayd; *Cyb. Hib.*
Derry—Only observed on Clontygearagh Mountain, parish of Ballynascreen; D.M.

First found in Ireland by Messrs. Templeton, Stokes, and Mackay, in the Mourne Mountains, but the statement in Flora Hibernica that the plant was abundant there would seem to be an error. Mr. Templeton's specimens in the museum herbarium are scanty, and stunted. The localities—Cave Hill, and near Ballyclare—mentioned in Flora of Ulster, are either errors, or they refer to specimens which had been planted out.

POLYPODIUM *Linn.* POLYPODY.

1. P. vulgare *Linn.* COMMON POLYPODY.
Banks, walls, rocks, and trees in shady places—common.
Var. SEMILACERUM.
Antrim—Redhall Glen; *Templeton.* Glenoe, and south end of Gobbins; *B.N.F.C.*, 1875, and 1879. Woodburn Glen—right branch; T.H.C. Knockagh, and Garron Point; R.Ll.P.
Confounded by Templeton with var. CAMBRICUM *Linn.*

2. P. phegopteris *Linn.* BEECH FERN.
Wet rocks, and damp banks—not common.
Down—Top of Slieve Bignian, Black Mountain above Tollymore Park, and rough ground two miles south of Slieve Croob; *Flor. Ulst.* Rostrevor; *Rev. Geo. Robinson.* By a stream on Luke's Mountain; *W. H. Phillips.* Slieve Commedagh; S.A.S. Slieve Bearnagh, Slievenaglough, and sparingly at Craigauntlet

above Holywood; R.Ll.P.
Antrim—Glen of the north branch of the Glenarve River; *Templeton*, 1809. Slievenanee, and Glendun; *Rev. S. A. Brenan*. Trostan Mountain, and also by Carnlough, Inver, and Linford rivers; R.Ll.P.
Derry—Ness Glen; *Templeton, Herb. Belf. Mus.* Dungiven; *B.N.F.C.*, 1872. By the Owenrigh River at Lignapeiste near Banagher; S.A.S.

The first Irish specimens were found by Robt. Brown, on Carlingford Mountain—1801.

3. P. dryopteris *Linn.* OAK FERN.

Rocky places—very rare

Antrim—North side of Knocklayd Mountain near Ballycastle; *Cyb. Hyb.*

It is about half a century since Dr. Moore observed a single plant on Knocklayd, and as no one has found it there since that time it is most probably extinct now in Antrim. The reported stations at Knockagh, and Tollymore Park are doubtless erroneous.

LASTREA *Presl.*

1. L. thelypteris (*Linn.*) *Presl.* MARSH FERN.

Wet places—very rare, and not found recently.

Antrim—Plentiful about Portmore Park and wood; *Templeton*, 1794. Banks of the Sixmilewater near Ballyclare; *Flor. Ulst.*

First found in Ireland by *Mr. Templeton*.

2. L. oreopteris (*Ehr.*) *Presl.* MOUNTAIN FERN.

Ditch banks, and mountain pastures—not common.

Down—About the birch wood, and farthest plantation in Tollymore Park; *Templeton*, 1805. Above Newcastle; *W. H. Phillips*. Moneyscalp west of Bryansford, and above Holywood Waterworks, sparingly; R.Ll.P.
Antrim—Above Carrickfergus; *Flor. Ulst.* Abundant in Glendun; *B.N.F.C.*, 1867. Glenarm, Culraney, Cushendall, Glenaan, and Glenariffe; R.Ll.P. Glenshesk; (*Dr. O'Connor*) *Proc. B.N.F.C.*, 1885-6, *appendix.*
Derry—Abundant in Lignapeiste, and Bennedy Glens near Dungiven; D.M. Moyola Park near Castledawson; *B.N.F.C.*, 1870. Cushcapel, south of Dungiven, and plentiful at base of Slievegallion above Desertmartin; S.A.S.

3. L. filix-mas (*Linn.*) *Presl.* MALE FERN.

Woods, and damp shady places—common.

4. L. spinulosa (*Roth.*) *Presl.*
Damp shady rocks and woods—very rare.
Derry—Frequent in rocky places by the Roe below Dungiven; D.M. Dr. Moore sent specimens from Derry to England, and had his own determination confirmed by Mr. Thomas Moore, of Chelsea. In Smith's English Flora there is a vague record "In the north of Ireland, Dr. Scott."

5. L. dilatata (*Willd.*) *Presl.* BROAD FERN.
Woods, glens, and damp shady rocks—common from sea level to over 2000 feet.

6. L. æmula *Brack.* (L. RECURVA *Bree*). BREE'S FERN.
Damp woods and banks, frequent, but not plentiful.
Down—Dundrum sandhills, and lower slopes of Slieve Donard; *Flor. Ulst.* Tollymore Park; T.H.C. Rademon demesne, and near Tyrella; *C. Dickson.* A single clump in Newtownards Glen, also sparingly at Dunbeg northwest of Ballynahinch; found also in glens near Dundonald, and in the Holywood Hills; R.Ll.P.
Antrim—Cushendall; *Moore's Brit. Ferns.* Murlough, and Runabay Head; R.Ll.P.
Derry—Coleraine, and Garvagh; *Moore's Brit. Ferns.*

POLYSTICHUM *Roth.*

1. P. aculeatum (*Linn.*) *Roth.*
Damp rocks and stony places—frequent, and often abundant.
Down—Rostrevor; *W. Thompson.* Tollymore Park, Crawfordsburn, and Holywood; *Flor. Ulst.* Above Holywood and Bangor; R.Ll.P.
Antrim—Colin Glen, Kilwaughter, Glenarm, and Glenariffe; *Templeton*, 1797 & 1804. Near Carrickfergus; *Flor. Hib.* Stonyford, and Larne; *W. H. Phillips.* Glens in Belfast Hills, Woods at Muckamore, and frequent around the Antrim coast; S.A.S. Glynn, Glendun, Carnlough, and Murlough; R.Ll.P.
Derry—By the Faughan below Gort bridge; *Templeton.* By the Roe above Limavady, and at the Ness waterfall; D.M. Lignapeiste Glen; S.A.S.
The variety *lobatum* occurs, but it is not easy to draw the line between the two forms, and we have not attempted to separate them here.

2. P. angulare (*Willd.*) *Newman.*
Woods and shady banks—very common.

FILICES.

CYSTOPTERIS *Bernhardt.* BLADDER FERN.

1. C. fragilis *(Linn.) Bernh.*
Damp shady rocks and walls—rare, but frequently abundant where it does occur.
Down—Mountain above Tollymore Park; *Flor. Ulst.* Slieve Donard; *Dr. Dickie.* Rademon demesne, Shanslieve, and Mountain west of Slieve Donard; R.Ll.P.
Antrim—Rocks by the waterfall in Carrickfergus River; *Templeton*, 1811. Sallagh Braes; *Flor. Ulst.* Glenariffe; *B.N.F.C.*, 1866. Walls of Retreat Castle; T.H.C. Glenarm deerpark, Trostan Mountain, and Murlough, also plentiful by the roadside near Retreat rail. station; R.Ll.P.
Derry—Sparingly on rocks by the Roe above Limavady; D.M.

The variety, C. DENTATA *Smith* is recorded as occuring at Woodburn, but cannot be found now.

ATHYRIUM *Roth.*

1. A. filix-fœmina *(Linn.) Roth.* LADY FERN.
By streams, and in wet places—common. The variety *rhæticum* is frequent.

ASPLENIUM *Linn.* SPLEENWORT.

1. A. adiantum-nigrum *Linn.* BLACK SPLEENWORT.
Damp shady rocks—frequent, less common on banks and old walls.
Down—Mourne Mountains; *Wade Rar.* Holywood, and Castlereagh; *Flor. Belf.* Crawfordsburn, and Copeland Isles; *Flor. Ulst.* Walls at Portavo; T.H.C. Mealough Hill, and walls between Greencastle and Rostrevor—occurs on slate rocks of Mourne, but seems not to ascend to the granite; S.A.S. Crossgar, Bryansford, and Warrenpoint; R.Ll.P.
Antrim—Colin Glen; *Flor. Belf.* Wolfhill, Cave Hill, Ballycarry, and cliffs at Glenarm; *Flor. Ulst.* On the Knockagh, and extends all along the basaltic escarpment of the coast line, frequent in Rathlin; S.A.S. Dunadry, Cushendall, and Portrush; R.Ll.P.
Derry—Common in many places, ascending to 1600 feet on the mountains; D.M. Lignapeiste Glen, walls of Banagher old church, and rocks of Benbradagh, and Benevenagh; S.A.S. Downhill, and Castlerock; R.Ll.P.

Var. β A. ACUTUM *Bory.*
Damp shady rocks and banks—very rare.

FILICES.

Down—"In a dark cave among the mountains of Mourne"; *Sherard, Herb. Oxon.* (*Filix minor longifolia*), also *Raii Syn.* ed. II. p. 51, 1696. On a ditch bank in an old road west of Newtownards—sparingly; R.Ll.P., 1883.
Derry—Benevenagh Mountain, growing in small quantity on precipitous cliffs of basalt; S.A.S., 1884.
First found in Britain by Sherard.

2. A. trichomanes *Linn.* MAIDENHAIR SPLEENWORT.
Damp shady rocks, and old walls—frequent.
Down—Rostrevor; *Moore's N. P. Ferns.* Walls and glens at Holywood; *Flor. Belf.* Old walls at Newtownards; *Flor. Ulst.* Tollymore Park, and walls at Donaghadee; T.H.C. Hillsborough Park; S.A.S. Dundonald, Crossgar, and Mourne Mountains; R.Ll.P.
Antrim—Under stones at the Knockagh; *M'Skimmin's Hist. C.fergus.* Colin Glen; *Flor. Belf.* Cave Hill, Wolfhill, Ballycarry, and Glenarm; *Flor. Ulst.* Glenariffe; *B.N.F.C.*, 1866. Woodburn, Dunadry, and Glynn; T.H.C. In glens all round the Antrim coast; S.A.S. Larne, Cushendall, and Fairhead; R.Ll.P.
Derry—Abundant on rocks, especially basalt, ascending to 2000 feet on the mountains; D.M. Slieve Gallion, and Benbradagh; S.A.S. Castlerock, and Downhill; R.Ll.P.

3. A. marinum *Linn.* SEA SPLEENWORT.
Rocks by the sea—frequent, but local.
Down—Rocks at Newcastle; *Miss. Hincks.* Cliffs by the shore south of Newcastle; *B.N.F.C.*, 1869.
Antrim—Rocks at Ballygally Head, also cave at Blackhead, and other caves along the sea coast; *Templeton*, 1794. On basalt rocks at Blackhead, 1863, and caves at Ballintoy, 1870; *B.N.F.C.* Very fine in a cave at south end of Gobbins; *W. Swanston.* Fairhead, Giants' Causeway, and abundant on Rathlin; S.A.S. Tor Head; T.H.C. Garron Point, Redbay, Cushendun, and Carrick-a-rede; R.Ll.P.
Derry—Clefts and caves in the rocks from Downhill to Portrush; D.M. Castlerock; R.Ll.P.

4. A. ruta-muraria *Linn.* WALL RUE.
Old walls—common, rarely, if ever, found on rocks in this district.

SCOLOPENDRIUM *Smith.*

1. S. vulgare *Symons.* HART'S-TONGUE.
On banks, rocks, walls, and old trees in damp shady places—very

common, ascending to 1200 feet in Derry mountains.

CETERACH *Willdenow*.

1. C. officinarum (*Linn.*) *Willd.* SCALY HART'S-TONGUE.
Old walls—rare.

Down—Rostrevor, Bryansford, and Donaghadee; *Flor. Ulst.* North of Newry on a wall at Treemount one mile from Sheepbridge; *Rev. Geo. Robinson.* Old wall near Newcastle; W. H. Phillips. Wall at Turner's Hill near Newry; *C. Dickson.* Rademon, sparingly; R.Ll.P. Hillsborough park wall; *Rev. C. H. Waddell.*

Antrim—Wall of Belfast Deerpark (still there), and on the wall of the old deerpark at Galgorm near Ballymena; *Templeton*, 1801. Walls at Castle Dobbs; *W. H. Phillips.* Walls of a bridge over the Cloghwater northwest of Broughshane; *Rev. H. W. Lett., Proc. B.N.F.C.*, 1885-6. *App.* Abundant on walls of old bridge a half mile east of Roughfort; *B.N.F.C.*, 1878. Parkmill bridge, and on walls in Glenarm Deerpark; R.Ll.P. Very sparingly on an old wall at Muckamore, and at Greenmount south of Antrim ; *D. Redmond.*

It seems remarkable that we have not any note of the occurrence of this fern in Derry.

BLECHNUM *Linn.*

1. B. spicant *Roth.* (B. BOREALE *Sw.*). HARD FERN.
Woods, banks, stony moors, and mountain slopes—abundant throughout the district. Ascends to 1500 feet in the mountains.

PTERIS *Linn.* BRAKES.

1. P. aquilina *Linn.* BRACKEN.
Sandy seashores, thickets, and on the lower slopes of mountains—very common.

HYMENOPHYLLUM *Smith.* FILMY FERN.

1. H. tunbridgense (*Linn.*) *Sm.*
On rocks in damp places—very rare.
Down—On Slieve Donard above Donard Lodge ; *Flor. Ulst.*
Antrim—Glendun ; *Rev. S. A. Brenan*, 1884.

2. H. unilaterale *Willd.* (H. WILSONI *Hook.*).
Amongst moss in damp woods, and on wet rocks—rare, ascends to 2000 feet in Derry.

Down—On the margin of Cove Lough southwest of Slieve Donard, also on rocks on the north side of Diamond Mountain, and by the waterfall in Tollymore Park; *Templeton*, 1796.* Slieve Lamagan, and between Lamagan and Commedagh; R.Ll.P.

Antrim—On a rock in Colin Glen, and by the river in Glenariffe; *Templeton*, 1809. Amongst stones below the cliffs of the Cave Hill; *Hyndman*, 1838, and subsequently *Phillips*. Slievenanee; *Flor. Ulst.* Glenarm Deerpark; *W. Darragh.* Sallagh Braes; S.A.S. Above the falls in Carulough River; R.Ll.P. Glendun; *Rev. S. A. Brenan.*

Derry—Not general, but occurs occasionally on rocks from Benbradagh to Clontygearagh, and also on the top of Dart; D.M.

* Noted as *H. tunbrigense*, *H. unilaterale* not having been distinguished at that time.

OSMUNDA *Linn.* FLOWERING FERN.

1. O. regalis *Linn.* ROYAL FERN.
By streams, and in boggy places—very rare, and now extinct in some of the old stations.

Down—Donard's cave in Mourne Mountains, and in Kirkiston bog, also in a bog beside Greyabbey, and margin of a lake southwest of Castlewellan; *Templeton*, 1793. Side of the river above Newcastle; *Flor. Ulst.* In a bog east of Kirkcubbin; *C. H. Brett.* Sparingly by the side of the river at Bloodybridge south of Newcastle, and in an old cave near the same place; *C. Dickson.* In the long valley above Newcastle; T.H.C. Sea cliffs south of Newcastle; R.Ll.P. By the Kilbroney River; *Turretin.* By a stream near Killinchy; *Jas. Wilson.*

Antrim—Shores of Lough Neagh near Shane's Castle; *Flor. Ulst.* (not found there now). Formerly grew in a bog near Glarryford; *Proc. B.N.F.C.*, 1885-6, *App.*

Derry—West side of the Bann a little below Portglenone, and margin of one of the small lakes above Kilrea; D.M.

BOTRYCHIUM *Swartz.*

1. B. lunaria (*Linn.*) *Sw.* MOONWORT.
Short pastures in damp moory ground—not common. Ranges from sea level to 1200 feet.

Down—Scrabo Hill; *Templeton*, 1806. Scanty and stunted on sandhills near Newcastle station, but abundant, and luxuriant on heath at Carngaver Hill above Holywood; S.A.S. Carrowreagh Hill, and at Craigauntlet; R.Ll.P.

Antrim—Meadow about 200 yards north of the second lock of Lagan canal, also by the stream below Hannahstown, and on Knocklayd; *Templeton*, 1802. Roughfort; *Moore's N. P. Ferns*. Belfast hill range; *Flor. Belf.* East side of Cave Hill below the caves, also above Wolfhill, and on the Knockagh, and in Altmore Glen near Cushendall; *Flor. Ulst.* By the Main River below Randalstown, 1863, and on Squires Hill near Belfast, 1881; *B.N.F.C.* In the hollow between Divis and Black Mountain, also south side of Cave Hill; S.A.S. Shore at Cushendun, and on Carnaneigh, and Crosslieve near Cushendall; R.Ll.P. Near Kells; *D. Redmond*.

Derry—Frequent along the basaltic range of hills; D.M. Magilligan; *Flor. Hib.* Benbradagh at 1200 feet; S.A.S.

OPHIOGLOSSUM *Linn.* ADDER'S-TONGUE.

1. O. vulgatum *Linn.*

Pastures and meadows—frequent.

Down—Plentiful in meadows near the old abbey at Greyabbey, 1864, and in Mountstewart grounds, 1877; *B.N.F.C.* Killinchy; *Jas. Wilson.* Field near Cultra; *Robt. Patterson.*

Antrim—Field north of second lock Lagan canal; *Templeton*, 1806. Banks of Lagan three miles above Belfast; *Newman.* Shane's Castle, 1863, and Cave Hill, 1871; *B.N.F.C.* Whitehead; *G. C. Hyndman.* By the Blackstaff River north and west of Malone Cemetery; *Flor. Ulst.* Wolfhill, and Knockagh; *W. Millen.* Springfield; *W. H. Phillips.* Colin Glen, and Rathlin; S.A.S. Abundant in some fields at Cairncastle; R.Ll.P. Birch Hill near Antrim; *Rev. W. S. Smith.*

Derry—Springhill, Garvagh, Somerset, Mountsandys, and damp sandy warrens at Magilligan; D.M.

SERIES II. ORDER III. **MARSILACEÆ.**

PILULARIA *Linn.* PILLWORT.

1. P. globulifera *Linn.*

In ditches, and by rivers and lakes—extremely rare. Fl. June—Aug. (*Cyb. Hib.*),

Antrim—At Selshan harbour; (*D. Moore*) *Flor. Ulst.* and *Cyb. Hib.*
Derry—Only observed by the side of the Bann below Coleraine, and near the same place in a ditch a little below Jackson Hall; D.M.
Noted in Flora of Ulster as occurring by the Lagan, near Belfast. Not found there now, and the note considered erroneous.

SERIES II. Order IV. **LYCOPODIACEÆ.**

ISOETES *Linn.* Quillwort.

1. I. lacustris *Linn.*
Under water in lakes—rare. Fl. July and Aug.
Down—Fragments washed up on the shore of Annsborough Lake east of Castlewellan; S.A.S., 1883. Bignian Lake at 1500 feet (*H. C. Hart.*); *Proc. R.I.A.*
Antrim—Plentiful in Lough-na-Cranagh near the summit of Fairhead, and found abundantly in the southeast corner of Loughmourne when that lake was drained in 1882; S.A.S.
Derry—In Lough Beg, and in Enagh Lough at the west side; D M.

LYCOPODIUM *Linn.* Clubmoss.

1. L. clavatum *Linn.*
Grassy heaths—not common. Fl. mid. June till mid. Aug.
Down—Mountains of Mourne; *Harris's Down.* Tollymore Park; *C. Dickson.*
Antrim—Top of Black Mountain, 1798, and on Agnew's Hill, 1804; *Templeton.* Divis, and Carrickfergus commons; *Flor. Belf.* Boughal Hill, banks of stream between Divis and Black Mountain, and heath at Loughmourne; S.A.S. Slievenanee; *Rev. S. A. Brenan.*
Derry—Heathy pastures near Ballyharrigan Glen, and other places; D.M. Slieve Gallion, and Benbradagh; S.A.S.

2. L. alpinum *Linn.*
Mountain heaths—rare. Fl. July and Aug. Ranges from 1200 to 2700 feet.
Down—On the mountains of Mourne; *Harris's Down.* Slieve Donard; *Templeton.* Slieve Commedagh, and Slieve Bernagh; *H. C. Hart, op. cit.*
Antrim—On the top of Black Mountain; *Templeton*, 1793 (still there).

On Divis Mountain ; *Dr. Rea.* Knocklayd, and Slievenanee ; *Cyb. Hib.* Garron Plateau ; R.Ll.P.

Derry—Only observed on the top of Dart, and on the north side of Slieve Gallion near the cairn ; D.M.

3. L. selago *Linn.* FIR CLUBMOSS.
Mountain heaths, and moory peat bogs—common. Fl. mid. May till mid. Sept. Ranges from near sea level to 2500 feet on the Mourne Mountains.

SELAGINELLA *Pal. de Beauvais.*

1. S. spinosa *Beauv.* (LYC. SELAGINOIDES *Linn.*).
Boggy heaths in rocky ground—frequent. Fl. July and Aug. At various elevations from sea level to 1500 feet.

Down—Crevices of rocks in the Mourne Mountains ; *Templeton.* At Balloch ; *Flor. Ulst.* Slieve Croob ; *C. Hickson.*

Antrim—In moist places among the sandhills near the Giants' Causeway, and on Agnew's Hill ; *Templeton,* 1804. Slemish Mountain ; *Flor. Ulst.* Knocklayd, summit of Fairhead, and sparingly on Rathlin ; S.A.S. Glenarm, Slievenanee, and Lurigethan ; R.Ll.P.

Derry—Near Derry ; *Templeton.* Moist places on the sandy warren between Downhill and the Bann ; D.M. Benevenagh ; *B.N.F.C.,* 1863. Sandhills at Magilligan ; *Flor. Ulst.* Benbradagh ; S.A.S.

SERIES II. ORDER V. **CHARACEÆ.**

CHARA *Vaillant.*

1. C. fragilis *Desv.*
Lakes, and ponds—frequent.

Down—Lough Aghery, and pond in the Ormeau Park ; S.A.S. Holywood (var. CAPILLACEA *Thuill.*) ; R.Ll.P.

Antrim—County Antrim ; *Grov. Brit. Char.* In several of the lakelets in the Island of Rathlin ; S.A.S. Lough Neagh (var. DELICATULA *A. Braun*) ; S.A.S. In County Antrim (var. BARBATA *Gant.*) ; *Wm. Thompson.*

2. C. aspera *Willd.*
Lakes—apparently not common.

Down—Loughinisland, and plentiful in Clandeboye Lake; S.A.S.
Antrim—County Antrim; *Grov. Brit. Char.* In Lough Neagh (var.
SUBINERMIS *Kutz.*); (*Rev. H. W. Lett*) Jour. Bot., 1886.

[**C. polyacantha** *A. Braun.*
Down—" *Equisetum muscosum sub aquis repens semine lithospermi.* This grows in the turf bogs of this county" *Harris's Down*, 1744.]
The above note doubtless relates to Sherard's plant, which has been referred to the present species. No County Down specimens of *C. polyacantha* are, however, known to exist, and it is probable that the plant recorded by Harris was in reality *C. hispida.*

3. C. hispida *Linn.*
Ditches, and lakes—not common. Fl. mid. June till end of July.
Down—Abundant in Loughinisland Lake (var. RUDIS *Braun*); S.A.S.
Derry—Abundant in a brackish ditch at Limavady Junction; S.A.S.
County Derry (var. GYMNOTELES *Braun*); *Grov. Brit. Char.*

4. C. vulgaris *Linn.*
Ditches, ponds, and lakes—frequent. Fl. mid. June till end of July.
Down—Loughinisland Lake, and in a brackish ditch in the Victoria Park; S.A.S.
Antrim—In Co. Antrim (var. LONGIBRACTEATA *Kutz.*) (*Rev. W. M. Hind*); *Grov. Brit. Char.*

TOLYPELLA *Leonhardi.*

[**T. nidifica** (*Sm.*) *Leon.*
Antrim—Lough Neagh near Langford Lodge (*D. Moore*, 1840); *A. Braun.*]
There would seem to be diversity of opinion regarding this, but we have preferred to rely on the high authority of Prof. Braun. Meanwhile it is to be hoped that some of our local botanists may succeed in re-finding the plant. Good specimens, with mature fructification would be decisive.

NITELLA *Agardh.*

1. N. translucens *Ag.*
Ponds, and lakes—rare. Fl. July (*Groves*).
Down—Derry Lough north of Ballynahinch; S.A.S.
Antrim—Sparingly in ponds and lakes in Rathlin Island; S.A.S.

2. N. opaca *Ag.*

Lakes, and ponds—frequent. Fl. July.

Down—Lakes at Loughinisland, Ballynahinch, and Clandeboye; S.A.S.

Antrim—Marsh at Whitehead, and plentiful in the ponds and lakelets of Rathlin; S.A S.

Derry—In Co. Derry; *W. Thompson.*

CLASS MUSCI.

SERIES II. ORDER VI. **ANDREÆACEÆ**.

ANDREÆA *Ehrhart.*

1. A. petrophila *Ehr.*
Shady rocks on the mountains—rather common. Fr. April and May. Ranges from 700 to 2000 feet.
Down—Granite rocks of Slieve Donard, and on slate near the summit of Slieve Croob; S.A.S. Rostrevor, and Shanslieve; C.H.W. Common on the Mourne Mountains; H.W.L.
Antrim—Abundant on Slemish, and other mountains in the county; D.M. Sparingly on trap rocks at base of the cliffs north of the caves of the Cave Hill, and on trap at 900 feet on north-east side of Divis; S.A.S. Basalt rocks of Sallagh Braes; H.W.L., and C.H.W.
Derry—On Clontygeragh Mountain; D.M. On trap at summit of Slievegallion, and on Mullaghmore Mountain; S.A.S.

Var. β ACUMINATA *Schimper.*
Down—Slievenabrock, and on Rocky Mountain near Hilltown; H.W.L.

Var. γ GRACILIS *Schimper.*
Granite rocks between Slieve Dermot and Hilltown; H.W.L.

2. A. alpina (*Linn.*) *Smith.*
Mountainous rocks—frequent. Fr. April and May. Ranges from 1000 to 2500 feet.
Down—Granite rocks of Slieve Donard; S.A.S. Slievenamady; H.W.L. Thomas Mountain, and Pigeonrock Mountain; C.H.W.
Antrim—Slemish Mountain; D.M. Trap rocks on north side of Divis, and abundant on Slemish; S.A.S.
Derry—On Clontygeragh, and Dart Mountains at 1100 to 2000 feet; D.M. Benbradagh; S.A.S.

3. A. crassinervis *Bruch.*
Exposed rocks—rare, the typical plant not seen.
  Var. HUNTII (*Limpricht*) *Braith.*
Down—Slievemartin, Slievedermot, Slieve Commedagh, and Slievenamaddy; H.W.L.
This var. has not previously been noted in Ireland.

4. A. rupestris (*Linn.*) *Roth* (A. ROTHII W. & M.).
Rocks on the higher mountains—frequent. Fr. April and May.
Down—On granite rocks, Slieve Donard, and on slate at summit of Slieve Croob; S.A.S. Slievemeel Beg, and Slieve Commedagh; H.W.L. Rocky Mountain near Hilltown, by the Yellow Water River, and above Tollymore Park; C.H.W.
Antrim—Abundant through Co. Antrim; D.M.
Derry—Rare, but abundant on Slievegallion from 1500 feet upwards; D.M.
  Var. FALCATA *Schimper.*
Down—Rocks at Pierce's Castle northeast of Rostrevor; H.W.L.
Antrim—Rocks by Lough-na-Cranagh near the summit of Fairhead; H.W.L.

SERIES II. ORDER VII. **GEORGIACEÆ.**

GEORGIA *Ehrhart.*

1. G. pellucida (*Linn.*) *Raben.* (TETRAPHIS *Hed.*).
Damp rocks—very rare. Fr. July—Sept. (B.M.F.).
Antrim—Glenariffe; *Moore, Proc. R.I.A.*, 1872.
Derry—Moist places in Co. Derry; *Templeton.*

2. G. Brownii (*Dicks.*) *Muell.* (TETRODONTIUM *Schw.*).
Damp shady glens—very rare. Fr. July (B.M.F.).
Antrim—In fine fruit in a glen near Ballycastle, July 1835; *Moore, Proc. R.I.A.*, 1872, and also Brit. Moss Flora. Glenmakerron, June 1836; D.M.

SERIES II. ORDER VIII. POLYTRICHACEÆ.

CATHARINEA Ehrhart (ATRICHUM Beauv.).

1. C. undulata (*Linn.*) *Web. et Mohr.*
Damp banks, and moist ground in shady places—very common. Fr. Nov. –March. Ranges from sea level to over 1000 feet.

OLIGOTRICHUM *Lamk. & D.C.*

1. O. incurvum (*Huds.*) *Lindb.* (O. HERCYNICUM *L. & D.C.*).
Wet gravelly places on mountains—rare, though locally abundant. Usually barren, but found in fruit in June by C.H.W. Ascends to nearly 2000 feet.
Down—Bencrom, Deers' Meadow, Slieve Donard, Slieve Commedagh, Tollymore Park, and by Bloodybridge River; C.H.W. Speltha Mountain, Slievemeel Beg, and Slievenamady; H.W.L. Plentiful on granite sand by the Spinkwee River above Tollymore Park; S.A.S.

POLYTRICHUM *Dillenius.*

1. P. nanum *Dill.*
On sandstone rocks, and shaded sandy banks—frequent. Fr. Oct.—Feb.
Down—Ditch banks on the hill above Holywood, and at the Manyburn Glen; *Templeton.* Slieve Donard, and Slieve Commedagh mountains, also sandy banks at Castlereagh, Shawsbridge, and Sydenham; S.A.S. Near Hilltown; C.H.W.
Antrim—Cranmore; *Templeton.* Malone; *W. Thompson.* Tor Head; *Moore, op. cit.* On red sandstone at Derriaghy, sandy banks at Stranmillis, and grassy heaths at Black Mountain, Slievetrue, and Sallagh Braes; S.A.S.
Derry—Maghera, and frequent in the county; D.M. Old quarry at Castledawson; S.A.S.

Var. β LONGISETUM *Hampe.*
Down—Ditch bank in Victoria Park, and old quarry at Ballymaghan; S.A.S. Mourne Mountains; C.H.W.
Antrim—Heathy and sandy bank on Slievetrue; S.A.S.

2. P. aloides *Hedwig.*
Dry, shady banks—common, especially in hilly districts. Fr. Oct.—March.

Var. β P. DICKSONI *Turner.*

Derry—Near Derry (*R. Brown*) *Templeton*, also *Musc. Hib.*, and *Brit. Moss Flora*.

3. P. urnigerum *Linn.*

Sandy and gravelly heaths, and wastes—frequent. Fr. Nov.—April. Ranges from sea level to 2000 feet.

Down—Rostrevor, Slieve Donard, Slieve Croob, also on quarry *debris* at Castlereagh Hill, and Cregagh Glen; S.A.S. Deers' Meadow, Slieve Commedagh, and Shanslieve; H.W.L.

Antrim—Side of a pond at Cranmore, and dry sandy field near Dunmurry; *Templeton*. Banks in Glendun and other places; D.M. Near Lisburn; *J. Creeth*. Sandpit at Newforge, also on Belfast Hills, Tardree, Whitehead, Knocklayd, and in Glenshesk; S.A.S. By the Lagan near Belfast; *J. J. Andrew*.

4. P. alpinum *Linn.*

On heaths—not common. Fr. May and June. Ranges from near sea level to 2600 feet.

Down—Mourne Mountains; *Templeton*. Deers' Meadow, Slieve Commedagh, and at the lesser cairn on Slieve Donard; H.W.L.

Antrim—On Slemish; *Templeton*, 1809. Black Mountain, Knockagh, and Glenariffe; S.A.S.

Derry—Lagnagapagh; *Templeton*. Top of Dart Mountain; D.M. Bogs at Toome, and abundant on the Sperrin Mountains; S.A.S.

5. P. gracile *Dickson.*

Peat bogs—not common, though in some places abundant. Fr. May and June.

Down—Woods at foot of the Mourne Mountains (*P. pallidisetum*); *T. Drummond*, *Musci Scotici*, vol. 3. Plentiful and luxuriant in Cotton moss; S.A.S. By the shore of Lough-na-shanagh; H.W.L. and C.H.W.

Antrim—Tardree, and Ballynoe near Antrim; S.A.S.

6. P. piliferum *Schreber.*

Heaths, and boggy ground on the hills—common. Fr. April—June. Ascends to 2000 feet.

7. P. juniperinum *Willd.*

Dry stony heaths—frequent. Fr. May and June.

Down—Cotton moss, Slieve Donard, and near Banbridge; S.A.S.

Antrim—Lambeg moss; *Templeton*. Frequent on banks in the glens; D.M. Near Lisburn; *J. Creeth*. Cave Hill, Ballygally Head, Skerry Hill, and Rathlin Island; S.A.S. Clough, H.W.L.

Derry—Very common in the county; D.M.

8. P. strictum *Banks.*
Mountain rocks—very rare, and no fruit seen.
Down—Slievemartin, and rocks on Slieve Donard called Blackstairs; H.W.L.
Mentioned by Templeton as found by him, but no definite locality stated.

9. P. commune *Linn.*
Woods, heaths, bogs, and stony banks—very common. Fr. April—June. From sea level to 2700 feet.

SERIES II. ORDER IX. **FISSIDENTACEÆ.**

FISSIDENS *Hedwig.*

1. F. exilis *Hedwig.*
Damp ditch banks—very rare. Fr. Feb. and March.
Antrim—Ditch banks near Belfast; *T. Drummond, Musci Scotici,* vol. 3. Bank of a lane leading from Ballygomartin road, south of the Forth River, to the Black Mountain. The moss was only observed, sparingly, on the south side of the lane near the foot of the mountain; S.A.S., March, 1880.
Not known to occur elsewhere in Ireland.

2. F. pusillus *Wilson.*
Sandstone, and limestone rocks—very rare. Fruiting in spring.
Antrim—On fallen sandstone rocks on the east face of the Black Mountain to the north of Whiterock, and chalks rocks in Redhall Glen; S.A.S.
The only Irish localities as yet known.

3. F. bryoides (*Linn.*) *Hedwig.*
Damp, shady clay banks—very common. Fr. Nov.—March.

4. F. osmundoides (*Swartz*) *Hedwig.*
Damp shady rocks, and wet stony heaths—rare. Fr. Feb.—April.
Down—Moist rocky places in Newtownards Park, quarter of a mile from

the glen; *Templeton*, 1803. Mourne Mountains; *T. Drummond, op. cit.* Wet rocks on Slieve Donard, and abundant on boggy heath at summit of Slieve Croob; S.A.S. Thomas Mountain; C.H.W. Frequent in Mourne Mountains; H.W.L.

5. F. taxifolius (*Linn.*) *Hedwig.*
Damp banks in woods and shady places—common. Fr. Nov.—March.

6. F. adiantoides (*Linn.*) *Hedwig.*
Damp shaded banks and rocks, and in woods—frequent. Fr. Dec.—Feb.
Down—Slieve Donard, Drumbo Glen, Holywood Hill, and Cregagh Glen; S.A.S. Moygannon Glen, and Tollymore Park; C.H.W. Frequent; H.W.L.
Antrim—Bog meadows; *Templeton.* Glenariffe, and Cushendall; D.M. Shady banks at top of Colin Glen, wet heath on Black Mountain, and in abundant fruit under trees at Kilwaughter, also Agnew's Hill, Sallagh Braes, Knocklayd, Fairhead, and Rathlin; S.A.S. By the Lagan near Belfast; *Dr. S. M. Malcomson.*
Derry—Frequent; D.M. Ditch banks near Moneymore; S.A.S.

SERIES II. Order X. **LEUCOBRYACEÆ.**

LEUCOBRYUM *Hampe.*

1. L. glaucum (*Linn.*) *Schimp.*
Damp shaded rocks, and wet woods—frequent, but without fruit.
Down—Birky moss, and other places; *Templeton.* Under granite boulders on Slieve Donard, and at Kilbroney, and on boggy heath at Conlig; S.A.S. Rostrevor wood, and Tollymore Park; C.H.W. Common, but not abundant in the Mourne Mountains; H.W.L.
Antrim—In bogs on the mountains, not rare in the county; D.M. Black Mountain, and Mount Aughrim; *J. Creeth.* Frequent on heaths in Rathlin Island; S.A.S. Larne; C.H.W.

DICRANACEÆ.

SERIES II. Order XI. DICRANACEÆ.

Sect. I.—Ditricheæ.

ARCHIDIUM *Bridel.*

1. A. alternifolium (*Dicks.*) *Schimp.*
Banks—very rare, and seems not to have been seen in Ireland during the past 50 years.
Antrim—" Belfast (*Drummond !*)" ; *Brit. Moss Flora.*
[Lane near the Black Mountain ; *Drummond, Musci Scotici,* vol. 3. The specimen mounted by Drummond in Belfast Museum copy is only *Ceradoton purpureus,* young.]

PLEURIDIUM *Bridel.*

1. P. axillare (*Dicks.*) *Lindb.* (P. NITIDUM *Hed.*).
Cultivated land and damp sandy ground—not common. Fr. July—Nov.
Down—In the garden of Aghaderg Glebe near Loughbrickland ; H.W.L.
Sparingly on damp bank in the sandhills at Newcastle ; S.A.S.
Antrim—In the drills of a bean-field at Malone, in fruit 1st Dec., 1802. *Templeton.* Ditch banks near Belfast ; *Musc. Brit.* Banks near Belfast ; *Flor. Hib.* Ditch banks near Belfast ; *T. Drummond, op. cit.* Rathlin Island ; D.M., 1836.

2. P. subulatum (*Huds.*) *Raben.*
"Bare, somewhat moist places on sandy lands, as ditch banks, and potato furrows ;" *Templeton.* Damp sandy banks—not frequent. Fr. Nov.—April.
Down—Gravelly bank at Ballymaghan, wet bank at Mealough Hill, wet sandy bank at Sydenham, and gravel pit at Dundonald ; S.A.S. Brickfield at Ballynafeigh ; *J. J. Andrew.*
Antrim—Grassy heath on White Mountain, and clay fields at Carnmoney Hill, and at Carnearny ; S.A.S. Woodburn ; H.W.L.
Derry—Dry banks in the Bennedy, and in Ballyharrigan glens ; D.M. Sandy banks near Moneymore ; S.A.S.

3. P. alternifolium *Rabenhorst* (P. SUBULATUM *Schreb*).
Sandy banks—very rare. Fr. May and June (B.M.F.).
Down—Dry bank near Warrenpoint ; C.H.W.

DITRICHUM *Linn.*

1. D. tenuifolium (*Schrad.*) *Lindb.* (TRICH. CYLINDRICUM *Hed.*).

Sandy banks—extremely rare. Fr. May and June (B.M.F.).
Antrim—Ditch banks near the Belfast Bot. Gardens, fruit extremely rare;
T. Drummond, Musci Scotici, vol. 3. Ditch bank at Cranmore (*Drummond*); *Flor. Hib.* On sandy banks near the Botanic Gardens; *Moore, Proc. R.I.A.,* 1872. Belfast (*Drummond*); *Brit. Moss Flora.*

First found in Britain, by Drummond, at Belfast, the only Irish locality, and has not been seen in Ireland for more than 50 years.

2. D. pusillum (*Hedw.*) *Timm.*
Sandy banks—very rare.
Antrim—Fields beyond the Botanic Gardens; *T. Drummond, Musci Scotici,* vol. 3. Near Belfast, Templeton, afterwards Drummond; *Flor. Hib.* In the late Mr. Templeton's garden at Cranmore; *Moore, Proc. R.I.A.,* 1872, (spec. in Herb. Coll. Sc. Dublin). Near Belfast (*Drummond*); *Brit. Moss Flora.*

First found in Ireland by Templeton, from whose collection Drummond, no doubt, derived his information.

3. D. homomallum (*Hedw.*). *Hampe.*
Wet rocky or gravelly banks, and by streams—frequent. Fr. July—Oct.
Down—Deers' Meadow, Slievemeel Beg, and Tullybranagan; H.W.L. Woods at Narrow-water and Rostrevor, also on Slieve Donard, and in Tollymore Park; C.H.W.
Antrim—Sparingly on sandy bank between Stranmillis road and the Lagan, but most abundant on decomposing mica schist by the stream on Brean Mountain above Ballycastle; S.A.S.
Derry—Ness Glen, and mountains and glens in Derry, and by the Cookstown road near the foot of Sawel; *Templeton,* 1809. Slievegallion, Mullaghmore, and Sluggada Mountain; S.A.S.

4. D. flexicaule (*Schleich.*) *Hampe.*
Dry basaltic rocks, and coast sandhills; rare and barren.
Down—Plentiful on sandhills at Ballykinler; S.A.S.
Antrim—Basalt rocks of Sallagh Braes; H.W.L., and C.H.W.
Derry—Abundant on sands at Portstewart; S.A.S. Sandhills at Magilligan (*G. V. Craig*); C.H.W.

Var. β DENSUM *Br. & Schimp.*

Antrim—Peaty heath on the northeast side of Divis Mountain; C.H.W.

Sect. II.—Dicranelleæ.

DICRANACEÆ.

DICRANELLA *Schimper*.

1. D. crispa (*Ehr.*) *Schimp.*
Sandy or gravelly banks—rare. Fr. Nov.—Feb.
Down—Banks of Manyburn River near Purdysburn; *Templeton*, 1801. Sparingly on damp rocks in Cregagh Glen; S.A.S.
Antrim—Dry banks near Belfast (*Templeton*); *Flor. Hib.* Belfast (*Temp.*); *Brit. Moss Flora.* On red sandstone at Derriaghy; *J. H. Davies.* Plentiful on sandy bank at a path from north side of Stranmillis road to the Lagan, and abundant on the mica schist of Brean Mountain; S.A.S.
Derry—Ditch bank near Derry (*R. Brown*); *Templeton.* White Mountain; S.A.S.

2. D. secunda (*Sw.*) *Lindb.* (D. SUBULATUM *Hed.*).
Arenaceous rocks, and sandy banks—rare. Fr. July—Sept.
Down—Side of the Manyburn Rivulet (*Temp.*); *Flor. Hib.*
Antrim—Belfast (*Miss Hutchins*); *Brit. Moss Flora.* In Glenariffe sparingly; D.M. Sparingly on Rathlin Island; S.A.S.

3. D. heteromalla (*Dill.*) *Schimp.*
Woods, and dry shady banks—frequent. Fr. Sept.—Feb. Ascends to near 2000 feet.
Down—Rostrevor, Bloodybridge, Slieve Croob, Portavo, Ballyalloley, Moneyreagh, Cregagh, and Ormeau Park; S.A.S. Abundant; H.W.L.
Antrim—Roots of trees at Cranmore, and on ditch banks in turfy soils; *Templeton.* Cave Hill, Larne, and Fairhead; S.A.S. Divis; C.H.W. By the Lagan near Belfast; *J. J. Andrew.*
Derry—Ness Glen, and other places; *Templeton.* Carndaisy Glen, and plentiful about Dungiven; S.A.S.

4. D. cerviculata (*Hed.*) *Schimp.*
Turfy banks in peat bogs, and heaths—frequent. Fr. June—Aug.
Down—Near the summit of Slieve Croob, and in the bog at Moneyreagh; S.A.S. Slievenabrock, and Deers' Meadow; H.W.L. Mourne Mountains; C.H.W.
Antrim—Observed in a bog near Bruce's Castle, Rathlin Island; D.M. Kings' Moss, Brean Mountain, and bogs in the northwest of Rathlin; S.A.S.

ANISOTHECIUM *Mitten.*

1. A. rubrum (*Huds.*) *Lindb.* (DIC. VARIUM *Hed.*).
Wet clay banks—frequent. Fr. Nov.—March.

Down—Holywood Hills, Cregagh, Sydenham, and Knock; S.A.S. Hillsborough; H.W.L. Warrenpoint, banks of Kilkeel River, and in Quarry, and on river bank at Magheralin; C.H.W.
Antrim—Found on the slip of a bank in Colin Glen, also field north of Lambeg, and side of road to Grogan's Glen; *Templeton*, 1801. Marshy ground at Black Mountain, and abundant in wet clay fields at Stranmillis, Cave Hill, and Islandmagee : S.A.S.
Derry—Curleyburn, and Ballyharrigan Glen, but not common; D.M.

Var. CALLISTOMUM (*Dicks.*) *Braith.*

Antrim—Colin Glen (*Dr. Scott*, 1802); *Brit. Moss Flora.*
Derry—Near Derry (*R. Brown*); *Templeton.* Near Derry (*Dr. Scott*); *Brit. Moss Flora.*

Var. TENUIFOLIUM *Bruch.*

Down—Wet clay bank at Stormount; S.A.S.
Antrim—On chalk rocks south of Larne; S.A.S. Larne (*Stewart*); *Brit. Moss Flora.*

2. A. rufescens (*Dicks.*) *Lindb.* (DIC. RUFESCENS *Sm.*).
Arenaceous rocks, and damp shady banks—rare. Fr. Aug.—Nov.
Down—On Crown Mount near Newry; C.H.W.
Antrim—Base of a little bridge at Cranmore; *Templeton*, 1801. On sandbanks near Belfast; *T. Drummond, Musci Scotici*, vol. 3. On mica schist in Glenshesk, and on Brean Mountain; S.A.S.
Derry—On bare clay banks near Derry (*R. Brown*); *Templeton.*

3. A. squarrosum (*Starke*) *Lindb.*
Wet stony places, and dripping rocks—not common, and usually barren.
Down—On dripping rocks by the riverside between Slieve Donard and Diamond Mountain; *Templeton.* By Bloodyburn River, and in several places on Slieve Donard, and Slieve Croob, also on Carngaver Hill; S.A.S. Slieve Commedagh; H.W.L. Rostrevor, Yellow Water, and Slievenabrock; C.H.W.
Antrim—Peaty marsh on Slemish, and wet stony places on Carrickfergus commons; S.A.S.
Derry—Rocks by stream on Meenard Mountain, south of Dungiven; S.A.S.

Sect. III.—Seligerieæ.

SELIGERIA *Br. & Schimp.*

1. S. Donii (*Sm.*) *C. Muell.*
Wet sandstone rocks—extremely rare.
Antrim—In good fruit on wet Greensand rocks by the side of the first

waterfall in upper Colin Glen ; S.A.S., 19th June, 1876.
Occurs here only on concealed sloping sides of rocks—not on exposed surfaces. The only Irish locality.

2. S. pusilla (*Ehr.*) *Br. & Schimp.*

Damp Chalk rocks, forming a bright green crust on their shaded sides —rare. Fr. April—Aug.

Antrim—In dense patches on the white limestone in the neighbourhood of Belfast (*Templeton*); *Musc. Brit.* Near Belfast (*Templeton*); *Flor. Hib.* Limestone rocks near Belfast ; *T. Drummond, op. cit.* Limestone of Sallagh Braes, and not infrequent on limestone in Co. Antrim; D.M. In considerable abundance on white limestone rocks above the mill at Wolfhill, fruiting in May, 1837, also on shady limestone rocks above Lisburn ; *Moore, Proc. R.I.A.*, 1872. Belfast (*Drummond*); *Brit. Moss Flora.* Limestone rocks at Windy Gap, and in several other places on Black Mountain north and south of Whiterock, also in Crow Glen, Carr's Glen, and Redhall Glen ; S.A.S. Chalk rocks at foot of Sallagh Braes ; C.H.W.

First found in Britain by Mr. Templeton, in the Belfast Hills.

3. S. calcarea (*Dicks.*) *Br. & Schimp.*

Damp Chalk rocks, but more often on detached stones where overgrown by grass—very rare. Fr. March—June. Occurs in much smaller patches than the preceding.

Antrim—On small pieces of white limestone on Black Mountain, both north and south of Whiterock, also on chloritic limestone by the stream above the waterfall in upper Colin Glen, and on Chalk rocks in Carr's Glen ; S.A.S.

Under Smith's name, *Grimma calcarea*, this was given as an Irish plant, by Dawson Turner in 1804, but as no definite locality was stated, and as the preceding species was omitted by Turner it seems to have been assumed that his record was erroneous, and it has been ignored by subsequent writers. In Mr. Templeton's copy of Muscologia Hibernica, now in the possession of Mr. R. M. Young, there is a M.S. note by Templeton which mentions the finding of this plant on 27th April, 1800, but no special station is assigned. As this, and the preceding species grow in close proximity it is highly probable that he found both, and was therefore also the first to discover this moss in Ireland.

Sect. IV.—Dicraneæ.

BLINDIA *Br. & Schimp.*

1. B. acuta (*Huds.*) *Br. & Sch.*

Shallow, rocky mountain streams, and wet stony heaths—frequent on
the higher mountains. Fr. April—July.

Down—On the hollow moist places about Slieve Donard; *Templeton*,
1805. *M.S. note in Musc. Hib.* On Slieve Donard, and
frequent on mountains above Newcastle, also wet heath on
Slieve Croob; S.A.S. Tollymore Park; C.H.W. Abundant
in the Mourne Mountains; H.W.L.

Antrim—Slemish Mountain, and on wet banks near waterfalls; D.M.
Knocklayd; S.A.S. Sallagh Braes; C.H.W.

DIDYMODON *Hedwig.*

1. D. denudatus (*Brid.*) *Lindb.* (DIC. LONGIROSTRE B. & S.).

Var. β ALPINUS (*Schimp.*) *Braith.*

Antrim—Moist rocks at Cushendall; *Moore, Proc. R.I.A.*, 1872.
Cushendall (*Moore*) *Brit. Moss Flora.*

CAMPYLOPUS *Bridel.*

1. C. pyriformis (*Schultz.*) *Brid.*

Peaty, and heathy ground—frequent. Fr. Feb.—April, and Aug.—Oct.

Down—Slieve Donard, and rocky shore near Groomsport; S.A.S. Frequent in the Mourne Mountains; H.W.L.

Antrim—Divis Mountain, and peaty heath on Slievetrue; S.A.S. Sallagh
Braes; C.H.W.

Derry—Fruiting abundantly on the peat bog at summit of Benbradagh;
S.A.S.

2. C. fragilis (*Dicks.*) *Br. & Sch.*

Stony heaths, and sometimes in turfy places on rocks—frequent. Fr.
Feb.—May.

Down—Peaty heaths at Slieve Croob, Conlig, and Braniel, also rocks by
the sea at Groomsport; S.A.S. Warrenpoint, Rostrevor, and
Dromore; C.H.W. Frequent in Mourne Mountains; H.W.L.

Antrim—Belfast Mountains; *Templeton*, 1801. Crow Glen, Slievetrue,
Agnew's Hill, Rathlin, and frequent on heaths; S.A.S.
Sallagh Braes; C.H.W.

Derry—In the mountains; *Templeton.*

3. C. flexuosus (*Linn.*) *Brid.*

Peaty, and stony heaths—frequent. Fr. May—July.

Down—Laghey bog, and Annahilt bog; *Templeton.* Slieve Commedagh,
and mountains above Bloodybridge; S.A.S. Bencrom, and
Teevedocharrah; C.H.W. Common in Mourne Mountains;
H.W.L.

Antrim.--Black Mountain; *T. Drummond, op. cit.* Fairhead; *Moore, Proc. R.I.A.* Boggy heaths on Slemish, and Knocklayd; S.A.S.
Derry—Common; D.M. Benbradagh, and frequent in the Sperrin Mountains; S.A.S.

Var. PALUDOSUS *Schimper*.

Down—Bog near Rocky River; C.H.W., 1883.

4. C. setifolius *Wilson*.

Damp rocks—very rare, and without fruit.

Down—Slieve Commedagh, and at Blackstairs on Slieve Donard; H.W.L.

5. C. atrovirens *De Notaris*.

Wet rocks—rare, though locally abundant. No fruit found.

Down—Wet boggy rocks on Slieve Donard; S.A.S. Rostrevor Mountain, and common on the Mourne Mountains; C.H.W. Slievemeel Beg, and frequent in Mourne; H.W.L.
Antrim—Frequent on wet rocks at the northeast of Rathlin; S.A.S. Sallagh Braes; C.H.W.

6. C. brevipilus *Br. et Schimp*.

Damp rocky heaths—rare, and barren.

Down—Slieve Commedagh, Shanslieve, and east side of Slieve Donard; H.W.L.

DICRANOWEISSIA *Lindberg*.

1. D. cirrata (*Linn.*) *Lindb*.

On rocks, and occasionally on wood—not common. Fr. Nov.—Mar.

Down—On a stone in a little glen at Ballyhanet, 1808; *Templeton, M.S. note in Musc. Hib*. Rocks east of Groomsport; *W. Thompson*, 1835. Boulder stone on mountain above Rostrevor; S.A.S. Bencrom, and Clandeboye; H.W.L. On trees at Warrenpoint, and on rocks at Shanslieve, Pigeonrock Mountain, and Rocky Mountain near Hilltown; C.H.W.
Antrim—In clefts of the timber of the beams of the second lock of the Lagan canal; *Templeton*, 1808. On the wood of the locks of the Lagan canal, May 1838; D.M. (not found there now).
Derry—On rocks at 1400 feet on Slievegallion; D.M. (*sub W. crispula*).

DICRANUM *Hedwig*.

1. D. majus *Smith*.

Damp shady rocks, and banks—frequent. Fruiting in spring, and

DICRANACEÆ.

early summer.
Down—At the waterfall in Tollymore Park; *Templeton*, 1805. Glenmaghan; S.A.S. Slieve Donard; H.W.L. Rostrevor wood; C.H.W.
Antrim—Colin Glen; *T. Drummond, op. cit.* Belfast hills, Blackhead, Cushendall, and amongst heath on Rathlin Island; S.A.S. Loughmourne, H.W.L.
Derry—Carndaisy Glen, and amongst rocks and heath on Slievegallion; S.A.S.

2. D. scoparium (*Linn.*) *Hedw.*
Heaths, banks, and rocks—very common. Fr. April—Aug. Ranges to 1500 feet on the mountains.
Var. β ALPESTRE *Huebens.*
Down—Slievemartin, Slieve Donard, Slieve Commedagh, and Slievenabrock; H.W.L.

3. D. Bonjeani *De Not.* (D. PALUSTRE B. & S.).
Wet boggy places—not common. Fr. July and Aug. (B.M.F.).
Down—Donard demesne; H.W.L. Moygannon Glen; C.H.W.
Antrim—In a little bog southwest of Seymour Hill; *Templeton.* Wood at Portglenone; D.M. Near Lisburn; *J. Creeth.* Sparingly on Rathlin Island; S.A.S.

4. D. fuscescens *Turner.*
Mountainous rocks—barren, and very rare.
Down—Shady rocks on the northeast side of Slieve Bignian, and rocks on Bencrom Mountain; H.W.L.

5. D. Scottii *Turner.*
Mountain rocks—rare. Fr. July and Aug.
Down—Mourne Mountains; *T. Drummond, op. cit.* Shanslieve, and Slieve Commedagh; H.W.L.
Named by Turner from specimens found near Swanlinbar, Co. Cavan, by Dr. Robert Scott.

DICHODONTIUM *Schimper.*

1. D. pellucidum (*Linn.*) *Schimp.*
Rocky banks of streams in hilly districts—frequent. Fr. Oct.—March.
Down—Slieve Donard, and mountain above Tollymore, also in a little glen west of Dundonald rail. station; S.A.S. Drumcro, Sheepfield, and Moygannon Glen; C.H.W.
Antrim—Stream near Larne; *Templeton* (*sub Dic. virens*). Spring by the

roadside at Derriaghy, also in Colin Glen, and Glendivis;
S.A.S.

Derry—Stream on north side of Sawel Mountain, also on Slievegallion, and in Carndaisy Glen; S.A.S.

Var. β PAGIMONTANUM *Brid.*

Antrim—Stream above Whitewell; C.H.W. Sallagh Braes; H.W.L.

2. D. flavescens (*Dicks.*) *Lindb.*
Rocky banks of streams—not rare. Fr. Oct.—Feb.

Down—Slievenabrock; H.W.L. Tollymore Park; C.H.W. Abundant, and very fine in Crawfordsburn, and in Drumbo Glen, also in the glen at Cregagh; S.A.S.

Antrim—On a stone near Grogan's Glen; *Templeton*. Colin Glen; S.A.S.

Sect. V.—Oncophoreæ.

ONCOPHORUS *Bridel.*

1. O. Bruntoni (*Sm.*) *Lindb.* (CYNODONTIUM B. & S.).
Extremely rare. Fr. July and Aug. (B.M.F.).
Antrim—Rare, only in one place in the Little Deerpark at Glenarm; D.M.

2. O. crispatus (*Dicks.*) *Lindb.* (RHAB. DENTICULATA B. & S.).
Mountain rocks—rare. Fr. June and July (B.M.F.).
Down—Diamond Mountain; *W. Thompson.* Amongst rocks on the ascent of Donard by the glen river; S.A.S. Slievenabrock; H.W.L.
Antrim—Slemish (*Moore*); *Phytologist, vol.* II. 1857. One perfect capsule, and several old ones, in rock crevices at summit of Slemish, Nov. 1875; S.A.S.

3. O. striatus (*Schrad.*) *Lindb.* (RHAB. FUGAX B. & S.).
Damp, shady rocks—very rare. Fr. June and July (B.M.F.).
Down—Mourne Mountains; *Templeton,* 1808.
Antrim—Slemish, and Collon, rare; D.M. Sallagh Braes; *Moore, Proc. R.I.A.,* 1872.
Derry—Benevenagh; *Moore, op. cit.*

CERATODON *Bridel.*

1. C. purpureus (*Linn.*) *Brid.*
Walls, banks, and waste ground—very common. Fr. March—May.

SERIES II. Order XII. TORTULACEÆ.

Sect. I.—Tortuleæ.

EPHEMERUM *Hampe.*

1. E. serratum *Hampe.* (PHASCUM *Schreb.*).
Sandy ground—rare. "In fructification nearly all the year" *Templeton.*
Down —Damp bank in the Newcastle sandhills; S.A.S.
Antrim—In the garden at Cranmore, and in a dry sandy pasture field near Seymour Hill, also in a field near Lambeg moss, 1801, and in clover ground at Malone, 1805; *Templeton.* Fields now occupied by the Belfast Botanic Gardens; *T. Drummond.* Frequent about Belfast; *Moore, op. cit.*

PHASCUM *Linn.*

1. P. acaulon *Linn.* (P. CUSPIDATUM *Schreb.*).
Sandy fields, and sandy or gravelly waste ground—not uncommon. Fr. Dec.—March.
Down—Gravel pits north of Giants' Ring, hedge banks at Belmont, and pasture field at Sydenham; S.A.S. Drumcro; C.H.W.
Antrim—Found on the flower beds at Cranmore, and on sandy soil in several places in Malone; *Templeton*, 1801.* Sparingly on Rathlin Island; S.A.S.

Var. β PILIFERUM *Schreb.*

Antrim—Very fine on crumbling basalt rocks at Blackhead; S.A.S. Sandy bank by the sea at Ballygally; C.H.W.

* Named *P. muticum* by Templeton, but as he does not record the present species, but remarks that he considers *P. cuspidatum*, and *P. muticum* to be one, it may be concluded that he did not distinguish between them.

POTTIA *Ehrhart.*

1. P. Heimii (*Hedw.*) *Fuern.*
Rocks, and banks near the sea—not common. Fr. April and May.
Down—Rocks by the shore south of Newcastle, and among rocks above high-water mark at Bangor, also on the sides of drains in Ormeau Park; S.A.S.
Antrim—Among stones by the quay at Carrickfergus, and on rocky seashore of Rathlin Island; S.A.S.

TORTULACEÆ.

2. P. truncatula (*Linn.*) *Lindb.*
Damp sandy fields, and banks—common. Fr. Nov.—Feb.

3. P. intermedia (*Turner*) *Fuern.*
Damp banks—frequent. Fr. Oct.—Feb.
Down—Common on ditches near Killyleagh; *Templeton*, 1804 (as *Gymnostomum obtusum*). Wet banks at Mealough Hill, and roadside at Belmont; S.A.S. Roadside near Crossgar; H.W.L.
Antrim—Banks of drains in Bog meadows, and basalt rocks at Ballygally Head; S.A.S. Common in the furrows of cornfields in autumn. It varies greatly in size; *Templeton*. The latter remark shows that there was more than one species.

4. P. Starkei (*Hedw.*) *C. Muell.*
Fields, and ditch banks—rare. Fr. Nov.—Jan.
Var. DAVALLII *Sm.* (P. minutula *Fuern.*).
Down—Fields below Holywood; *T. Drummond, Musci Scotici*, vol. III. Pasture field at Sydenham, also on Holywood Hill, and banks near the shore at Crawfordsburn; S.A.S.

5. P. asperula *Mitten.*
Sandy banks—very rare. Fr. Feb. and March.
Down—Banks by the sea between Donaghadee and Millisle, and in an old sandpit by the roadside at Donaghadee warren; S.A.S.

6. P. viridifolia *Mitten.*
Basalt rocks—very rare. Fr. Jany. and Feb.
Antrim—Ledges of dry, crumbling rocks by the shore at Blackhead; S.A.S. Blackhead, Belfast (*Stewart*, 1884); *Brit. Moss Flora*.
This, and the preceding species are not, as yet, known to occur in Ireland except in the stations mentioned above, but no doubt they will be found elsewhere when carefully sought for.

7. P. crinita *Wilson.*
On amygdaloidal rocks—very rare. Fr. Jany. and Feb.
Antrim—Steep dry rocks by the shore at Blackhead, growing sparingly with the preceding species, but distinguished by its larger size, longer hair points, and longer lid; S.A.S. Blackhead, Belfast (*Stewart*, 1882); *Brit. Moss Flora*.

TORTULA *Hedwig.*

1. T. lamellata *Lindberg.*
Sandy banks—very rare.

Down—Sparingly on sandy ditchbank between Donaghadee and Millisle, in good fruit on 2nd Feb, 1874 ; S.A.S.

2. T. stellata (*Schreb.*) *Lindb.* (T. RIGIDA *Schultz*).
Damp rocks—very rare. Fr. Jany. and Feb.
Down—Three capsules on rocks by the waterfall in Cregagh Glen, Feb., 1877, and sparingly at same locality in Jany., 1880 ; S.A.S.
Antrim—In Colin Glen a little above the new bridge ; D.M.

3. T. ericæfolia (*Neck.*) *Lindb.* (T. AMBIGUA *B. & S.*).
Rocks, and clay banks—rather rare. Fr. Jany.—March.
Down—Walls at Portavo, and Newtownbreda, also on clay banks at Scrabo ; S.A.S.
Antrim—Plentiful on red sandstone rocks by the roadside north of Derriaghy church ; S.A.S.

4. T. aloides (*Koch*) *De Not.*
Damp clay banks, wall tops, and earthy ledges of rocks—frequent. Fr. Nov.—Feb.
Down—Walls at Ballylesson, and Dundonald, also clay banks above Groomsport church, ditch bank at Sydenham, and on grit rocks in Cregagh Glen ; S.A.S. Narrow-water, and old quarry at Magheralin ; C.H.W.
Antrim—On a moist clay bank among some firs opposite Newforge (*Dr. Stokes*, 1802) ; *Templeton.* In the crevices of the stonework at the lower end of the west side of the lock on the Lagan canal, and on the north, and south walls of the Belfast Deerpark ; *Templeton* (as *B. rigidum*, *E. B.*, 180). Colin Glen ; *T. Drummond.* Damp banks at Drumbridge, and on ledges of limestone rocks all along the Belfast hills, as Black Mountain, Cave Hill, and Whitewell, also on ditch bank near Troopers' Lane, and basalt rocks north of Whitehead ; S.A.S. Sallagh Braes ; C.H.W.

5. T. atrovirens (*Sm.*) *Lindb.* DID. NERVOSUS *H. & T.*).
Rocks near the sea—extremely rare. Fr. March—May (*B.M.F.*).
Antrim—One small tuft collected on Rathlin Island ; S.A.S.

6. T. muralis (*Linn.*) *Hedw.*
Walls, and occasionally on dry rocks—very common. Fr. Jan.—June.
Var. β RUPESTRIS *Schultz.*
Dry rocks—frequent, especially on limestone, and basalt.

7. T. subulata (*Linn.*) *Hedw.*
Dry shady banks, and occasionally on rocks, and trees—frequent. Fr.

March—June. Ranges to over 1000 feet.
Down—Banbridge, Donaghadee, Warrenpoint, and coast intermediate, also Comber, Castle Espie, and Mealough Hill; S.A.S. Loughbrickland. Hilltown, and Annalong; H.W.L.
Antrim—Banks in shady places; *Templeton*. Lisburn, Belfast hills, Woodburn, Islandmagee, Rathlin Island, and on trees at Templepatrick; S.A.S. Sallagh Braes; C.H.W.
Derry—Common through the county; D.M. Toome, and Moneymore; S.A.S.

8. T. mutica (*Schultz*) *Lindb.* (T. LATIFOLIA *Hartm.*).
At roots of trees in damp places—rare, and usually barren.
Down—On trees by the Lagan at Shawsbridge; S.A.S. Plentiful by the Lagan at Drumcro; C.H.W.
Antrim—On trees in the Bog meadows, and by Lagan canal between 1st and 2nd locks, also at Drumbridge, and by the Sixmilewater above Templepatrick; S.A.S.
First found in Ireland by S.A.S. in 1874.

9. T. papillosa *Wilson.*
Rugged bark of trees—frequent, and widely distributed but not plentiful. Barren here, as elsewhere.
Down—Greyabbey demesne, Purdysburn, and near Knock; S.A.S. Sparingly at Magheralin; C.H.W.
Antrim—On apple trees at Glenavy, and trees by Sixmilewater near Antrim, also on trees between Dunadry and Crumlin, and at Glencairn near the Ballygomartin road; S.A.S.

10. T. lævipila (*Brid.*) *Schwg.*
Basaltic, and limestone rocks, and on old walls, rarely on trees—common throughout the district. Fr. Jan.—April.

11. T. montana (*Nees*) *Lindb.* (T. INTERMEDIA *Berk.*).
Limestone rocks, and occasionally on basalt—rare. Fr. March and April.
Down—On a slate roof in Ballywalter, barren; S.A.S.
Antrim—East side of Black Mountain, and on the basaltic cliffs o Knockagh, and Blackhead, also on limestone at Waterloo near Larne; S.A.S. Sallagh Braes; C.H.W.

12. T. ruralis (*Linn.*) *Ehr.*
Frequent on old roofs, rare on rocks or walls. Fr. April and May, but more often barren.
Var. β ARENICOLA *Braithwaite.*

Abundant on sands by the sea all around the coast, usually barren, but fruiting at Ballycastle.

MOLLIA *Schrank.*

1. M. crispa *Lindb.* (PHASCUM CRISPUM *Hedw.*).
Damp banks—very rare. Fr. April and May (*B.M.F.*).
Antrim—Banks near Belfast (*Templeton*) ; *Flor. Hib.*, and *Moore, Proc. R.I.A.*, 1872.

2. M. microstoma *Lindb.* (GYM. MICROSTOMUM *Hedw.*).
Gravelly banks, and waste ground, and on earthy ledges of rock—not rare. Fr. Feb.—May.
Down—Ditch banks at the Newcastle sandhills, gravel bank at Glenmachan, and rocks, and wet bank by the shore at Groomsport ; S.A.S. Near Newcastle ; H.W.L.
Antrim—Frequent in Antrim ; D.M. On gravel at Whiterock, and foot of Black Mountain, and on ledges of basaltic rocks at Cave Hill, Blackhead, and Rathlin ; S.A.S. Woodburn Glen ; H.W.L.

3. M. viridula (*Linn.*) *Lindb.* (WEISSIA CONTROVERSA *Hedw.*).
Hedge banks, and ledges of rocks—very common. From sea level to 1500 feet. Fr. Oct.—Feb.

4. M. rutilans (*Hedw.*) *Lindb.* (WEISSIA MUCRONATA *Schp.*).
Banks—very rare. Fr. March and April (*B.M.F.*).
Down—Dry banks near Warrenpoint ; C.H.W.
The only Irish station known for this plant.

5. M. tenuis *Lindb.* (GYMNOSTOMUM *Schrad.*).
Sandstone rocks—very rare. Fr. July—Sept.
Down—On New Red Sandstone by the Lagan at Glenmore ; J. H. Davies, 1880. Sparingly on sandstone blocks of a bridge at upper end of Crawfordsburn, 1885, and abundant on rocks in the freestone quarries at Scrabo, 1886 ; S.A.S.
Antrim—Belfast (*Drummond*) ; *Brit. Moss Flora.*

6. M. æruginosa (*Sm.*) *Lindb.* (GYM. RUPESTRE *Schleich.*).
Shady glens—rare. Fruit not seen.
Antrim—In the glens of Antrim, not rare ; *Moore, op. cit.*
   Var. RAMOSISSIMA *B. & S.*
Down—Tollymore Park ; H.W.L.

7. M. verticillata (*Linn.*) *Lindb.* (WEISSIA *Brid.*).
In dense compact tufts on limestone rocks—rare, not found in fruit.
Antrim—Dripping limestone rocks north of the Black cove near Larne, and on limestone cliffs in Rathlin; S.A.S. Limestone rocks at upper end of Colin Glen, and on Cave Hill; C.H.W. Murlough Bay; H.W.L.

8. M. crispula *Lindb.* (TRICH. CRISPULUM *Bruch*).
Basaltic rocks—very rare. Fr. May and June (*B.M.F.*).
Antrim—Sparingly at the Black cove near Larne; D.M. In small quantity on stone fence at Sallagh Braes, in fruit Oct., 1874; S.A.S.
Var. ELATA (*Schimper*) *Braith.*
Antrim—Sparingly on rocks at the west end of Rathlin; S.A.S. Rathlin Island (*Stewart*, 1882); *Brit. Moss Flora.*
The only Irish locality known, at present, for this var.

9. M. litoralis (*Mitten*) *Braith.*
Rocks by the shore, and on mountains—rare, or perhaps overlooked.
Down—Rocks near Groomsport; H.W.L.
Antrim—On rocks in several places by the shore on Rathlin Island; S.A.S., 1882. Cave Hill; C.H.W. Sallagh Braes; H.W.L.
Derry—Benevenagh (*Hart*); *Jour. Bot.*, 1886.

10. M. brachydontia *Lindb.* (TRICH. MUTABILE *Bruch*).
Rocks, especially limestone and basalt—rare. Fr. April—June (*B.M.F.*).
Down—On the shady side of "the Rock" at Newcastle; H.W.L.
Antrim—In considerable abundance on basalt rocks, but generally barren; *Moore, Proc. R.I.A.*, 1872. Sparingly on basalt at Fairhead, and abundant on Chalk at Whitepark Bay near Ballintoy; S.A.S. Cliffs at Sallagh Braes; H.W.L., and C.H.W.

11. M. tenuirostris (*Hook. & Tayl.*) (DID. CYLINDRICUS *B. &S.*).
On rocks—rare, and barren.
Down—Near Rockport below Holywood; *T. Drummond.* Pigeonrock Mountain (*Lett* and *Waddell*); *Brit. Moss Flora.*
Other localities have been given, but not verified by authentic specimens. This is probably less rare than appears from the above, but is often confounded with other mosses to which it bears a general resemblance.

12. M. inclinata *Lindb.* (TORTULA *Hedw.*).
Sandy banks, and crevices of rocks—very rare.

Down—Rostrevor, and rocks by the shore at Groomsport; C.H.W.
Rocks by the shore at Annalong; H.W.L., and C.H.W.
Rocks by the sea at Groomsport (*Waddell*); *Brit. Moss Flora.*
Antrim—Basalt rocks at Blackhead; S.A.S., March, 1884.
First found in Ireland by Rev. C. H. Waddell, in 1883 at Omeath Co. Louth.

13. M. tortuosa (*Linn.*) *Schrank.* (TORTULA *Ehr.*).

In dense cushions on rocks in mountain districts, and occasionally on walls—frequent and often abundant; Fr. July (*B.M.F.*).
Down—Slieve Donard, and Newtownards Glen; S.A.S. Tollymore Park, and Slievenabrock; H.W.L. Narrow-water, Rostrevor Mountain, Pigeonrock Mountain, and Rocky Mountain near Hilltown; C.H.W.
Antrim—Rocks in Colin Glen; *Templeton.* Belfast hills, Knockagh, Slemish, and Sallagh Braes; S.A.S. Fairhead; H.W.L.
Derry—Abundant on the basalt; D.M. Trap rocks of Benevenagh, with immature fruit in July; S.A.S.

Var. β ANGUSTIFOLIA (*Juratz.*) *Braith.*
Down—Wall west of Bryansford (*Lett*); *Brit. Moss Flora.*

BARBULA *Hedwig.*

Sect. I.—Hymenostylium.

1. B. curvirostris (*Ehr.*) *Lindb.* (GYMNOSTOMUM *Hedw.*).
Damp shady rocks—rare. Fr. Sept. (*B.M.F.*).
Down—Wet rocks at the head of Moygannon Glen; H.W.L., and C.H.W.
Antrim—Rocks at Fairhead (*Templeton*); *Flor. Hib.* Glen at Cushendall; *Moore, op. cit.*
Derry—Sandy places near Derry (*R. Brown*); *Templeton.* On Clontygeragh at 1100 feet, and by the road from Coleraine to Portrush; D.M.

Sect. II.—Erythrophyllum.

2. B. rubella (*Hoff.*) *Mitt.* (DIDYMODON *B. & S.*).
Walls, rocks, gravelly waste ground, and especially abundant on old quarry heaps. Fr. Sept.—Dec.

Sect. III.—Eubarbula.

3. B. brevifolia (*Dicks.*) *Lindb.* (TRICH. TOPHACEUM *Brid.*).
Wet rocks, and dripping stony banks—frequent. Fr. Oct.—Jan.
Ranges from sea level to 1500 feet.

Down—Wet rocks by Spinkwee River, and at Groomsport, also near the shore below Bangor, and in Cregagh Glen; S.A.S. Moygannon Glen, Annalong, Bloodybridge, and Tollymore Park; H.W.L. Wet rocks in railway cutting at Dromore, and on clay at Magheralin; C.H.W.

Antrim—Rocks at Redbay cave, and on the "organ" at Giants' Causeway; *Templeton*, 1814. Moist banks near Dunluce; D.M. The Glens of Antrim; *Moore, op. cit.* Colin Glen, Crow Glen, Glendivis, and Woodburn, also on vertical limestone rocks at Whitehead, and rocks by streams at Glynn, Glenariffe, and Rathlin; S.A.S.

Derry—Rocks by the sea at Portstewart, and boggy rocks at summit of Benbradagh; S.A.S.

4. B. fallax *Hedwig.*
Damp clay banks, and waste ground in moist places—very common. Fr. Oct.—Feb.

Var. γ B. BREVIFOLIA (*Sm.*) *Brid.*
Down—Wet places by stream in Cregagh Glen; S.A.S. Railway cutting near Dromore; H.W.L.

5. B. spadicea *Mitten.* (DIDYMODON RIGIDULUS *Brid.*).
Wet banks, and rocky margins of streams—rare. Fr. Sept.—Nov. (*B.M.F.*).

Down—Newcastle (*Lett*); *Brit. Moss Flora.*

Antrim—Near Belfast; *Templeton*, 1807. Moist banks near Belfast; *T. Drummond, Musci Scotici,* vol. III. On stones in stream at lower end of Carr's Glen, basalt rocks in south Woodburn, and in Rathlin Island; S.A.S. Belfast (*Stewart*), and Fairhead (*Lett*); *Brit. Moss Flora.* Colin Glen; C.H.W. Mr. Waddell's specimen is probably the right plant, but wants mature fruit to give entire certainty.

Derry —[Occurs, no doubt, in this county, but has not yet been collected. Mr. Lett finds it in Loughery demesne, which is over the border in Co. Tyrone.]

6. B. rigidula (*Hedw.*) *Mitten.*
Damp rocks, especially limestone—not common. Fr. Nov.—Feb.
Ranges from near sea level to 1500 feet.

Down—On the rocks by the pen weir above Belvoir, and walls of Belvoir demesne; *Templeton*, July, 1807, *Herb. Belf. Mus.*

Antrim—Limestone rocks near Belfast; *T. Drummond, op. cit.* Banks by the sea near Dunluce; D.M. The Glens of Antrim; *Moore, Proc. R.I.A.*, 1872. Castle Robin; *J. Creeth.* By the stream at Glendivis, and on limestone at Black cove near Larne; S.A.S.

Derry—Plentiful on the Chalk rocks near summit of Slievegallion, also on a wall near Moneymore; S.A.S. Slievegallion (*Stewart*); *Brit. Moss Flora.*

7. B. cylindrica (*Tayl.*) *Schp.* (TORT. INSULANA *De Not.*).

Rocks, and banks—frequent in the barren state. Fr. March and April, but not common in fructification.

Down—On slate rocks in Cregagh Glen; S.A.S. Rocks in railway cutting near Dromore, and roadside near Aghaderg schoolhouse; H.W.L. Side of the Lagan at Drumcro; C.H.W.

Antrim—Crumlin waterfoot, and about roots of trees at Derriaghy, also on basalt rocks in Woodburn, on stones in stream at Kilroot, and in Rathlin Island; S.A.S. Divis Mountain; C.H.W. Kilroot (*Stewart*, 1874); *Brit. Moss Flora.*

Var. β B. VINEALIS *Bridel.*

Old walls, and on dry rocks—rare, apparently.

Antrim—Plentiful in the barren state on the walls of Carrickfergus Castle, and on the basaltic rocks on which the castle is based; S.A.S.

Recorded, and probably does occur in other localities, but only from the above station have satisfactory specimens been seen.

This moss seems, with us, so different in aspect, and *habitat* as to merit being kept distinct.

[B. HORNSCHUCHII *Schultz.*

Recorded by Moore as occurring on the walls of Carrickfergus Castle. Subsequent botanists have failed to find this plant at Carrickfergus, and as it is not known elsewhere in Ireland the record is probably erroneous. Some form of *B. vinealis* may have been taken for it.]

8. B. revoluta *Schrader.*

Old walls—common. Fr. April and May.

Sect. IV.—Leptopogon.

9. B. convoluta (*Huds.*). *Hedwig.*

Walls, waste ground, quarry heaps, and occasionally on rocks—frequent, and usually in profusion where it occurs. Fr. April—June.

Down—Flags of Drumbridge weir; *Templeton*, 1803. Slieve Croob, and walls about Banbridge, also on walls of bridge over the Ravernet River, and between that place and Hillsborough,

quarry heaps near Moira, walls on Castlereagh Hill, waste ground in Cregagh Glen, and abundant on walls of Belvoir Park ; S.A.S. Common in the county ; H.W.L.
Antrim—Old walls at Carr's Glen, and on basalt rocks at the Knockagh ; S.A.S.

10. B. unguiculata (*Huds.*) *Hedw.*
On ditch banks, shady rocks, waste ground, and quarry *debris*—common. Fr. Nov.—Feb.
Var γ B. APICULATA *Hedwig.*
Down—Dromore (*Waddell*) ; *Brit. Moss Flora.*
Antrim—Wall of Ballymoney bridge ; *Templeton,* 1814 (as *T. aristata*).

CINCLIDOTUS *Pal. de Beauv.*

1. C. fontinaloides (*Linn.*) *Beauv.*
On submerged stones in streams and lakes—common. Fr. March—May.

LEERSIA *Hedwig.*

1. L. exstinctoria (*Linn.*) *Leyss.* (ENC. VULGARIS *Hedw.*).
Basaltic rocks—very rare. Fr. March—May (*B.M.F.*).
Antrim—On Cave Hill near the northern termination of the rocks of MacArt's Fort ; *Templeton.* Lurigethan, rare ; D.M.
Derry—On Benbradagh at 1100 feet ; D.M.

2. L. laciniata *Hedw.* (ENC. CILIATA *Hoff.*).
Ledges of basaltic rock—rare. Fr. July and Aug.
Antrim—Agnew's Hill, Sallagh Braes, and Lurigethan ; D.M., 1836. Sparingly on cliffs of Sallagh Braes ; S.A.S., 1873.
Derry—Plentiful on Benbradagh at 1300 feet ; D.M.

3. L. contorta (*Wulf.*) *Lindb.* (ENC. STREPTOCARPA *Hedw.*).
Old walls, and sometimes on rocks—frequent, but without fruit.
Down—Slieve Donard, and Bryansford ; H.W.L. Walls at Narrowwater, and Tollymore Park, also Thomas Mountain, Rocky Mountain near Hilltown, and Dromara ; C.H.W.
Antrim—Glenarm, and wall of the Belfast Deerpark ; D.M. Wall near Castle Robin ; J. *Creeth.* Plentiful on limestone rocks in Crow Glen, and at Whitewell quarries, and abundant on the wall of the deerpark at the top, and again at the lower part near the Antrim road, also in great abundance on the wall of

the bridge over Clady river south of Dunadry station ; S.A.S. Glarryford ; H.W.L.

SERIES II. ORDER XIII. **WEBERACEÆ.**

WEBERA *Ehrhart.*

1. W. sessilis (*Schm.*) *Lindb.* (DIPHYSCIUM FOLIOSUM *Linn.*).
Damp rocks in the mountains—rare, and barren.
Down—Slieve Bignian, Slieve Commedagh, Slievenamady, and Slievenabrock ; H.W.L. By the Bloodybridge River, also on Pigeonrock Mountain, and Thomas Mountain ; C.H.W.

SERIES II. ORDER XIV. **GRIMMIACEÆ.**

GRIMMIA *Ehrhart.*

Sect. I.—Schistidium.

1. G. pruinosa *Wilson.*
Rocks—rare, but locally abundant on amygdaloidal trap. Fr. Nov.—March. Ranges from 400 to over 1000 feet.
Down—Slieve Donard ; H.W.L.
Antrim—Cave Hill ; *W. Thompson*, 1836. On rotten trap rocks near Belfast ; *Moore, op. cit.* Basaltic rocks of Cave Hill, Knockagh, and Rathlin, and on mica schist at Brean Mountain; S.A.S. Sallagh Braes ; C.H.W.
Derry --Cliffs of Benbradagh ; S.A.S.

2. G. apocarpa (*Linn.*) *Hedw.*
Wet stony or rocky places by streams, and on mountains—very common. Fr. Nov.—April.
Var. β G. RIVULARE *Bridel.*
Frequent in wet places, Holywood Hill, Carr's Glen, etc. ; S.A.S. Magheralin, and other places ; C.H.W. Between Rostrevor and Hilltown ; H.W.L.

3. G. maritima *Turner.*
Rocks by the shore about high-water mark—common and abundant all around the coast. Fr. Nov.—March. First discovered by Dr. Robt. Scott.

Sect. II.—Eugrimmia

4. G. funalis *Schwaeg.* (G. SPIRALIS *H. & T.*).
On rocks—rare, though locally abundant. Barren.
Down—Granite rocks on Slieve Donard; S.A.S. Slievenabrock, and Thomas Mountain; C.H.W.
Antrim—Slemish (*Templeton*); *Flor. Hib.* (still abundant there). East side of Slemish Mountain; *Musc. Brit.* Agnew's Hill; D.M. On trap boulders at Sallagh Braes; S.A.S.

5. G. microcarpa *Gmel.* (RAC. SUDETICUM *Funck*).
Rocks and stones on the mountains—rare, though occasionally abundant. Fr. April and May. Ranges from 1000, to 2700 feet.
Down—On Slieve Donard, and on the summit of Slieve Croob; S.A.S. Thomas Mountain, and Shanslieve; H.W.L.
Antrim—Abundant on the top of Slemish; *Templeton*, 1809. Sparingly on Slemish; D.M., 1837.
Derry—On Dart Mountain at 2000 feet; D.M.

6. G. pulvinata (*Linn.*) *Smith.*
Walls, stones, and rocks—very common. Fr. Oct.—Feb.
A form with shorter capsule, and short blunt lid, occurs (*var. obtusa ?*), but it merges into the type.

7. G. trichophylla *Greville.*
Rocks, and stones—very rare. Fr. Nov.—Jan. (*B.M.F.*).
Down—Plentiful, but barren on a large glacial boulder between Giants' Ring and the Lagan; S.A.S. In fruit on stone fence near Bryansford, and on rocks by the coast at Annalong, also (*forma robusta*) on wall at Kinnehalla; C.H.W. On the grit rocks of Whitewater Glen, and on Slieve Donard; H.W.L.

8. G. robusta *Fergusson.*
Rocks and stones on mountains—very rare, and without fruit.
Down—Top of stone fence on Spalga Mountain, rocks by stream on Slievenamady, and above the ice-house on Slieve Donard; H.W.L.
Antrim—Rocks at Fairhead, 1862; *Moore, Proc. R.I.A.*, 1872.

First found in Ireland by Dr. Moore, at Fairhead, in 1862.

9. G. decipiens *Schultz* (G. SCHULTZII *Brid.*).
Mountainous rocks—rare. Fr. Nov.—Feb.
Down—Rocks at Bloodybridge, and plentiful on granite of Slieve Donard;
    S.A.S. On slate, and granite at Annalong, Tollymore Park,
    and Tievedocharrah; H.W.L.
Antrim—On the top of Fairhead; *Moore, op. cit.*

10. G. elatior *Schimper.*
On the Silurian grit rocks—extremely rare.
Down—Fruiting on stones in Ballagh Park, Slieve Donard; H.W.L.,
    Oct., 1884.
The only Irish station yet known.

11. G. Donii *Smith.*
Rocks, and stones on the mountains—very rare. Fr. late in autumn,
and again early in spring.
Down—Slievenamady, and Slieve Donard; H.W.L. On the walls of
    Tollymore Park; C.H.W.
Antrim—Very sparingly on stone fence at the summit of Sallagh Braes,
    in good fruit 27th Oct., 1873; S.A.S.
First found in Ireland by S.A.S. as above.

12. G. leucophæa *Greville.*
Basaltic rocks—very rare. Fr. April (*Hobkirk*).
Antrim—On basalt at Fairhead; *Musc. Brit.* Trap rocks near the
    Giants' Causeway, and on similar rocks in Rathlin, 1837;
    Moore, Proc. R.I.A., 1872.

Sect. III.—Dryptodon.

13. G. elliptica *Arnott.* (RAC. ELLIPTICUM *Turner*).
Cliffs and ledges of rock—rare. Fr. winter and spring (*Berk.*).
Down—Mourne Mountains; *Templeton, Herb. Belf. Mus.* Mountains
    above Newcastle; S.A.S. Slieve Donard, and by the Spink-
    wee River; H.W.L. Slievenabrock; C.H.W.
Antrim—Slemish, and top of Fairhead; *Templeton,* 1815 (still plentiful
    in both places).
Derry—Only seen on Clontygeragh; D.M. Basaltic cliffs of Ben-
    bradagh; S.A.S.

14. G. patens (*Dicks.*) *Br. et Schimp.*
Mountainous rocks—rare, ascends to 2700 feet. Fr. April (*Moore*).

Antrim—Slemish; D.M., 1836, *Phytologist* ser. II., vol. II. In large barren tufts on Slemish; S.A.S.

Derry—At 1100 feet on Clontygeragh; D.M.

There is a note by Templeton "Found, mixed with *Hyp. filicinum*, on the left branch of Colin Glen river, beside the second fall, 1803." This is doubtfully correct, but if right it was the first discovery of this species in Ireland. Templeton says " Dickson's plant appears more leafy, leaves when highly magnified are slightly serrate, a brownish red medrib, and finely reticulated." Probably the Colin Glen plant was merely a form of the next species.

Sect. IV.—Trichostomum.

15. G. acicularis (*Linn.*) *C. Muell.*

Rocky river banks, and stones in wet places—common. Fr. Nov.—April.

Var. β DENTICULATA *Wilson.*

Of frequent occurrence.

Cave Hill; *Templeton.* Colin Glen, and Carr's Glen; *J. H. Davies.* By streams in Glendivis and Windy Gap; S.A.S.

16. G. aquatica (*Brid.*) *C. Muell.* (RAC. PROTENSUM *Braun*).

Wet rocky places—rare. Fr. March—May.

Down—Mourne Mountains; *T. Drummond.* In several places between Donard and Tollymore; S.A.S. Slieve Bignian, Tievedocharrah, Lough Shanagh, Slievemeel Beg, and Slieve Donard; H.W.L. In Rostrevor Wood, and rocks near Hilltown; C.H.W.

17. G. heterosticha *C. Muell.* (RACOMITRIUM *Hedw.*).

Stony heaths—very common on the mountains, ascending to 2000 feet. Fr. Nov.—April.

Var. ALOPECURUM *Br. et Schimp.*

Down—On Thomas Mountain, and on the lesser cairn of Slieve Donard; H.W.L.

Antrim—On rocks of Slemish; S.A.S.

Var. GRACILESCENS *Br. et Schimp.*

Down—Mourne Mountains; *T. Drummond.* Slievenabrock, and Slievenamady; H.W.L.

18. G. fascicularis (*Schrad.*) *C. Muell.* (RACOMITRIUM *Brid.*).

On rocks and stones in hilly and heathy places—common. Fr. Nov.—March.

GRIMMIACEÆ.

19. G. hypnoides (*Linn.*) *Lindb.* (RAC. LANUGINOSUM *Brid.*).
Stony mountain heaths—common. Fr. Nov.—March. Ranges from 500 to 2500 feet.

20. G. canescens (*Dicks.*) *Lindb.* (RACOMITRIUM *Hedw.*).
Sandy waste ground, and gravelly heaths—frequent. Fr. Nov.—March. Ranges from near sea level to 2000 feet.

Down—Sandhills at Newcastle, and stony heath near Newtownards; S.A.S. Common in the Mourne Mountains; H.W.L.
Antrim—Malone, Divis, and Cave Hill; *Templeton*, 1806. Belfast hills, Carrickfergus commons, and by the river at the top of Woodburn Glen; S.A.S. Sallagh Braes; C.H.W.
Derry—Common in Ballynascreen; D.M. Gravelly ground at Drumcormick wood near the base of Slievegallion; S.A.S.

GLYPHOMITRIUM *Bridel.*

1. G. polyphyllum (*Dicks.*) *Mitt.* (PTYCHOMITRIUM *B. & S.*).
Rocks and stones, sometimes on bushes—very common. Fr. Nov.—Feb. From sea level to 1500 feet.

2. G. Daviesii (*Dicks.*) *Bridel.*
Basalt, grit, and slate, also micaceous, and granitic rocks—rare, but locally abundant on the basalt. Fr. April—June.

Down—Mourne Mountains; *T. Drummond, op. cit.* Sparingly on granite rocks of Slieve Donard; S.A.S. Frequent on slate rocks at east face of Slieve Donard; H.W.L. One tuft on trap dyke above the waterfall on Donard; C.H.W.
Antrim—On the columns of Fairhead (*R. Brown*); *Templeton.* Common on the columns of the Giants' Causeway; *Musc. Brit.* On basalt at Giants' Causeway, Fairhead, and Rathlin; *Moore, op. cit.* Abundant on the Grayman's path on Fairhead, and plentiful on the basaltic rocks at summit of Ballygally Head, also sparingly on Knocklayd, and Rathlin Island; S.A.S.
Derry—Not common, but occurs on the Mica Schist of Dart Mountain, and on Trap rocks at Benevenagh; D.M. Basalt rocks of Mullaghmore, at 1300 feet; S.A.S.

First found in Ireland by Robt. Brown.

SERIES II. ORDER XV. ORTHOTRICHACEÆ.

ANŒCTANGIUM *Hedwig.*

1. A. Mougeottii *Lindb.* (ZYGODON *B. & S.*).
Rocks in damp shady places—frequent, but usually barren.
Down—Slieve Donard, Slieve Bignian, Pigeonrock Mountain, and Cove Mountain; H.W.L. Rostrevor Mountain, and Tollymore Park; C.H.W. Mountains above Newcastle; S.A.S.
Antrim—Trap rocks at upper end of Colin Glen; *J. H. Davies, Phytologist,* 1859. At the head of Glenballyeamon, a single stem in fruit in June 1863; D.M. Sallagh Braes; C.H.W. Plentiful on rocks below MacArt's fort on Cave Hill; S.A.S.
Derry—Basaltic cliffs of Benevenagh; S.A.S.
First found in Ireland by Isaac Carroll.

PLEUROZYGODON *Lindb.*

1. P. æstivus (*Hed.*) *Lindb.* (AN. COMPACTUM *Schw.*).
Damp rocks—rare and barren.
Down—Steep wet rocks on Slieve Commedagh, and Slievenamady; H.W.L.
Antrim—Carnlough, and frequent in northern glens; D.M.

ZYGODON *Hook. & Tayl.*

1. Z. Stirtoni *Schimper.*
Damp shady places on rocks and walls—rare, and usually only in small quantity. Fr. Feb. and March.
Down—Barren in crevices of rocks on the shore at Portavo, and fruiting on old wall by the roadside opposite Portavo demesne; S.A.S. Annalong, and Bloodybridge; H.W.L.
Antrim—Sparingly on basaltic rocks by the waterfall in Gleno, and rather more plentiful on the inner side of the southwest wall of the bridge on the Larne road at Kilroot, also on Rathlin Island; S.A.S.
Derry—On the trunk of a tree brought from Co. Derry to Glenmore; *J. H. Davies.*

2. Z. viridissimus (*Dicks.*) *Hook. & Tayl.*
Trunks of trees—frequent, and often abundant. Fr. Feb. and March, but more commonly sterile.

Down—Abundant on trees at Rostrevor, and on the ground, forming a carpet on a path in the wood ; S.A.S. Drumcro, and by the waterworks at Rostrevor ; C.H.W. Frequent ; H.W.L.
Antrim—On blocks of limestone near Grogan's Glen ; *Templeton.* At Cherryvalley, and the waterfoot near Crumlin, also Shane's Castle, Shawsbridge, Cave Hill, Kilroot, and roadside north of Magheramorne ; S.A.S.

There is no specimen of Templeton's plant, but it is probable that it should be placed under the preceding species.

3. Z. conoideus *(Dicks.) Hook. & Tayl.*
On trees—very rare, and not found recently. Fr. May *(Berk.).*
Down—Belvoir Park ; *T. Drummond, op. cit.*
Antrim—On pear and apple trees in Cranmore orchard ; *Templeton,* 1800 *(spec. in Belfast Mus. Herb.).* Rare, only observed on trees in Glenarm Deerpark ; D.M.

First found in Ireland by Templeton.

ORTHOTRICHUM *Hedwig.*

Sect. I.—Gymnoporus.

1. O. rupestre *Schleich.*
Rocks—very rare. Fr. early summer.
Antrim—On rocks at Doneygregor Head near Ballycastle ; *Colby's Surv. Londonderry.* Rocks near Ballycastle, and beside the larger lake on Fairhead ; D.M. On basaltic rocks near the Giants' Causeway ; *Moore, Proc. R.I.A.,* 1872.

Var. β O. STURMII *Hopp. et Hornsch.*
Rocks, and stones—very rare. Fr. April—June.
Antrim—Fairhead ; *Moore, op. cit.* Abundant on stones in a boggy moor on Knocklayd, and on a large boulder by Lough Neagh at Sandy Bay south of Glenavy ; S.A.S.

2. O. affine *Schrader.*
Trunks of trees and bushes, occasionally on stones—very common. Fr. April and June.

3. O. Sprucei *Mont.*
On trees and bushes overhanging water—extremely rare. Fr. April—June.
Down—Plentiful on trees at the bye-wash of the canal immediately above Drumbridge ; *J. H. Davies,* 1878.
The only Irish locality.

4. O. striatum (*Linn.*) *Hedw.* (O. LEIOCARPUM *B. & S.*).
On trees—frequent, but very rare on rocks. Fr. Feb.—June.
Down—On trees, and stones; *Templeton.* On trees in Belvoir Park;
S.A.S. Loughbrickland, and Hillsborough demesne; H.W.L.
Beech trees in Tollymore Park, and Finnebrogue; C.H.W.
Antrim—Glenarm Park, and near Ballycastle; D.M. On trees at Crumlin waterfoot, Antrim, Dunadry, Cave Hill, Redhall, and Sallagh Braes, also found in abundance on basaltic rocks at Knocklayd, and on Rathlin; S.A.S.
Derry—Trunks of old ash trees near Dungiven; D.M. On beech trees at Drumcormick above Moneymore near the foot of Slievegallion; S.A.S.

5. O. Lyellii *Hook. & Tayl.*
On trunks of old trees—rare, and usually sterile.
Down—Trees near Banbridge, also Rademon demesne, Crawfordsburn, and Belvoir Park; S.A.S. Loughbrickland, Waringstown, and Finnebrogue; H.W.L. Gillhall, Drumcro, and Tollymore Park; C.H.W.
Antrim—By the Lagan above Shawsbridge, also Crumlin, Kilroot, and Glenarm Park; S.A.S.

Sect. II.—Calyptoporus.

6. O. diaphanum *Schrader.*
On trees, stones, and walls—frequent. Fr. Jany.—April.
Down—On trees at Drumbo, and Newtownbreda, and on stones at Moira S.A.S. Annalong; H.W.L. By the Lagan at Drumcro, and on thatched roof, and walls at Warrenpoint; C.H.W.
Antrim—Trees at Leslie Hill near Ballymoney (*R. Brown*); *Templeton*, and stones at Drumbridge, and Derriaghy; *Templeton*, 1803. Stones by the shore of Lough Neagh, and at Whitehouse, and Rathlin Island, also on trees at Colin Glen, Cave Hill, Kilroot, and Blackhead; S.A.S.

7. O. cupulatum *Hoffman.*
On stones in streams—frequent, but only in the northern half of the district, rarely found on rocks. Fr. Jany.—May.
Our plant is mainly the var. *nudum* of Dickson.
Antrim—Rocks at Fairhead; D.M. Stones in Crumlin River, and at Trench beyond Andersonstown, Glendivis, Kilroot, Gleno, and Glynn in profusion, also on limestone rocks at Black Mountain, north Woodburn, and Rathlin Island, and on basalt cliffs at the Gobbins; S.A.S.

Derry—On rocks, and trunks of trees, and plentiful on the headstones in Dungiven Churchyard; D.M.

8. O. saxatile *Bridel.*
Dry rocks, and on stones and walls—frequent. Fr. Feb.—May. Ranges from near sea level to over 1000 feet.
Down—Warrenpoint; C.H.W.
Antrim—Frequent on white limestone rocks in the county; D.M. On basalt and limestone rocks in the Belfast hills, also Woodburn, Carrickfergus commons, and Larne; S.A.S. Megabbery, and Islandmagee; H.W.L.
Derry—Bennedy Glen; D.M. Benbradagh, and frequent in the Sperrin Mountains; S.A.S.

9. O. stramineum *Hornsch.*
Trees—very rare. Fr. June (C.H.W.).
Down—On a beech tree in Tollymore Park; C.H.W.

10. O. rivulare *Smith.*
Rocks, and trees beside streams—rare. Fr. May—Aug.
Down—Trees by the canal at Drumbridge; *J. H. Davies.* Plentiful on trees in Gillhall demesne, and by the Annacloy river at Rademon; S.A.S.
Antrim—On a stone in a rivulet near Derriaghy (*Templeton*); *Musc. Hib.*, 1804. Abundant on stones and bushes by the Glenavy River near the waterfoot; S.A.S. In the Braid near Broughshane, and near a waterfall at Drumnasole; H.W.L.

11. O. tenellum *Bruch.*
Trees—very rare.
Down—Sparingly on trees in Gillhall demesne; S.A.S. In fruit 29th June, 1884.
First found in Ireland by Miss Hutchins.

12. O. pulchellum *Smith.*
On trunks and branches of trees in shady places—frequent, widely diffused but rarely abundant. Fr. March—May.
Down—Trees in Gillhall demesne, and on guelder rose in Dundonald Glen; S.A.S. Hillsborough demesne; H.W.L. Drumcro; C.H.W.
Antrim—On trees at Glenarm, and other places; D.M. Colin Glen; *J. H. Davies, Phytologist,* 1859. Near Lisburn; *J. Creeth.* Crumlin waterfoot, several places about Antrim, also at Cave Hill, Glendivis, Kilroot, Castle Chichester, and on a boulder at Gleno; S.A.S.

Derry—Side of the Faughan in Ogilby's demesne ; D.M. Beech trees growing on rath at Drumcormick at foot of Slievegallion ; S.A.S.

WEISSIA *Ehrhart.* (ULOTA *Bridel*).

[W. DRUMMONDII *Hooker.*
As the tuft, so named, found in Colin Glen in 1869 by S.A.S. is too far advanced to be entirely satisfactory, and as there has been no confirmation of the record, it is better to leave this species out of our lists until re-found, and in a more satisfactory condition.]

*1. **W. Bruchii** (*Hornsch.*) *Lindb.*
Usually on trees, rarely on stones—frequent. Fr. June—Sept.
Down—On oak trees by the Quoile below Downpatrick, Rademon demesne (very fine), and in Crawfordsburn ; S.A.S. Rostrevor wood, and Gillhall demesne ; C.H.W.
Antrim—Plentiful in Colin Glen, and very fine on bushes by stream at Black Mountain ; S.A.S.
Derry—Woods at Castledawson, and Lignapeiste ; S.A.S.

2. **W. ulophylla** (*Ehr.*) *Lindb.* (ULOTA CRISPA *Hedw.*).
On trees, and occasionally, but very rarely on stones—frequent. Fr. June—Sept.
Down—Wood by the Quoile below Downpatrick ; S.A.S.
Derry—Carndaisy Glen ; S.A.S.

Var. β W. CRISPULA (*Bruch*) *Lindb.*
On trees—rare ? Fr. July—Sept.
Down—Oak trees by the Quoile below Downpatrick, and on apple trees in Belvoir Park ; S.A.S.
Antrim—Colin Glen ; S.A.S.
Derry—Lignapeiste Glen ; S.A.S.

3. **W. phyllantha** (*Brid.*) *Lindb.*
On trees, and on stones in bushy places—very common, but always sterile.

4. **W. vittata** (*Mitt.*) *Braith.* (ULOTA CALVESCENS *Wils.*).
On trees—very rare.

* As regards this and the succeeding species, with its variety many other localities could have been enumerated, but no stations have been given except such as have yielded mature and satisfactory specimens. These have been confirmed by Mr. Holt, and the records though meagre are reliable. *Ulota intermedia* is scarcely distinguishable even as a variety.

Antrim—Glenshesk above Ballycastle; S.A.S. In good fruit in July, 1882.
The label which was attached to the specimen having been lost there is not entire certainty as to the exact locality. There is, however, but little doubt that the station given above is correct.

SERIES II. ORDER XVI. **SPLACHNACEÆ.**

SPLACHNUM *Linn.*

1. S. ampullaceum *Linn.*
On the droppings of cattle in wet mountain pastures, and heaths—rare. Fr. June and July.
Down—Turf and peat bogs, always on old cowdung; *Templeton*. Slieve Bernagh, Deers' Meadow, and Rocky Mountain near Hilltown; H.W.L.
Antrim—Rasharkin, and Glenravel; D.M. Bog on top of Knockagh, and wet bog on Colinward Mountain; S.A.S.

2. S. pedunculatum (*Huds.*) *Lindb.*
Var. SPHÆRICUM *Hedwig.*
On cowdung in boggy ground—frequent. Fr. June and July.
Down—Donaghadee bog; *Templeton*, 1797. Abundant on Slieve Croob; S.A.S. Tievedocharragh, Cratlieve, and Pierce's Castle; H.W.L. Bog between Hilltown and Rostrevor, also by Kilkeel River, and on Eagle Mountain; C.H.W.
Antrim—Top of Black Mountain, and on Cave Hill; *Templeton*, 1806 (still there). Colinward west of Cave Hill, and wet heath above Carnlough; S.A.S.
Derry—Bogs, and mountains in Derry; *Templeton*. Craignashoke; D.M.

TETRAPLODON *Br. et Schimp.*

1. T. bryoides (*Zoeg.*) *Lindb.* (T. MNIOIDES *Hedw.*).
On cattle droppings in the higher mountains—rare. Fr. June and July. Ranges from 800 to 2700 feet.
Down—On mountains near Belfast (*Templeton*); *Flor. Hib.* Close to the great cairn on summit of Slieve Donard; *B.N.F.C.*, 1860. Wet places near summits of Donard and Bignian; S.A.S.

By the lesser cairn on Donard; H.W.L. At 2000 feet on Slieve Donard; *J. J. Andrew*. Rocky Mountain near Hilltown, and Crocknafeola plantation at 800 feet; C.H.W.

TAYLORIA *Hooker*.

1. T. tenuis (*Dicks.*) *Schimper.*
On old cowdung in mountain bogs—extremely rare. Fr. July.
Derry—Wet peaty and heathy pastures on the top of Benbradagh Mountain; S.A.S. Commencing to fruit 13th June, 1868, and, again, with ripe capsules on 8th July, 1884.
The only station in Ireland. The elevation is about 1300 to 1400 feet, which is the lowest known for this plant in Britain.

SERIES II. ORDER XVII. **FUNARIACEÆ**.

PHYSCOMITRELLA *Schimper.*

1. P. patens *Schimp.* (PHASCUM PATENS *Hedw.*).
Damp banks—rare. Fr. Autumn (*Berk.*).
Antrim—Found on the bottom and sides of a dry drain in the Bog meadows, and in the lower drain of Allen's meadow between first and second lock of the Lagan canal; *Templeton*, 28th Aug., 1800. Bog meadows near Belfast; *T. Drummond, Musci Scotici*, vol. III. Banks near Belfast (*Temp.*); *Flor. Hib.* Moist banks near Belfast; *Moore, Proc. R.I.A.*, 1872.

PHYSCOMITRIUM *Bridel.*

1. P. pyriforme (*Linn.*) *Br. et Schimp.*
Damp sandy, or gravelly banks—not common. Fr. April—June.
Down—Dromore, Portavo, Ballyalloley, and Ormeau Park; S.A.S. Loughbrickland; H.W.L. Warrenpoint, and abundant at Drumcro; C.H.W. Cregagh Glen; *J. J. Andrew*.
Antrim—Ditch banks at Cranmore; *Templeton*. Glenarriffe, and not uncommon in furrows of wet sandy fields; D.M. Shore of the lough north of Crumlin waterfoot, also near Lisburn, and on the Black Mountain, the Plains near Belfast, Kilroot, and Loughmourne; S.A.S.

FUNARIACEÆ.

FUNARIA *Schreber.*

Sect. I.—Entosthodon.

1. F. obtusa (*Dicks.*) *Lindb.* (E. ERICETORUM *De Not.*).
Wet rocky places by streams—rare. Fr. March—May. Ranges from 600 to near 2000 feet.
Down—Rocky banks of streams on Slieve Donard at 1500 to 1800 feet; S.A.S. Tollymore Park; H.W.L. Moygannon Glen; C.H.W.

2. F. fascicularis (*Dicks.*) *Schimp.*
Pastures, and banks—rare. Fr. April and May.
Down—Ballymaghan, and Sydenham; S.A.S. Warrenpoint; C.H.W.
Antrim—Giants' Causeway, the Glens, and near Belfast; *Moore, op. cit.* Grazing fields at Kilroot, and Blackhead; S.A.S.
Derry—Found in crevices of rocks at the side of the Faughan two miles below Cumber; *Templeton.*

3. F. Templetoni (*Hook.*) *Smith.*
By mountain streams, and in wet rocky places—not rare. Fr. May—Aug.
Down—Moist hollows of rocks at Bangor Bay; *Templeton.* Mourne Mountains near Newcastle; *T. Drummond, op. cit.* Plentiful, and fine, by the Causeway Water, and by streams on Slieve Donard; S.A.S. Mourne Mountains at Rostrevor, and by the Yellow-water River; C.H.W. Pigeonrock Mountain, and frequent on the Mourne range; H.W.L.
Antrim—Abundant by the sides of rivers in northern glens; D.M. Sparingly on rocky margins of streams in Rathlin; S.A.S. Sallagh Braes; C.H.W.

Sect. II.—Eufunaria.

4. F. calcarea *Wahlenberg.*
On limestone—extremely rare.
Antrim—Limestone soil in the Belfast Deerpark; D.M.

5. F. hygrometrica (*Linn.*) *Sibth.*
Heaths, walls, waste ground, and rubbish heaps—common. Fr. Feb.—June.
This cosmopolitan moss is quite as abundant here as elsewhere.

AMBLYODON *Pal. de Beauv.*

1. A. dealbatus (*Dicks.*) *Beauv.*

Damp banks—extremely rare. Fr. Summer (*Berk.*).
Down—Hollow banks at the south end of the fir grove near Holywood, April, 1803, and southeast bank of the canal about 200 yards above Blaris bridge, June, 1803; *Templeton.*
Antrim—On a flow bog in the parish of Rasharkin, 1837; *Moore, op. cit.*
First found in Ireland by Robert Brown, in Co. Donegal.

SERIES II. ORDER XVIII. **BRYACEÆ.**

LEPTOBRYUM *Schimper.*

1. L. pyriforme (*Linn.*) *Schimp.*
Shady banks—extremely rare. Fr. early Summer (*Berk.*).
Down—Solitude near Banbridge; H.W.L. Near Newtownbreda; *J. J. Andrew.*

POHLIA *Schreber* (WEBERA *Hedwig*).

1. P. elongata (*Dicks.*) *Hedw.*
Mountainous rocks—extremely rare. Fr. Summer (*Berk.*).
Down—Crevices of rocks at the top of Slieve Commedagh; C.H.W. At the lesser cairn on Slieve Donard, and on slate rocks of Slieve Commedagh; H.W.L.
Derry—Found in crevices of the Eagle Rock by the side of the Cookstown road at the foot of Sawel; *Templeton*, 1809.

2. P. cruda (*Linn.*) *Lindb.*
Damp rocks, and boggy places—very rare. Fr. Summer (*Berk.*).
Down—Lambeg moss; *Templeton.* Slieve Donard; H.W.L.
Antrim—Carnlough Glen near the waterfall; D.M. Sallagh Braes; C.H.W.

3. P. nutans *Schreber* (WEBERA *Hedw.*).
Bogs, and peaty heaths—frequent. Fr. April—June. Ranges from sea level to over 2000 feet.
Down—Laghey bog, and on hillocks in the peat bogs; *Templeton.* Slieve Donard, Slieve Croob, Ballynahinch, Dundonald Glen, and abundant on damp banks in Cotton moss; S.A.S. Warrenpoint; C.H.W. Frequent in Mourne Mountains; H.W.L.

Antrim—Slemish; D.M. Lisburn, Cave Hill, and southward along the whole range, also Woodburn, and Carrickfergus Commons; S.A.S.
Derry—Benevenagh, Benbradagh, and White Mountain; S.A.S.

4. P. carnea (*Linn.*) *Schreb.*
Wet banks, and sides of drains—frequent. Fr. March—May.
Down—Ballymaghan, and Cregagh Glen; S.A.S. Magheralin; H.W.L. and C.H.W.
Antrim—At the base of the lower wall of the Belfast Deerpark, and by the side of a watercourse where it crosses the road to Grogan's Glen; *Templeton.* Banks of a small rivulet at Ballintoy; D.M. Lisburn, Colin Glen, Glendivis, Ballysillan, and on stones by the stream at Woodburn; S.A.S.

5. P. annotina *Linn.*
Wet rocky places—very rare. Fr. early Summer (*Berk.*).
Down—Sparingly on Slieve Donard; H.W.L.
Antrim—In fruit in Colin Glen in March, 1806, and fruiting plentifully in Carrickfergus Glen in April, 1809; *Templeton.*

6. P. albicans (*Wahl.*) *Lindb.*
Wet rocks, and by streams—frequent. Fr. April—June.
Down—Roadside near Banbridge, and wet quarry at Ballymaghan; S.A.S. Tollymore Park, and Slieve Donard; H.W.L. Deers' Meadow, Yellow-water River, and in Cregagh Glen; C.H.W.
Antrim—The Glens near Cushendall; *Moore, op. cit.* Prospect Hill near Lisburn; J. *Creeth.* Rocks at the weir in Crow Glen, and bank of stream on Black Mountain; S.A.S. Colin Glen, and Sallagh Braes; C.H.W.
Derry—Wet rocky places on White Mountain; S.A.S.

PLAGIOBRYUM *Lindberg.*

1. P. Zierii (*Dicks.*) *Lindb.*
Damp rocks—rare. Fr. Oct. and Nov. (*Hobk.*).
Antrim—Near Cushendall; *Moore, Proc. R.I.A.*, 1872.
Derry—Clontygeragh; *Moore, op. cit.*

BRYUM *Dillenius.*

1. B. concinnatum *Spruce.*

Damp shady rocks—extremely rare.
Antrim—Mixed with *B. filiforme* on rocky banks at foot of the cliffs of Sallagh Braes; H.W.L. and C.H.W.

2. B. filiforme *Dickson.* (B. JULACEUM *Schrad.*).
Wet rocks—very rare. Fr. Autumn (*Berk.*).
Down—Tollymore Park; H.W.L. By a stream on Slieve Donard; H.W.L. and C.H.W.
Antrim—Abundant in Glendun, in fruit July, 1836; D.M. Sallagh Braes; *Lett and Waddell.* Rocky river bank on the moor above Colin Glen; S.A.S.

3. B. inclinatum (*Swartz*) *Bland.*
Damp banks and walls—rare. Fr. June—Sept.
Down—Old walls at Portavo (seems correct, but not absolutely certain), and wet sandy bank in the unfinished Victoria Park; S.A.S. Peaty ground on Slieve Donard; H.W.L.

4. B. pendulum *Hornsch.*
Walls, and waste places—perhaps not rare. Fr. May and June.
Antrim—Walls near Lisburn, and abundant on old brick and mortar rubbish at Woodburn; S.A.S.
Derry—Walls at Toome; S.A.S.

5. B. intermedium *Web. et Mohr.*
Damp banks—rare? Fr. March—Aug.
Down—Gravel bank at Ballymaghan, and brickfield at Ballynafeigh; S.A.S.
Antrim—Gravelly shores of Lough Neagh; S.A.S.

6. B. bimum *Schreber.*
Wet boggy places—rare. Fr. Summer (*Berk.*).
Down—Slieve Donard, and old bog drains in Deers' Meadow; H.W.L.
Antrim—In the black bog at Cranmore (obsolete), and bog between Divis and Black mountain; *Templeton.* Fairhead, and Wolfhill; S.A.S. By the Lagan near Belfast; *J. J. Andrew.*

7. B. cæspiticium *Linn.*
Walls, rocks, quarry heaps, and waste places—common. Fr. April—June.

8. B. argenteum *Linn.*
Walls, roofs, waysides, and rubbish heaps—very common. Fr. Oct.—Feb.

Very variable. The var. *lanatus* occurs on crumbling basalt at Blackhead, and the var. *majus* on walls at Antrim, and damp rocky cliffs of Sallagh Braes.

9. B. bicolor *Dicks.* (B. ATROPURPUREUM *W. & M.*).
Waste ground, walls, and quarry debris—frequent. Fr. April—June.
Down—Walls at Dundonald; *T. Drummond, op. cit.* Dundrum, Millisle, and quarry heaps at Newtownards; S.A.S. Near Magheralin; C.H.W.
Antrim—Quarry waste at Kilcoreg, and walls at Lisburn, also on Cave Hill, several places about Belfast, and at Fairhead; S.A.S.

10. B. alpinum *Hudson.*
Rocks, and heaths—frequent. Usually barren, but found in fruit, by Moore, in June, 1836.
Down—Plentiful at various elevations on Slieve Donard; S.A.S. Rostrevor Mountain, and Tollymore Park; C.H.W.
Antrim—Fairhead, and Carrickfergus commons (fruit); D.M. Summit of Divis, also on Agnew's Hill, Slemish, and Rathlin; S.A.S.
Derry—Near Craignashoke; *Templeton.* Slievegallion; S.A.S.

11. B. murale *Wilson.*
Old walls, and limestone rocks—rare. Fr. May and June (*Hobk.*).
Down—Limestone rocks near Moira; H.W.L., 1882.

12. B. pallens *Swartz.*
Wet rocky, or gravelly places—frequent. Fr. June and July.
Down—Deers' Meadow; H.W.L. Kirkcassock; C.H.W.
Antrim—Banks of stream above Colin Glen; *J. H. Davies, Phytologist,* 1859. Colin Glen, Wolfhill, Woodburn, and bog at Duneane; S.A.S.
Derry—White Mountain, also banks of stream on Meenard Mountain, and by Sluggada Burn near Dart Mountain; S.A.S.

13. B. ventricosum *Dicks.* (B. PSEUDO-TRIQUETRUM *Hedw.*).
Wet rocks, and by mountain streams—rather rare. Fr. May and June.
Down—Rocks on Slieve Donard, and near Dundrum, also on slate rocks in Newtownards Glen; S.A.S. Annalong, and Moygannon Glen; H.W.L. Knockbarragh, and Tollymore Park; C.H.W.
Antrim—Bogs in Co. Antrim (*Orr*); *Phytologist,* 1857.

14. B. capillare *Linn.*
Walls, rocks, banks, heaths, and waste ground—very common. Fr. March—June.

BARTRAMIACEÆ. 237

15. B. roseum *Schreber.*
On rocks—rare, barren.
Antrim—Dry rocks in the Little Deerpark at Glenarm ; D.M.

SERIES II. ORDER XIX. **BARTRAMIACEÆ.**

BARTRAMIA *Hedwig.*

Sect. I.—Oreadella.

1. B. Œderi (*Gunn.*) *Swartz.*
Shady rocks—very rare. Fr. Summer (*Berk.*).
Antrim—Colin Glen, and Glenarm Deerpark ; *Moore, op. cit.* (not found recently).

Sect. II.—Eubartramia.

2. B. pomiformis (*Linn.*) *Hedw.*
Rocks, and shady banks—not rare. Fr. March—May. Ascends to nearly 2000 feet on the mountains.
Down—Slieve Donard, Drumbo Glen, and woody bank near Ballylesson ; S.A.S. Knockbarragh; C.H.W. Loughbrickland, and frequent on Bignian, and the northern slopes of the Mourne Mountains ; H.W.L.
Antrim—Rocks at the upper end of Colin Glen, and on a very dry sandy bank beside the upper road going to Stranmillis; *Templeton.* Grassy slopes of Black Mountain, rocks of Sallagh Braes, and very luxuriant on the basaltic cliffs of Slemish ; S.A.S.
Derry—Abundant near Dungiven ; D.M.

Sect. III.—Vaginella.

3. B. ithyphylla *Bridel.*
Rocks, and dry shady banks—rare. Fr. March—May.
Down—Crevices of rocks on Slieve Donard, and on mountain above Bryansford ; S.A.S. Leitrim Hill near Hilltown, and on Slievenabrock; H.W.L. Rostrevor woods ; C.H.W.
Antrim—Rocks above Carrickfergus ; D.M. Basaltic rocks on Black Mountain, and basalt cliffs at Knockagh, and in North Woodburn ; S.A.S. Cave Hill; H.W.L.

4. B. norvegica *Gunner.* (B. HALLERIANA *Hedw.*).
Shady rocks—very rare. Fr. Summer (*Berk.*).
Antrim—At Colin Glen near the top; D.M. Colin Glen; *Flor. Hib.* Glenarm, and Carnlough; *W. Thompson* (not recently seen in any of above stations).

Sect. IV.—Philonotis.

5. B. fontana (*Linn.*) *Swartz.*
By springs, and in wet rocky places—common. Fr. June—Aug.

6. B. calcarea *Br. et Schimp.*
Swampy ground in the hills—not common. Fr. July and Aug.
Down—Old quarry at Kilwarlin, with stems nine inches in length; H.W.L. and C.H.W.
Antrim—Upper end of Colin Glen, and damp places on Belfast hills; S.A.S.

BREUTELIA *Schimper.*

1. B. chrysocoma (*Dicks.*) *Schp.* (BART. ARCUATA *Hedw.*).
Heaths, and stony places—common; ascending to 2000 feet. Fruit very rare.
"Abundant in fruit by the river which flows through Glenchain, Co. Derry"; D.M. Found also in fruit near Carrickfergus; D.M. (*spec. in herb.*). A single stem in fruit, in November, near Rostrevor; C.H.W.

SERIES II. ORDER XX. **MNIACEÆ.**

SPHÆROCEPHALUS *Necker.*

1. S. palustris (*Linn.*) *Lindb.* (AULACOMNION *Schag.*).
Wet heathy, and peaty ground—frequent, and locally abundant. Fr. May and June. Ranges from near sea level to 1600 feet.
Down—Slieve Croob, Moneyreagh, and Conlig; S.A.S. Slieve Commedagh, and Deers' Meadow; H.W.L. Rostrevor wood, Kilbroney bog, and on Slievenabrock; C.H.W.
Antrim—In the bogs; *Templeton.* Bog at Duneane, marshy heaths on Black Mountain and Cave Hill, also plentiful at Kings' moss, Carrickfergus commons, Sallagh Braes, Slemish, and Rathlin; S.A.S.

[ORTHOPYXIS ANDROGYNUM (*Linn.*) (Aulacomnion *Schwaeg.*).
"Found in the Manyburn, at Purdysburn, by Mr. J. Drummond";
Templeton.
Not in Moore's list, and very doubtful. The only certain Irish station is Montiaghs, near Lurgan, where Rev. Mr. Lett found it on weathered stumps of bog-wood. Mr. Lett's locality is in Co. Armagh close to the border of Down, and the plant may be hoped for in that county.]

MNIUM *Linn.*

1. M. hornum *Linn.*
At the roots of trees and shrubs in woods, and hedges—common. Fr. March—May.]

2. M. stellare *Hedw.*
Wet, rocky banks of streams—very rare, and without fruit.
Antrim—Sparingly by the stream below waterfall at top of Colin Glen; S.A.S., April, 1885.
Found also by Mr. G. A. Holt on wet rocks by the stream above Torc Cascade, Killarney, in June, 1885, the only two Irish stations being at extreme points of the island, and discovered almost simultaneously.

3. M. cuspidatum *Linn.* (M. AFFINE *auct.*).
Wet places—very rare, and no fruit found.
Down—In a marsh by the Lagan at Magheralin; H.W.L.

4. M. undulatum (*Schreb.*) *Hedw.*
Damp shady places—frequent. Fr. Feb. and March.
Usually barren, but fruiting abundantly at Narrow-water; C.H.W. Often in fruit; H.W.L.

5. M. rostratum (*Schrad.*) *Schwaeg.*
Wet rocky banks—not common. Fr. March—May.
Down—Cregagh Glen; S.A.S. Wall of a bridge in Tollymore Park; H.W.L. Rostrevor wood, Moygannon Glen, and Drumcro; C.H.W. Purdysburn; *J. J. Andrew.*
Antrim—Carr's Glen; *Templeton*, 1805. Colin Glen; *J. H. Davies, Phytologist*, 1859. Windy Gap, Crow Glen, Whitewell quarries, and Tardree; S.A.S.

6. M. pseudo-punctatum *Br. et Schp.* (M. SUBGLOBOSUM *Bry. Eur.*).
In very wet stony bogs—rare. Fr. Feb.—March.

240   *HYPNACEÆ.*

Antrim—Wet peaty bog at back of Cave Hill, marshy margin of Loughmourne, and peaty heath on Carrickfergus commons; S.A.S.

7. M. punctatum (*Schreb.*) *Hedw.*
Dripping rocks, wet banks, and margins of mountain streams—common. Fr. Feb.—March. Ranges from sea level to 2500 feet.

SERIES II. ORDER XXI. **HYPNACEÆ.**

THUIDIUM *Br. et Schimp.*

1. T. tamariscifolium (*Neck.*) *Lind.* (HYP. PROLIFERUM *Linn.*).
On stones, and banks in damp woods and shady glens—common. Fr. Oct.—Jan.

LESKEA *Hedwig.*
1. L. polycarpa *Ehrhart.*
On trees by rivers and lakes—not uncommon. Fr. May—July.
Down—Manyburn rivulet; *Templeton*, 1805. On trees by the Lagan at Drumcro; C.H.W. By the Annacloy at Rademon; S.A.S.
Antrim—Colin Glen; *Templeton.* By the Lagan near Belfast; *T. Drummond, Musci Scotici*, vol. III. Massereene Park, riverside above Templepatrick, and on bushes in the ditch by the Lagan between first and second lock; S.A.S. Drumbridge, and Woodburn; H.W.L.

ANOMODON *Hook. & Tayl.*

1. A. viticulosus (*Linn.*) *H. & T.*
Vertical rocks of limestone, and basalt, on walls occasionally—frequent, but almost always barren.
Antrim—Deerpark wall, and limestone rocks about the mountains of Belfast; *Templeton.* In fruit at Ballygally Head, Feb., 1837; D.M. Belfast hills, Knockagh, Gobbins, Gleno, Larne, Sallagh Braes, and Ballintoy; S.A.S. Kilwaughter; H.W.L. Killyglen; C.H.W.

HYPNUM *Dillenius.*

Sect. I.—Amblystegium.

1. H. filicinum *Linn.*

In wet rocky places, and by streams—frequent. Fr. Feb.—May. Ascends to 2500 feet.

Down—Holywood hills; S.A.S. Magheralin, Kilwarlin, Rostrevor Mountain, Moygannon Glen, and by the Yellow-water River; C.H.W. Frequent; H.W.L.

Antrim—Colin Glen; *Templeton*, 1803. Sparingly near the head of Glenariffe; D.M. Glenavy River, Stoneyford, Belfast hills, Woodburn, Agnew's Hill, and Rathlin; S.A.S. Kenbane Head; H.W.L. Marsh by Lagan canal near Belfast; *J. J. Andrew.*

Derry—Lignapeiste; D.M. Slievegallion, and Carndaisy Glen; S.A.S.

2. H. serpens *Linn.*

Damp banks, tree trunks, and on the ground in shady places—very common. Fr. Feb.—May.

3. H. riparium *Linn.*

River banks, lake shores, and other wet places—rare. Fr. Aug. and Sept.

Down—Abundant in several places by the canal above Moira, and plentiful, but sparingly in fruit, on peaty lake shores of Derry Lough, and other lakes about Ballynahinch; S.A.S. In Ballymaginn bog near Magheralin; C.H.W.

Antrim—On the stones of Shawsbridge weir; *Templeton.*

Sect. II.—*Campyliadelphus.*

4. H. stellatum *Schreber.*

Wet boggy land—not uncommon. Fruit rare.

Down—Pigeonrock Mountain, Slievenamady, and near Groomsport; H.W.L. Copeland Islands; *J. H. Davies.* Above Tollymore Park, and at Ballymaginn near Magheralin; C.H.W. Wet places by the river in Tollymore Park; S.A.S.

Antrim—In moss holes, and old peat pits at Cranmore; *Templeton*, 1805. On the moors above Carnlough; D.M. Wolfhill, and by the Lagan below the second lock; H.W.L.

Derry—Marshy places in Bennedy Glen above Dungiven; D.M.

Sect. III.—*Drepanocladus.*

5. H. glaucum *Lamk.* (H. COMMUTATUM *Hedw.*).

Wet banks, and dripping rocks—frequent. Fr. April—June.

Down—Gillhall demesne, Slieve Donard, Slieve Croob, Holywood Hills, and Drumbo Glen; S.A.S. Deers' Meadow, Tollymore Park,

and Groomsport; H.W.L. Moygannon Glen ; C.H.W.
Antrim—Frequent on moist springy bogs throughout the county ; D.M.
    Donegore Hill, Colin Glen, Black Mountain, Cave Hill, Car-
    rickfergus commons, Agnew's Hill, and Knocklayd ; S.A.S.
Derry—Slievegallion Mountain ; S.A.S.

6. H. falcatum *Bridel.*
Wet rocks and banks—not uncommon. Fr. May and June.
Down—Slieve Donard, and Annalong ; H.W.L.
Antrim—Colin Glen ; *J. H. Davies, Phytologist*, 1859. Cave Hill, and
    several places on Belfast hills, also in the rocky swamps of
    Rathlin ; S.A.S. Marsh by the Lagan near Belfast ; *J. J.
    Andrew.*
First noted in Ireland by J. H. Davies.

7. H. aduncum *Linn.* (H. UNCINATUM *Hedw.*).
Moist rocks and banks—not common. Fr. May—July.
Down—Rocks on lower slopes of Slieve Donard, stones in Rademon
    wood, and on margin of Lough Aghery ; S.A.S. Tollymore
    Park, and on stones in the river at Drumcro ; C.H.W.
Antrim—Wet places by stream at top of Colin Glen, and of Crow Glen,
    also wet rocks in north Woodburn ; S.A.S. On stone fence
    by the roadside near Clough ; H.W.L.

8. H. exannulatum *Gumbel.*
Wet boggy places—rare Fr. April—June.
Down—Glenaveagh, and Deers' Meadow ; H.W.L. Old quarry near
    Holywood, a form which may be the var. *stenophyllum* of
    Wilson ; *J. J. Andrew.*
Antrim—Bogs near Ballycastle, and other places ; D.M. Boggy banks
    of stream on Black Mountain, and rocks by the waterfall in
    Colin Glen ; S.A.S.
Derry—Not uncommon in the mountains ; D.M.

9. H. fluitans *Linn.*
Lakes, ditches, and pools—rare. Fruit not found.
Down—Margin of Lough Aghery, and abundant in brackish drains at
    Castle Espie quarries ; S.A.S. In pools on Slievedermot,
    and on Rostrevor Mountain ; H.W.L.
Antrim—Black Mountain (very slender form) ; *J. J. Andrew.*

10. H. Kneiffii *Bry. Eur.* (H. ADUNCUM var. KNEIFFII *Sch. Syn.*).
Wet places—very rare ? Fr. June (*Berk.*).
Antrim—Wet rocks by the stream in Glendivis ; S.A.S.

11. H. intermedium *Lindb.*
Boggy, and wet heathy places—rare. Fr. not found.
Antrim—Rathlin Island; S.A.S.
Derry—Church Island near Toome; *J. J. Andrew.*

12. H. revolvens *Swartz.*
Wet stony and peaty places—not common. Fr. May and June, but usually sterile.
Down—Sparingly on Slieve Donard, and Slieve Croob; S.A.S. Rostrevor Mountain, Pigeonrock Mountain, and Tollymore Park; C.H.W. Frequent on the Mourne range; H.W.L.
Antrim—Bog on Cave Hill, wet heath on Carrickfergus commons, and on Slemish Mountain; S.A.S.
[H. HAMIFOLIUM *Schimper.*
In pools, and marshes by the shore of Lough Neagh at Kinnego, Co. Armagh; C.H.W., 1882. Not yet found in any other British locality, but may be expected on the Antrim shores of Lough Neagh. A prize worth searching for.]

13. H. lycopodioides *Necker.*
In bogs—very rare.
Antrim—Fruiting in a bog in the parish of Rasharkin; *Moore, op. cit.*

Sect. IV.—Scorpidium.

14. H. scorpioides *Linn.*
Wet boggy places—rare. Fr. Spring (*Berk.*).
Down—Boggy heath on Slieve Croob; S.A.S. Rostrevor, and Pigeonrock Mountain; H.W.L. Mountains above Tollymore Park; C.H.W.
Antrim—Turf bogs; *Templeton.* Bogs in the north, but not in fruit; D.M. Luxuriant, but barren on boggy places in the northwest of Rathlin; S.A.S.
Derry—Abundant in Glenchain; D.M.

Sect. V.—Hygrohypnum.

15. H. ochraceum *Turner.*
Rocky river banks—rare. Fr. May and June (*Berk.*).
Down—Slieve Commedagh, and Slievenabrock; H.W.L. Abundant at the upper waters of the Bann; C.H.W. By the river in

Tollymore Park; S.A.S.
Antrim—By the stream in Colin Glen; C.H.W.
Derry—By streams on Mullaghmore, and White Mountain; S.A.S.

16. H. palustre *Huds.*
On stones in, and beside flowing water, especially abundant on the rocky margins of streams in the hills—common. Fr. May—Aug.

Var. β H. SUBSPHÆRICARPON *Schleich.*

[Near Carrickfergus, and Cushendall (*Johns*); Moore, Proc. R.I.A., 1872. May be correct, but specimens are wanting, and for the present should be considered as doubtful.]

Sect. VI.—Calliergon.

17 H. giganteum *Schimper.*
Wet stony heaths—rare, and barren.
Down—Lower slopes of Slieve Croob at the west side; S.A.S. Ballymaginn moss near Magheralin; C.H.W.
Antrim—Wet moory pastures on Carrickfergus commons; S.A.S.
First found in Ireland by S.A.S., in 1875. *Vide Proc. Belf. Nat. Field Club*, Ser. II., vol. I. pt. I., app. 1875. Elsewhere in Ireland it has only been met with at Howth, by Mr. Orr, *vide Jour. Bot.*, 1881.

18. H. cordifolium *Hedwig.*
Bogs, and marshes—not common. Fr. May and June.
Down—Peaty heath west of Moneyreagh; S.A.S. Magheralin, and sandhills at Newcastle; H.W.L. Bog near Rathfriland; C.H.W.
Antrim—Common in the Bog meadows; *Templeton.* In bogs amongst *Sphagnum* at Ballycastle; D.M. Boggy heath on Carrickfergus commons; S.A.S. By the Lagan near Belfast; J. J. Andrew. Between Larne and Glenarm; C.H.W.

19. H. sarmentosum *Wahl.*
Rocks, and banks—very rare. Fr. Summer (*Berk.*).
Down—By the Spinkwee River above Tollymore, and in White-water Glen; H.W.L. Slieve Commedagh; C.H.W.

20. H. stramineum *Dickson.*
Gravelly banks of mountain streams—very rare. Fr. early summer (*Berk.*).
Down—Glenaveagh; H.W.L. Deers' Meadow; C.H.W.

Antrim—Glenmakerron, June 1836; D.M.
Derry—By a rocky mountain stream on Mullaghmore, barren; S.A.S.

Sect. *VII.—Scleropodium.*

21. H. purum *Linn.*
Grassy, and heathy banks—common. Fr. Nov.—Jan., but usually barren.

22. H. illecebrum *Brid.*
Rocks, and stones—very rare. Fr. Autumn (*Berk.*).
Antrim—On the top of the Black Mountain; *Templeton*, Aug. 1804. Rocks at foot of Sallagh Braes; C.H.W.

Sect. *VIII.—Panckowia.*

23. H. striatum *Schreber.*
Woods, and shady banks—frequent. Fr. Nov.—Jan.
Down—Slieve Croob, Crawfordsburn, Ballylesson, and Glenmachan; S.A.S. Gilford, Hillsborough, and Bryansford; H.W.L. Narrow-water, Rostrevor, Kilwarlin, and Castlereagh; C.H.W.
Antrim—Banks, and woods through the county; D.M. Colin Glen; *J. H. Davies.* Derriaghy, Shawsbridge, Cave Hill, Woodburn, Redhall Glen, Glynn, and Rathlin; S.A.S. Portrush; H.W.L.
Derry—Errigal banks; D.M. Portstewart, and Carndaisy Glen; S.A.S.

24. H. prælongum *Dillenius.*
Damp shady banks—frequent. Fr. Nov.—Jan.
Down—By the river in Tollymore Park; *Templeton*, 1805. Hillsborough Park, Bangor, Crawfordsburn, Holywood Glen, Scrabo Hill, Drumbo Glen, Dundonald Glen, and Victoria Park; S.A.S. Loughbrickland, and Tullybranagan; H.W.L. Warrenpoint, Slieve Commedagh, Kilwarlin, and Drumcro; C.H.W. Cregagh Glen; *J. J. Andrew.*
Antrim—Belfast hills, Knockagh, and Ballymena; S.A.S. Fairhead, and Kenbane; H.W.L.
Derry—Frequent in the county; D.M. Carndaisy Glen, and old quarry at Springhill; S.A.S.

25. H. Swartzii *Turner.*
Damp shady banks, and rocks—not rare. Fr. Oct.—Jan.

Down—Rostrevor Glen; *Herb. Belf. Mus.* Crawfordsburn; S.A.S.
Under trees at Drumcro; C.H.W. Loughbrickland, and
Tollymore Park; H.W.L.
Antrim—Basaltic rocks at upper end of Colin Glen, Carr's Glen, sides of
drains at Newforge, under trees in Borough cemetery, rocks
in Woodburn Glen, and woody places at Broughshane; S.A.S.

26. H. pumilum *Wilson.*
Shady places—rare, and barren.
Down—Aghaderg Glebe; H.W.L. On ground inside the tower of
Dundrum Castle; C.H.W.
Antrim—In Co. Antrim (*Templeton in herb. Turner*); *Moore, op. cit.*

27. H. speciosum *Bridel.*
Damp rocks and stones—rare, and without fruit.
Down—Rocks by the stream in Drumbo Glen; S.A.S.
Antrim—Sparingly on damp rocks in Rathlin Island; S.A.S.

28. H. crassinervium *Taylor.*
On rocks—very rare. Barren.
Antrim—Cave Hill; *T. Drummond, Musci Scotici, vol. III.*

29. H. Teesdalii *Smith.*
Damp rocks and stones—rare. Fr. March—May.
Down—On a high rocky bank about 100 yards above Manyburn bridge
at Purdysburn; *Templeton.*
Antrim—Colin Glen (*Templeton*); *Flor. Brit.*, 1804. Moist rocks of
Colin Glen; *Templeton.* Colin Glen; *J. H. Davies*, 1859.
On rocky banks of the stream in Colin Glen at various spots
between the lower and the upper waterfalls, also on rocks by
the stream in Redhall Glen, and by the waterfall in south
Woodburn; S.A.S. Whiterocks at Portrush; H.W.L.

By a singular fatality this interesting moss, which was first found in
Ireland by Templeton, has been doomed to be overlooked. Described as
British in Flora Britannica, 1804, and as Irish by Turner, later in the
same year it, nevertheless, was not noticed in Muscologia Britannica,
1827, nor in Flora Hibernica, 1836. Dr. Moore, also, was unaware of
its history, and when making his list for the Royal Irish Academy, in
1872, recorded it from the south of Ireland only. Nevertheless this
plant was published 68 years previously as occurring in Colin Glen, and
Mr. Davies' confirmation of the original record was 13 years prior to the
publication of Dr. Moore's List of Irish Mosses.

Sect. *IX.—Rhynchostegium.*

30. H. piliferum *Schreber.*
Moist, shady banks and rocks—frequent. Fruit very rare.
Down—Banbridge, Newtownards, Cregagh Glen, Stormount Glen, and Glenmachan; S.A.S. Moira; H.W.L. Drumcro; C.H.W.
Antrim—Fruiting near Belfast (*Templeton*); *Flor. Hib.* Carr's Glen, Whitewell quarries, Woodburn, Kilroot, Ballycarry, Gleno, and Glynn; S.A.S. Sallagh Braes; H.W.L., and C.H.W.

31. H. rusciforme *Necker.*
Dripping rocks, and on stones in streams—very common. Fr. Oct.—March.

32. H. murale *Necker.*
Roadsides, and walls—rare. Fr. Nov.—Feb.
Antrim—Walls near Belfast; *T. Drummond, op. cit.* Tops of walls, and on stones near Ballycastle; D.M. In the water-table of the Falls road beyond Andersonstown, and on a stone dyke on the old road to Antrim above Ballysillan; S.A.S.
Derry—Common on walls, and stones; D.M.

33. H. confertum *Dickson.*
Banks, walls, stones, and at the roots of trees—very common. Fr. Nov.—Jan.

34. H. tenellum *Dickson.*
Limestone, or occasionally basaltic rocks, and on walls—not common. Fr. Feb.—April.
Down—Old walls of Dundrum Castle; S.A.S. Plentiful on the walls of a bridge in Tollymore Park; H.W.L., and C.H.W.
Antrim—Plentiful on the rocks in Redhall Glen; *Templeton*, 1809 (still there). On the southwest side of the wall of Belfast Deerpark, and on limestone in Carr's Glen, Trap rock at Sallagh Braes, and Chalk in Rathlin Island; S.A.S.

Sect. X.—Brachythecium.

35. H. velutinum *Linn.*
Damp banks, stones, and sometimes on trees—common. Fr. Nov.—Jan.

36. H. pseudoplumosum *Brid.* (H. PLUMOSUM *Auct. Brit.*).
Wet rocky banks, and on stones in streams—frequent. Fr. Nov.—March.
Down—On stones in the oak wood at Belvoir; *Templeton*, 1804. Slieve

Donard, Slieve Croob, Ballyoran, and Cregagh Glen; S.A.S.
Narrow-water, Rostrevor, Slievenabrock, and Tollymore
Park; C.H.W. Common in the Mourne Mountains; H.W.L.
Antrim—On rocks, and trees; *Templeton*. Moist banks and stones;
D.M. Belfast hills, Woodburn, Magheramorne, Gleno, and
Agnew's Hill; S.A.S.
Derry—Near the Ness waterfall; D.M. Near Moneymore; S.A.S.
Near Toome; *J. J. Andrew*.

37. H. viride *Lamk*. (H. POPULEUM *Hedw*.).
On stones, and banks, sometimes on tree trunks—common. Fr. Nov.
—Feb.

38. H. rutabulum *Linn*.
Damp rocks and walls and fields, occasionally on trees—very common.
Fr. Nov.—Feb.

39. H. rivulare *Bruch*.
On trees and stones by streams, and in wet places—not common, and usually sterile.
Down—Magheralin, Tollymore Park, and Struel Well; H.W.L. Kirkcassock, and Rostrevor Mountain; C.H.W. Rocky banks of stream in Drumbo Glen; S.A.S.
Antrim—Colin Glen; *J. H. Davies*, 1859. By the stream in Carr's Glen, and abundant on a willow tree by the riverside at Muckamore; S.A.S.
Derry—Mullaghmore Mountain; S.A.S.
First found in Ireland by Isaac Carroll.

40. H. albicans *Necker*.
Dry banks, and sandy shores—frequent, but seldom met with in fructification.
Down—On the sandy warren at Donaghadee; *Templeton*. Sandhills at Newcastle, sandy shore at Millisle, and fruiting sparingly in sandpit at Donaghadee, 20th Feb., 1876; S.A.S.
Antrim—Glenarm, and through the county; D.M. Dry ditch bank to the south of Carr's Glen; S.A.S.

41. H. glareosum *Br. et Schimp*.
Shady banks—rare. Fr. late autumn (*Berk*.).
Down—Disused quarry at Kilwarlin, with old fruit in March, 1884; C.H.W.
Antrim—Moist meadow in Glenaan; D.M. Ditch bank by roadside west of Carrick Junction station, barren; S.A.S.

Sect. XI.—*Pleuropus*.

42. H. lutescens *Hudson.*
Sandy banks, and limestone rocks—not rare. Fr. March and April.
Down—Railway bank at Newtownards, upper end of Holywood Glen, and in Cregagh Glen; S.A.S. Rostrevor Mountain, and Moira wood; C.H.W.
Antrim—Limestone rocks on Black Mountain above Whiterock; *Templeton* (still there). Belfast Deerpark wall, and moist boggy places; D.M. Quarry spoil-bank at Cave Hill, and sparingly on limestone rocks in Rathlin Island; S.A.S. Wall near Glarryford; H.W.L.
Derry—In old limestone quarry at Springhill near Moneymore; S.A.S. Church Island near Toome; *J. J. Andrew.*

43. H. sericeum *Linn.* (LESKEA *Hedw.*).
Walls, rocks, banks, and trunks of trees—very common. Fr. Oct.—Dec.

Sect. XII.—*Isothecium*.

44. H. myosuroides *Linn.*
Rocks, banks, and trunks of trees—frequent. Fr. Nov.—Jan.
Down—Rocks on Slieve Donard, wood and wet banks in Hillsborough Park, and at roots of trees in Portavo demesne; S.A.S. Woods at Rostrevor, Tollymore Park, Moira, and Groomsport; H.W.L.
Antrim—Rocks at Whiterock; *Templeton.* Frequent; D.M. Shawsbridge, Knockagh, Woodburn, Kilwaughter, Ballygally Head, and Glenshesk; S.A.S. Colin Glen; H.W.L.
Derry—Common; D.M. Carndaisy Glen; S.A.S.

45. H. viviparum *Necker.* (ISOTH. MYURUM *Poll.*).
On trees, stones, and rocks in shady places—common. Fr. Oct.—Dec.
Var. ELONGATUM *Schimp.*
Down—Slievenabrock; H.W.L.

46. H. ornithopodioides *Dill.* (PTER. GRACILE *Sw.*).
On rocks—very rare. Fr. Nov. (*Berk.*).
Antrim—Limestone rocks below Grogan's Glen; *Templeton,* 1802 (not found there now). In considerable quantity, but without fruit, on Trap rocks near the summit of Ballygally Head; S.A.S.

47. H. decipiens (*Web. et Mohr*).
(PT. FILIFORME var. heteropterum (*Brid.*) *Sehp.*).
On rocks—rare, and barren.
Down—Abundant in Tollymore Park; *T. Drummond, Musci Scotici, vol. III.*
Antrim—Rocks near Ballygally Head, and other places in the county; D.M.

Sect. XIII.—Heterocladium.

48. H. heteropterum (*Bruch*) (H. ATROVIRENS *Turn.*).
Damp rocks by streams—rare, and barren.
Down—In a little glen about a mile south of Holywood, 1804, and in hollows of rocks at foot of Diamond Mountain, 1805; *Templeton, Herb. Belf. Mus.* Wet rocks on Slieve Commedagh, and rocky banks of streams in Drumbo Glen, and Killeen Glen; S.A.S.

Sect. XIV.—Hylocomium.

49. H. brevirostrum *Ehrhart.*
Woods, and glens—not uncommon. Fr. Winter (*Berk.*).
Down—Slieve Croob, Newtownards Glen, Glenmachan, and fruiting sparingly in Holywood Glen; S.A.S. Moygannon Glen, and fruiting abundantly at Narrow-water; C.H.W. Tollymore Park, and Rademon wood; H.W.L.
Antrim—Colin Glen; *T. Drummond, op. cit.* Glenarm, and Glenariffe; D.M. Carr's Glen, Woodburn, Glynn, and Broughshane; S.A.S.
Derry—Carndaisy Glen; S.A.S.

50. H. proliferum *Linn.* (H. SPLENDENS *Hedw.*).
Damp banks in woods and shady places—common. Fr. March and April, but usually sterile. Ascends to 2200 feet on the mountains.

51. H. parietinum *Linn.* (H. SCHREBERI *Ehr.*).
Heaths, and damp sandy ground—common but always sterile. Ranges from sea level to 2500 feet.

52. H. triquetrum *Linn.*
Damp woods—common. Fr. Oct.—Feb., but only sparingly in fructification.

53. H. squarrosum *Linn.*
Hedge banks, fields, and damp waste ground—very common. Fr. Nov.—Jan., but only occasionally fertile. Ascends to 2700 feet.

54. H. loreum *Linn.*
Woods, and wet rocks—frequent. Fr. Nov.—Feb. Ranges from near sea level to about 2000 feet.
Down—Glen near Holywood; *Templeton.* Gillhall, Rostrevor wood, Slieve Commedagh, and Slieve Croob; S.A.S. Eagle Mountain, Slievenabrock, and Tollymore Park; C.H.W. Common in the Mourne Mountains; H.W.L.
Antrim—Glenariffe, and frequent on shady banks, and under trees; D.M. Sparingly at Whitewell quarries, and on Ballygally Head; S.A.S.
Derry—Not uncommon; D.M. Abundant in bog at summit of Slievegallion; S.A.S.

Sect. XV.—Ctenidium.

55. H. molluscum *Hedwig.*
Rocks, and shady banks—very common, especially abundant on the basalt and limestone. Fr. Jan.—March, but capsules rarely plentiful.
Var. β GRACILE *Boulay.*
Antrim—Sparingly on limestone rocks in Crow Glen; S.A.S.

Sect. XVI.— Stereodon.

56. H. cupressiforme *Linn.*
Banks, rocks, walls, trees, and damp ground—very common. Fr. Nov.—March.
Var. β LACUNOSUM *Wilson.*
Down—Narrow-water, and wall at Warrenpoint; C.H.W.
Antrim—Sparingly on stones on Black Hill above Hannahstown; S.A.S.
Var. γ FILIFORME *B. & S.*
Down—Tollymore Park; C.H.W. Warrenpoint, Slievemeel Beg, and Clandeboye; H.W.L. Seems to be common throughout the district; S.A.S.
Var. δ ERICETORUM *B. & S.*
Down—Slieve Donard; H.W.L. Slievenabrock; C.H.W.

57. H. resupinatum *Wilson.*
Banks, stones, and trunks of trees—frequent. Fr. Nov.—Feb.

Down—Ballyholme, Craigauntlet, and Cregagh Glen; S.A.S. Tievedocharrah, and Loughbrickland; H.W.L. Moygannon Glen, and on trees at Drumcro; C.H.W.
Antrim—On trees in Glendun; D.M. On trees near Antrim, and at Kilroot, also found on limestone blocks at Knockagh, and on rocks at Ballygally Head, Fairhead, and Rathlin; S.A.S.
Derry—Occurs in abundance on fallen timber, and on stones in, and about Killymoon where the counties of Derry and Tyrone meet; S.A.S.

Sect. XVII.—Isopterygium.

58. H. Patientiæ *Lindb.* (H. ARCUATUM *Sch. Syn.*, II).
Gravelly, or sandy banks—rare, and barren.
Down—Rostrevor; C.H.W. Very fine in Moygannon Glen; C.H.W., and H.W.L.
Antrim—Dry bank at Wolfhill near Belfast; C.H.W.

59. H. depressum *Br. et Schimp.*
On stones in damp shady places—rare, and barren.
Down—On stones by the stream at the upper end of Killeen Glen; S.A.S.

60. H. Borreri *Spruce.*
Shady places—rare, and always barren. Ascends to near 2000 feet.
Down—Under granite blocks on Slieve Donard; S.A.S. Shady banks in Tollymore Park; H.W.L. Eagle Mountain, and cave in Cove Mountain; C.H.W.

61. H. pulchellum *Dickson.*
Damp rocks—very rare. Fr. Summer (*Berk.*).
Antrim—Abundant in shady crevices of rocks on Sallagh Braes; D.M.
Derry—Ballyharrigan Glen, and south side of Meenard Mountain at 1500 feet; D.M.

Sect. XVIII.—Plagiothecium.

62. H. undulatum *Linn.*
Grassy heaths, and in woods—frequent. Fr. June and July. Ranges to 2000 feet.
Down—Slieve Donard; S.A.S. Tollymore Park, and Slieve Bignian; C.H.W. Frequent in the Mourne Mountains; H.W.L.
Antrim—Fruiting among heath on the north side of Divis; *Templeton.*

Frequent on moist banks in the northern glens; D.M. Black Mountain, Agnew's Hill, and Rathlin Island; S.A.S.
Derry—Ness Glen, and Bond's Glen, and on a boggy mountain between Cumber and Lagnagappagh; *Templeton.* Not uncommon; D.M. Among heath on Slievegallion, and under stones in several of the Sperrin mountains; S.A.S.

63. H. sylvaticum *Linn.*
Rocks, and shady banks—rare. Fr. July—Sept.
Down—Hedge bank at Tullygarnet south of Knock, and abundant in fruit on stones in Rostrevor wood; S.A.S. In a cave on Cove Mountain; C.H.W. Slievenabrock; H.W.L. Cregagh Glen; *J. J. Andrew.*
Antrim—Near Lisburn; *J. Creeth.* Stones on Black Mountain, and under fallen rocks on the Knockagh, and Ballygally Head; S.A.S.

64. H. denticulatum *Linn.*
Damp rocks, and woods—frequent. Fr. June and July. Ranges from near sea level to over 2000 feet.
Down—Eagle Mountain; C.H.W. Cove Mountain, Slieve Donard, Slievenabrock, and Tollymore Park; H.W.L.
Antrim—Woods, and moist shady places; *Templeton.* Ballygally Head, and frequent in shady crevices of rocks; D.M. Sparingly in hedge bank at Stockman's lane, and in abundance on basaltic cliffs of Slemish; S.A.S. Colin Glen; C.H.W.
Derry—At 2000 feet on Sawel Mountain; D.M.

Sect. XIX. Acrocladium.

65. H. cuspidatum *Linn.*
Swampy ground, especially on heath—very common. Fr. April—June.

SERIES II. ORDER XXII. **PTERYGOPHYLLACEÆ.**

PTERYGOPHYLLUM *Bridel.*

1. P. lucens (*Linn.*) *Brid.* (HOOKERIA *Smith*).
Damp clay banks near streams—frequent. Fr. Dec.—Feb.
Down—On rocks by the lower fall in Tollymore Park, and in a glen near

Holywood; *Templeton.* Crawfordsburn, Holywood wood, Dundonald, Drumbo, Killeen, and Cregagh Glen; S.A.S. Narrow-water, Rostrevor Mountain, and Rocky Mountain; C.H.W. Tievedocharrah, and Donard demesne; H.W.L.
Antrim—Carr's Glen; *Templeton.* On moist shady banks in northern glens; D.M. Woodburn, and Gleno; S.A.S. Sallagh Braes; C.H.W.
Derry—Ness Glen, and Bond's Glen; *Templeton.* Abundant by the Roe above Limavady; D.M.

SERIES II. ORDER XXIII. **NECKERACEÆ.**

Sect. I.—Neckereæ.

PAROTRICHUM (*Brid.*) *Mitt.* (THAMNIUM *Schimp.*).

1. P. alopecurum (*Linn.*) *Mitten.*
Damp shady places, and banks of streams—common. Fr. Nov.—Feb.

HOMALIA *Bridel.*

1. H. trichomanoides (*Schreb.*) *Bridel.*
On trees, rocks, and stones—very common. Fr. Nov.—Feb., but seldom found in fructification.

NECKERA *Hedwig.*

1. N. complanata (*Linn.*) *Hueb.*
Rocks, trees, and damp banks—common, but less so than the preceding species. Fruit rare, but occasionally found from December till end of February.

2. N. crispa (*Linn.*) *Hedw.*
Damp rocks, and on stones in shady places—frequent. Fruit very rare.
Down—Slieve Donard, and Newtownards Glen; S.A.S. Tollymore Park; C.H.W.
Antrim—With fruit in Colin Glen; *Flor. Hib.* In fruit at the head of Glenariffe; D.M. With over-ripe capsules on rocks at top of Colin Glen 11th April, 1885, also plentiful in the barren state on Black Mountain, Sallagh Braes, and Glenarm; S.A.S. Drumnasole; H.W.L.
Derry—Benbradagh Mountain; S.A.S.

Sect. II.—Meteorieæ.

CLIMACIUM *Web. et Mohr.*

1. C. dendroides (*Linn.*) *Web. et Mohr.*

Wet heathy pastures, lake shores and by the margin of rivers—common. Fr. Oct. and Nov., but usually sterile. Mr. J. H. Davies finds abundant fructification near Lisburn, and Mr. J. J. Andrew has fertile plants from the Lagan canal near Belfast. "In luxuriant fruit on the shores of Loughbrickland"; H.W.L.

FONTINALIS *Dillenius.*

1. F. antipyretica *Linn.*

On submerged stones, usually in running water, sometimes in lakes—common. Fr. Summer.

2. F. squamosa *Linn.*

On stones in streams—very rare. Fr. Summer (*Berk*).
Down —On stones in stream at Rostrevor Mountain; H.W.L., and C.H.W.
Derry—In the Faughan about two miles below Cumber; *Templeton*, 1800.

Sect. III.—Cryphæeæ.

LEUCODON *Schwaegrichen.*

1. L. sciuroides (*Linn.*) *Schwaeg.*

Trunks of trees—very rare. Fr. Spring (*Berk.*).
Antrim—On apple trees in Mrs. Barclay's orchard at Lambeg; *Templeton*, 1804.

Templeton's specimen in the Belfast Museum Herbarium is quite right, but the plant does not seem to have been found by any subsequent botanist.

CRYPHÆA *Web. et Mohr.*

1. C. arborea (*Huds.*) *Lindb.* (C. HETEROMALLA *Brid.*).
Trunks of trees, frequent. Fr. April—Sept.
Down—Gillhall, Ballywalter, Greyabbey, Rademon, Ballyalloley, and Clandeboye; S.A.S. Drumcro, and Kirkcassock,; C.H.W. Loughbrickland, and Annalong; H.W.L.
Antrim—On an apple tree at Cranmore; *Templeton*. Abundant on trees

in Glenarm demesne; D.M. On poplar trees at Derriaghy;
J. H. *Davies*. At Muckamore, and by the lake shore at
Antrim and Crumlin waterfoot, also on trees south of Clady
bridge, and by the stream at upper end of Magheramorne Glen;
S.A.S. Woodburn, and Clough; H.W.L.
Derry—Mountsandel wood near Coleraine; S.A.S.

HEDWIGIA *Ehrhart.*

1. H. imberbis (*Sm.*) *Spruce.*
On trap rocks—very rare. Not found with fruit.
Antrim—Very fine on Fairhead, May, 1854; *Moore, op. cit.*, and *Lett*,
    1884. Greenstone rocks at Ballygally Head, and on Basalt
    at Sallagh Braes; S.A.S.
The discoverer of this plant was our accomplished Irish lady botanist,
Miss Hutchins.

2. H. albicans (*Web.*) (H. CILIATA *Hedw.*).
On rocks, and stones in heathy places—common. Fr. April—June.
Ranges from 100 to 1600 feet.

SERIES II. ORDER XXIV. **SPHAGNACEÆ.**

* **SPHAGNUM** *Dillenius.*

1. S. papillosum *Lindb.*
Rocky and heathy bogs—not uncommon.
Down—Slieve Donard, and plentiful on wet rocks above Bloodybridge,
    and in Cotton moss; S.A.S. Slievenabrock; H.W.L. By
    Yellow-water River; C.H.W.
Antrim—On Black Mountain, and in the boggy heaths of Rathlin Island;
    S.A.S.
First recorded as an Irish plant by S.A.S. in Proc. Royal Irish Acad.,
Jany., 1884.

* The *Sphagna* of the district have not received sufficient attention, and
in consequence the list here given is, to some extent, incomplete, and the
notes of their distribution scanty. Additional stations might have been
assigned the species here enumerated, and other plants recorded, but only
such notes have been accepted as are based on rigidly authenticated
specimens.

Var. β CONFERTUM *Lindb.*
Down—Tollymore Park; C.H.W., Sept., 1884.
Antrim—In County Antrim; D.M. *Herb. Lindb.* (G.A.H.).

2. S. cymbifolium *Ehrhart.*
Bogs—frequent?
Down—Rostrevor Mountain; C.H.W. Rademon wood, and bog at Moneyreagh; S.A.S.
Antrim—Frequent in Co. Antrim; D.M.

3. S. tenellum *Ehrhart.*
Bogs—very rare.
Down—Hen Mountain; H.W.L.

4. S. subsecundum *Nees.*
Mountain bogs—frequent.
Down—Plentiful on mountains above Newcastle, and in bog at Moneyreagh; S.A.S. Shanslieve, Yellow-water River, and Tollymore Park; C.H.W. Common and abundant; H.W.L.
Antrim—In County Antrim; *Moore, op. cit.*

Var. CONTORTUM *Schultz.*
Down—Wet rocks of Slieve Commedagh, and mountains above Tollymore; S.A.S. Thomas Mountain; C.H.W. Shanslieve; H.W.L.
Derry—Marsh at Coleraine; S.A.S.
First recognised as an Irish plant by J. H. Davies.

Var. OBESUM *Wilson.*
Down—Bencrom; H.W.L. (scarce, and specimen not well marked).

5. S. molle *Sullivant.*
Var. MUELLERI *Schimp.*
Bogs—very rare.
Down—Rocky Mountain near Hilltown; H.W.L.
New to the Irish Flora.

6. S. rigidum *Schimper.*
Peaty, and stony marshes—rare.
Down—Kinnehalla, Chimney Rock Mountain, and Slieve Donard; H.W.L.
Not on record as occurring elsewhere in Ireland.

Var. COMPACTUM *Bridel.*
Down—Slieve Donard, and Deers' Meadow; H.W.L.

Antrim—Divis Mountain; *J. H. Davies, Phytologist,* 1859.

7. S. squarrosum *Persoon.*
Peat bogs—frequent?
Down—Moneyreagh, and Gillhall demesne; S.A.S.
Antrim—In County Antrim, but not common; D.M.

8. S. acutifolium *Ehrhart.*
Bogs, and marshes—very common.
   Var. ASCENDENS *Braith.*
Down—Shanslieve; H.W.L.
Var. new to the Irish Flora.
   Var. RUBELLUM *Wilson.*
Frequent on mountain bogs.
   Var. VERSICOLOR *Warnstorff.*
Down—Slieve Donard; *J. J. Andrew.*
New to the Irish Flora.
   Var. LÆTEVIRENS *Braith.*
Down—Boggy places by stream above Bloodybridge; S.A.S.
   Var. LURIDUM *Hueb.*
Down—Mourne Mountains (*Rev. H. W. Lett*); *Boswell, Jour. Bot.,* 1887, p. 111.
New to the Irish Flora.

9. S. intermedium *Hoffman.*
   Var. PULCHRUM *Lindb.*
Down—Deers' Meadow; H.W.L. (specimen not well marked).
New to Irish Flora.

10. S. cuspidatum *Ehrhart.*
Peat bogs—frequent?
Down—Peaty marsh at Hillsborough; S.A.S. On Speltha Mountain, and by Cove Lake; H.W.L.
Antrim—Flow bogs near Cloughmills; D.M. Rathlin Island; S.A.S.

CLASS HEPATICÆ.

SERIES II. Order XXV. **JUNGERMANIACEÆ.**

Sect. I.—Jubuleæ.

FRULLANIA *Raddi.*

1. F. tamarisci (*Schmid.*) *Dumort.*
Stones, and rocks—frequent.
Down—On Slieve Donard in Ballagh Park and other places; H.W.L. Rostrevor wood, and abundant in Tollymore Park; C.H.W.
Antrim—Rathlin Island; S.A.S. Abundant on basalt at Sallagh Braes; H.W.L., and C.H.W.
Var. CORNUBICA *Carrington.*
Antrim—On stones at Fairhead; H.W.L.
Mr. Lett's station is the only locality in Ireland where this var. has been, as yet, noticed.

2. F. dilatata (*Linn.*) *Dumort.*
On trees, and sometimes on rocks—common.

3. F. fragilifolia *Taylor.*
Rocks—very rare.
Down—On granite rocks in Cove Mountain; H.W.L.

4. F. germana *Taylor.*
Damp rocks—very rare.
Antrim—Rathlin Island; S.A.S.

JUBULA *Dumortier.*

1. J. Hutchinsiæ (*Hook.*) *Dumort.*

JUNGERMANIACEÆ.

Damp rocks—very rare.
Down—At the waterworks on Rostrevor Mountain, and rocks by the river in Tollymore Park; C.H.W.

LEJEUNEA *Libert.*

1. L. hamatifolia (*Hook.*) *Dumort*
Damp rocks—rare.
Down—On rocks by the stream at Blackstairs, Slieve Donard; H.W.L, and C.H.W.
Antrim—Glenarm, and Colin Glen; D.M. In the latter locality recently; C.H.W.

2. L. calcarea *Libert.*
Damp rocks, and stones—rare, apparently.
Down—Wall at the base of a bridge over the Shimna River in Tollymore Park; C.H.W.

3. L. ovata (*Dicks.*) *Tayl.*
On tree trunks—rare.
Down—Creeping over holly by the stream at Blackstairs, Slieve Donard; C.H.W. Slieve Donard, but rare; H.W.L.
Antrim—Near Belfast (*Dr. Dickie*); *Moore, Ir. Hep.*

4. L. ulicina *Tayl.* (L. MINUTISSIMA *Hook. non Sm.*).
On trees, and mosses—rare.
Down—On hawthorn in the paddock at Gillhall; C.H.W.
Antrim—Colin Glen; D.M.

5. L. serpyllifolia (*Dicks.*) *Libert.*
Damp banks, and tree trunks—common, and abundant.

Sect. II.—Jungermanieæ.

RADULA (*Dum.*) *Spruce.*

1. R. complanata (*Linn.*) *Dumort.*
Trees, and rocks—very common.

2. R. aquilegia. *Taylor.*
Damp rocks—rare.
Down—Rocks called Blackstairs on east side of Slieve Donard; C.H.W,

JUNGERMANIACEÆ.

PORELLA (*Dill.*) *Lindb.*

1. P. platyphylla (*Linn.*) *Lindb.*
Rocks, and moist banks—common.

2. P. rivularis *Nees.*
On trees—very rare.
Down—Trees by the Lagan at Drumcro; C.H.W.

PLEUROZIA *Dumortier*

1. P. purpurea (*Light.*) *Dum.* (J. COCHLEARIFORMIS *Weiss*).
Wet heaths—very rare.
Down—Rostrevor Mountain, Rocky Mountain, and Hen Mountain above Hilltown; H.W.L.
Derry—Abundant, but always barren; D.M.

HERBERTA *Gray.*

1. H. adunca (*Dicks.*) *Gray.*
Damp rocks—very rare.
Antrim—On the fallen rocks (undercliff) of Fairhead; *Templeton Herb. Belf. Mus.*

ANTHELIA *Dumort.*

1. A. julacea (*Linn.*) *Dumort.*
Mountain rocks—rare.
Down—Slieve Donard; H.W.L. Bloodybridge River; C.H.W.
Derry—Clontygeragh Mountain; D.M.

TRICHOCOLEA *Dumortier.*

1. T. tomentella (*Ehrh.*).
Shady banks in woods and glens—rare.
Down—By the Spinkwee River at the top of Tollymore Park; S.A.S. Tollymore Park, but rare; H.W.L. Rostrevor Wood, and Moygannon Glen; C.H.W. and H.W.L.
Derry—By the Roe above Limavady; D.M.

BLEPHAROSTOMA *Dumortier.*

1. B. trichophyllum (*Linn.*) *Dumort.*

Wet rocky places—rare.
Down—Amongst *Sphagnum* in Tollymore Park, and on Slievenabrock; H.W.L.
Antrim—Near Belfast (*Templeton*); *Moore, Ir. Hep.* Sparingly in Colin Glen; C.H.W., and S.A.S.

LEPIDOZIA *Dumortier.*

1. L. reptans (*Linn*) *Dumort.*
Damp banks, and stones in bushy places—not common.
Antrim—By the stream in Colin Glen; *J. H. Davies; Phytologist*, 1859.
Derry—In Bond's Glen; *Templeton.* Side of the Roe above Limavady; D.M.

2. L. setacea (*Web.*) *Mitten.*
Damp banks in boggy places—rare.
Down—Amongst the *Sphagnum* in Annahilt bog; *Templeton.* By the Yellow-water River; C.H.W. Slievenamady; H.W.L.
Antrim—By a drain on the north side of Divis; *Templeton*, 1803. Moist banks in the parish of Rasharkin; *Moore, op. cit.*

BAZZANIA *Gray.*

1. B. trilobata (*Linn.*) *Gray.*
Heathy hills—not common.
Down—Amongst heather on the side of Thomas Mountain; C.H.W. Slieve Donard; H.W.L.
Antrim—Slemish; D.M.
Derry—Dart Mountain, but rare; D.M.

CEPHALOZIA (*Dumort.*) *Spruce.*

[C. FRANCISCI (*Hook.*) *Dum.*
In the late Dr. Moore's MS. notes, this species is mentioned as common in Derry. In his published list the only station assigned it is Bantry, and it may be assumed that his early note was an error.]

1. C. divaricata (*Sm.*) *Dumort.*
Rocks, and banks in heathy and mountainous places—not common.
Antrim—Slemish, and Fairhead (*Moore*), and near Glenarm (*Dickie*); *Moore, op. cit.* Stones on Slemish; S.A.S. Colin Glen, and Fairhead; C.H.W.
Derry—Mullaghmore, and Benbradagh; S.A.S.

2. C. bicuspidata (*Linn.*) *Dumort.*
Wet banks, and boggy places—very common.

3. C. connivens (*Dicks.*) *Spruce.*
Wet mossy banks— rare, or perhaps often overlooked.
Derry—Bogs between Swatragh and Kilrea ; D.M.

SACCOGYNA *Dumortier.*

1. S. viticulosa (*Linn.*) *Dumort.*
Damp rocks, and banks—frequent.
Down— Rostrevor Mountain, Donard demesne, White-river Glen, Spink-wee River, and Tollymore Park ; H.W.L. Rostrevor wood ; C.H.W.
Antrim—In Co. Antrim ; *Moore, op. cit.* Woodburn, and rocks at summit of Ballygally Head ; S.A.S. Sallagh Braes ; H.W.L.

KANTIA *Gray.*

1. K. trichomanis (*Linn.*) *Gray.*
Damp, shady banks—frequent.
Down—Glenmachan, and Castlereagh Glen ; S.A.S. Narrow-water, and Tollymore Park ; C.H.W. Slievenamady ; H.W.L.
Antrim—By the river at Drumbridge, and wet banks in Rathlin Island ; S.A.S.
Derry—Abundant by the Roe above Limavady ; D.M.

2. K. arguta (*Nees et Mont.*) *Lindb.*
Damp clay banks—frequent?
Down—Mountains above Newcastle, and in Ballymaghan Glen ; S.A.S. Slievenamady; H.W.L. By the Yellow-water River; C.H.W.

SCAPANIA *Dumortier.*

1. S. undulata (*Dill.*) *Dumort.*
Wet rocks, and boggy places in the mountains—common. Fruiting in April.
  Var. β PURPURASCENS *Huben.*
Down—Mourne Mountains ; H.W.L. Tollymore Park ; C.H.W.

2. S. nemorosa (*Linn.*) *Dumort.*
Woods, and shady places—not common.

Down—Trees at Purdysburn, and Castlereagh Glen; *Templeton*, 1803.
Rostrevor wood, and Tollymore Park; C.H.W. Slieve
Donard; H.W.L.
Antrim—Moist shady places about the pond in the orchard at Cranmore; *Templeton*. Sallagh Braes; H.W.L.
Derry—Abundant near Lignapeiste; D.M.

3. S. resupinata (*Linn.*) *Dumort.*
Woods, heaths, and damp rocky places—frequent.
Down—Slieve Donard; H.W.L. Rostrevor wood, and rocks on Eagle
Mountain, and Thomas Mountain; C.H.W.
Antrim—Cave Hill, and Colin Glen; S.A.S. Sallagh Braes; H.W.L., and C.H.W.
Derry—Ballyharrigan Glen; D.M.

4. S. curta (*Mart.*) *Dumort.*
Damp banks—very rare.
Antrim—Sallagh Braes, and Slemish Mountain; D.M.

DIPLOPHYLLUM *Dumortier.*

1. D. albicans (*Linn.*) *Dumort.*
Damp banks—very common. Fruiting in April. Ranges from sea level to 2700 feet.

LOPHOCOLEA *Dumortier.*

1. L. bidentata (*Linn.*) *Dumort.*
Moist rocks, damp banks, and boggy places—very common. Fruiting in April.

CHILOSCYPHUS *Corda.*

1. C. polyanthos (*Linn.*) *Corda.*
Wet banks in shady places—frequent.
Down—Rostrevor Mountain; C.H.W. Glenmachan; S.A.S.
Antrim—Colin Glen, Carr's Glen, and Rathlin Island; S.A.S.
Derry—Side of the Faughan below Clady; D.M.

Var. β RIVULARIS *Nees.*
Antrim—In a spring near Clough; H.W.L.

PLAGIOCHILA *Dumortier.*

JUNGERMANIACEÆ.

1. P. asplenioides (*Linn.*) *Dumort.*
Woods, and shady banks—very common.

2. P. punctata *Taylor.*
Damp heaths—very rare.
Antrim—Colin Glen, and Loughmourne; H.W.L.

3. P. spinulosa (*Dicks.*) *Dumort.*
Damp shady banks—rare?
Antrim—Sallagh Braes in tufts with *Dicranum;* H.W.L.

MYLIA *Gray.*

1. M. Taylori (*Hook.*) *Gray.*
Mountain heaths, and rocks—frequent.
Down—Shanslieve, and Deers' Meadow; C.H.W. Slieve Donard; S.A.S. White-river Glen, Moygannon Glen, and by the Spinkwee River; H.W.L.
Antrim—Top of Divis Mountain; *Templeton, Belf. Mus. Herb.*
Derry—Abundant at Glenedra, and near Lignapeiste; D.M.

JUNGERMANIA *Ruppius.*

1. J. cordifolia *Hooker.*
River banks, and wet places—rare.
Antrim—By the river three-quarters of a mile above Cushendun; *Moore, Ir. Hep.*
Derry—At 1600 feet on Sawel; D.M.

2. J. riparia *Taylor.*
Wet rocks, and margins of streams—rare.
Antrim—Rathlin Island; S.A.S.

3. J. turbinata (*Raddi*) (J. AFFINIS *Wils.*).
Limestone rocks—not rare in suitable places.
Antrim—On white limestone at Glenarm; *Moore, op. cit.* Chalk rocks at Carr's Glen, and Springfield Glen; S.A.S. Limestone by the stream in Colin Glen, fruiting in March; C.H.W.

4. J. sphærocarpa *Hooker.*
Damp rocks and banks—frequent.
Down—Pigeonrock Mountain, and Tollymore Park; H.W.L. Eagle

Mountain ; C.H.W.
Antrim—Carr's Glen ; S.A.S.
Derry—By mountain rivulets above Maghera ; D.M.

5. J. bantriensis *Hooker.*
Moist shady places—very rare.
Down—Annahilt bog (*anon.*) ; *Herb. Belf. Mus.* In a glen on the shore of Belfast Lough ; *Templeton, Herb. Belf. Mus.*
Antrim—Colin Glen ; S.A.S.
[J. BARBATA *Schmidel.*
Dr. Moore noted this plant as of common occurrence in Ireland, but most abundant in the north, especially the counties of Antrim and Donegal, while, at the same time, giving only one Irish station for *J. Lyoni* (*quinquedentata*). This distribution is the reverse of that observed by other botanists, and not having seen specimens of the true plant we must doubt the correctness of the record in the Report on Irish Hepaticæ.]

6. J. quinquedentata *Huds.* (J. LYONI *Tayl.*).
Rocks, and stones on mountains—frequent.
Down—Rostrevor Mountain, Moygannon Glen, and Slieve Donard ; C.H.W.
Antrim—Divis Mountain ; *William Thompson.* Cave Hill, and Sallagh Braes ; S.A.S.

7. J. exsecta *Schmidel.*
On rocks—very rare.
Antrim—Sallagh Braes ; *Moore, Ir. Hep.*

8. J. porphyroleuca *Nees.*
Damp banks and rocks—rare.
Down—Ballyvally near Rostrevor ; C.H.W.

9. J. ventricosa *Dickson.*
Heathy, and rocky banks—frequent.
Down—Moygannon Glen, and Slievemartin ; H.W.L. Peat at summit of Slieve Commedagh ; C.H.W.
Antrim—In County Antrim ; *Moore, op. cit.*

NARDIA *Gray.*

1. N. crenulata (*Smith*).
Damp banks, and heaths—frequent.
Down—Warrenpoint, Tollymore Park, and frequent in the Mourne Mountains ; C.H.W. Slieve Commedagh ; H.W.L.

Var. GRACILLIMA *Smith.*
Down—Frequent; C.H.W. Moygannon Glen; H.W.L.

2. N. hyalina (*Lyell.*) *Carrington.*
By streams, and in damp shady places—rare.
Down—Banks of stream on Rostrevor Mountain, Tollymore Park, and Spinkwee Glen; C.H.W. Slievenabrock, and Slievenamady; H.W.L.

3. N. obovata (*Nees*) *Carrington.*
Wet rocks—very rare.
Down—Rocks in stream on Slieve Donard; H.W.L.

4. N. compressa (*Hook.*) *Gray.*
Damp rocky places—rare.
Down—By the Blue Lake, and on stones at foot of Eagle Mountain, and at Windy Gap; C.H.W. Spinkwee River; H.W.L.

5. N. scalaris (*Schrad.*) *Gray.*
Moist banks—very common.

6. N. emarginata (*Ehrh.*) *Gray.*
Wet rocks, and by streams—common.
Var MINOR *Carrington.*
Down—Slievenabrock above Newcastle; H.W.L., 1881.

7. N. Funckii (*Web. et Mohr.*) *Carring.*
Moist rocky banks—very rare.
Antrim—Black Mountain, 1837; *Moore, op. cit.*

CESIA *Gray.*

1. C. crenulata *Carrington.*
Mountain rocks—frequent.
Down—Granite rocks on Slieve Donard; S.A.S. Slieve Donard, Slieve Commedagh, and Slievenamady; H.W.L.
Derry—West side of Clontygeragh; D.M. Metamorphic rocks on Mullaghmore; S A.S.

2. C. obtusa (*Lindb.*).
Rocks in subalpine situations—not uncommon.
Down—Slieve Donard, Slieve Commedagh, and Slievenamady; H.W.L. Thomas Mountain, Hen Mountain, and Hare's Gap; C.H.W.

JUNGERMANIACEÆ.

FOSSOMBRONIA Raddi.

1. F. pusilla (*Linn.*) *Dumort.*
Wet places—very rare.
Antrim—On the shores of lakes in Rathlin Island; *Moore, op. cit.*

PALLAVICINIA Gray.

1. P. hibernica (*Hook.*) *Gray.*
Wet places in sandy seashores—very rare.
Down—In small quantity on a damp bank in the sandhills north of Newcastle railway station; S.A.S., July 1887.

BLASIA Micheli.

1. B. pusilla *Linn.*
Wet banks—not uncommon.
Down -Tollymore Park, and Victoria Park; S.A.S. Moygannon Glen, and Narrow-water; C.H.W. Slievenamady; H.W.L.
Antrim—Banks of a stream on Knockagh; S.A.S.

PELLIA Raddi.

1. P. epiphylla (*Linn.*).
Wet banks, and sides of streams—very common. Fr. April.

2. P. calycina (*Tayl.*) *Nees.*
Wet banks— frequent.
Down—Dundonald Glen; S.A.S. Moygannon Glen, and Tollymore Park; H.W.L.
Antrim—Glenballyemon; *Moore, op. cit.* Colin Glen, and Carr's Glen; S.A.S.

ANEURA Dumortier.

1. A. pinguis (*Linn.*) *Dumort.*
Wet rocks and banks—frequent.
Down—Slieve Donard, and Bencrom; H.W.L. Rostrevor Mountain; C.H.W.
Antrim—Wet rocks by stream in Falls Park; S.A.S.

Other species occur, but definite localities cannot at present be stated with certainty. Further observations are required.

MARCHANTIACEÆ.

METZGERIA *Raddi.*

1. M. furcata (*Linn.*) *Dumort.*
On trees, rocks, and stones—very common.
   Var. ÆRUGINOSA *Hook.*
Down—Kirkcassock, and Gillhall; C.H.W.

2. M. pubescens (*Schrank*) *Raddi.*
Limestone rocks—very rare, and always sterile.
Antrim—Mountains near Belfast; *Templeton.* Limestone rocks by the roadside near Larne, and at Sallagh Braes; *Moore, op. cit* On limestone in Carr's Glen, and at base of Sallagh Braes; S.A.S. Sallagh Braes; H.W.L. and C.H.W.

3. M. conjugata *Lindberg.*
Wet rocks—very rare.
Antrim—Colin Glen; S.A S. Sallagh Braes; C.H.W.

SERIES II. ORDER XXVI. **MARCHANTIACEÆ.**

HEPATICA *Micheli* (FEGATELLA *Raddi*).

1. H. conica (*Linn.*) *Lindb.*
Wet banks—common.

MARCHANTIA *March.-fil.*

1. M. polymorpha *Linn.*
Woods, and damp mossy banks—very common.

ASTERELLA *Pal. de Beauv.*

1. A. hemisphærica (*Linn.*) *Beauv.*
Damp rocks and stones—frequent?
Down—Shady bridge-wall in Tollymore Park; H.W.L.
Antrim—Sallagh Braes; *Moore, op. cit.* Rathlin Island; S.A.S.
Derry—Basaltic rocks of Benevenagh; D.M,

PREISSIA *Corda.*

1. P. commutata *Nees.*
Damp rocks—rare?
Down—Cregagh Glen ; S.A.S.
Antrim—Sallagh Braes ; *Moore, op. cit.* Carr's Glen ; S.A.S.

TARGIONIA *Micheli.*

1. T. hypophylla *Linn.*
Dry rocks—rare.
Antrim—Cave Hill ; *Templeton.* Basalt rocks in the Deerpark at Glenarm ; *Moore, op. cit.*

SERIES II. Order XXVII. **RICCIACEÆ.**

RICCIA *Micheli.*

1. R. fluitans *Linn.*
Floating in still water—not common.
Down—Ditches near Lough Neagh at the entrance of the Lagan canal ; *Moore, op. cit.* Several places in the Newry canal ; S.A.S.

SERIES II. Order XXVIII. **ANTHOCEROTACEÆ.**

ANTHOCEROS *Micheli.*

1. A. punctatus *Linn.*
By streams, and ditch banks—very rare.
Antrim—Bottom of a ditch bank, shaded with grass, on the Stranmillis road ; *Templeton,* 1803. Glendun ; *Moore, op. cit.*
Derry—Rare, only observed near Lignapeiste ; D.M.

PLANTS EXCLUDED.

α Plants which are not indigenous, and not naturalised.*
β Plants erroneously recorded.

α *Non-native plants.*

1. Ranunculus parviflorus *Linn.*
In a cornfield at Newforge near Belfast in 1846 ; *Flor. Ulst.*
A casual introduced with seed.

2. Ranunculus arvensis *Linn.* CORN CROWFOOT.
Mr. Templeton, who met with this plant at Agnew's Hill, and in Mr. Barclay's shrubbery at Inver near Larne, rightly considered it introduced in those places It is not now found there.

3. Helleborus viridis *Linn.* GREEN HELLEBORE.
Near Dundrum; *Flor. Ulst., Supp.* By the Lough near Antrim; *Smith, Gossip About L. Neagh.*
Garden escapes.

4. Aquilegia vulgaris *Linn.* COLUMBINE.
Shores of Lough Neagh at Salterstown ; *Flor. Hib.* Crawfordsburn ; *Flor. Ulst., Supp.* Shane's Castle, and other places ; S.A.S.
Escaped from cultivation in all the stations, and scarcely naturalised.

5. Berberis vulgaris *Linn.* BARBERRY.
Hedges at Conlig, Ballyhackamore, etc. ; S.A.S.—planted.

6. Corydalis lutea *D. C.* YELLOW FUMITORY.
Hedges between Whitehouse and Carrickfergus ; *Templeton.* Wall

---

* By a naturalised plant is understood not merely an exotic which is nature-sown, but one which in addition has become *permanently* established as a member of our flora.

of a bridge at Crumlin; *B.N.F.C.*, 1864. Wall at Purdysburn; *Flor. Ulst.* Old walls at Hillsborough Park; S A.S.

Has had a garden origin though now growing spontaneously in some places.

7. Cheiranthus cheiri *Linn.* WALLFLOWER.

Old walls at Carrickfergus; *Templeton.* Old walls at Lisburn; *Flor. Belf.* Walls at Strangford; *Flor. Ulst.* Old walls at Newtownards, and Greyabbey; S.A.S.

Introduced from gardens, but probably naturalised in some of the above localities.

8. Arabis perfoliata *Lamk.* TOWER MUSTARD.

In a small field at the foot of the mountain at Whiterock, 1847 (*Orr*); *Cyb. Hib.*

Not found since.

9. Cardamine impatiens *Linn.* BITTER CRESS.

Near the shore of Lough Neagh at Shane's Castle; *Rev. W. M. Hind, Phyt. O. S. vol.* 5.

This is a cultivated portion of the grounds embellished with many native, as well as exotic, plants. No plant found there, and there only, could be received as a member of the indigenous flora.

10. Hesperis matronalis *Linn.* DAMES' VIOLET.

Ballylesson, and Castlereagh Hill; *Flor. Belf.* Shane's Castle grounds; T.H.C.

Escapes from cottage gardens.

11. Brassica oleracea *Linn.* CABBAGE.

Among stones on the beach at Rathlin Island (*Gage*); *Flor. Ulst.*

Escaped from cultivation.

12. Diplotaxis tenuifolia *D.C.*

Rubbish heaps at the foot of the Milewater close to Belfast; *W. Millen.*

One of the large group of casuals which appeared at that spot when the railway to Ballymena was made.

13. Alyssum maritimum *Linn.*

Sandy shore near Groomsport church; S.A.S., 1866.

A garden outcast which established itself there, and maintained its place until lately.

14. Draba muralis *Linn.*
Old walls about Belfast; *Flor. Ulst., Supp.*
A garden escape.

15. Camelina sativa *Linn.*
Fields at Malone; *Templeton*, 1799. Flax fields about Ballinleg ; *Ir. Flor.* Fields about Castlereagh, Belfast, and Ballycastle ; *Flor. Ulst., Supp.* Ballyholme, and Troopers Lane ; T.H.C. About Antrim ; *Smith, op. cit.* Cushendun ; *Rev. S. A. Brenan.*
An immigrant imported with flaxseed.

16. Lepidium sativum *Linn.*
Shore near Belfast; *C. C. Babington*, 1837.
Escape from cultivation.

17. Senebiera didyma *Persoon.*
Waste heaps on the shore at the harbour of Magheramorne, 1865, and subsequently; S.A.S.
Ballast is emptied there, and thus the plant has been imported.

18. Reseda lutea *Linn.*
Banks of the river at Ballycastle ; *Flor. Ulst., Supp.* Ballast heaps, and quarry spoil-bank at Magheramorne, and gravelly waste ground close to the railway station at Larne ; S.A.S. Sandhills at Portrush ; R.Ll.P.
Possibly more than a casual, perhaps should be considered as naturalised.

19. Reseda suffruticulosa *Linn.* MIGNONETTE.
On the commons at Glenarm ; *Ir. Flor.* At Bangor, and on shores of Strangford Lough near Comber ; *Flor. Belf.* Shore at Newcastle ; *Cyb. Hib., Supp.*
Garden escapes.

20. Helianthemum vulgare *Gaertner.* ROCKROSE.
Glendarragh near Crumlin ; *Flor. Ulst., Supp.*
Planted, of course.

21. Tilia europæa *Linn.* LIME.
Skirts of the mountains in Co. Down ; *Ir. Flor.*
Lime trees are frequent, but none of these is truly wild.

22. Dianthus deltoides *Linn.* MAIDEN PINK.
Pastures at Sandymount near Belfast; *Flor. Ulst., Supp.*

Not to be found; most probably Sandymount near Dublin was intended. In any case a garden escape.

23. Silene armeria *Linn.* PINK.
Rathlin Island; *Dr. Marshall.* Sandy cornfield by the Roe in Co. Derry; *Flor. Hib.*
Escapes from cottage gardens.

24. Saponaria officinalis *Linn.* SOAPWORT.
Near Belfast, but an outcast; *Templeton.* Shores of Lough Neagh at Massereene Park, and glen of the Roe; *Flor. Ulst.* Rathlin Island; S.A.S.
In all cases an escape from floriculture.

25. Hypericum elatum *Aiton.*
Plantation above Donard Lodge; *Flor. Ulst., Supp.*
An introduced ornamental plant.

26. Acer campestre *Linn.* MAPLE.
Comber, and Ballyholme; *Flor. Belf.* Counties of Antrim and Derry; *Flor. Ulst.* Shores of Lough Neagh, and wild on north bank of Agivey (sic) River (? Moyola) near Castledawson, also hedge in Craigywarren; *Cyb. Hib.*
Apparently wild in some places, but at best descended from ancestors that were planted.

27. Acer pseudo-platanus *Linn.* SYCAMORE.
By several rivers in Antrim and Derry; *Cyb. Hib.*
Common in plantations, and by roadsides, but never really wild.

28. Geranium phæum *Linn.*
Roadside one mile south of the church in Islandmagee; *Cyb. Hib., Supp.* Plentiful, and fine, amongst dense vegetation by a little stream that cuts deep into the Boulder Clay, and empties into Knock Glen near the lower end; S.A.S.
Quite wild in appearance, but not admitted as an Irish native.

29. Geranium striatum *Linn.*
Ditch bank near Lisburn; *Cyb. Hib.* Banks near Templepatrick, and in an old orchard at the foot of Keady Hill; S.A.S. In Crawfordsburn, and on a hedge bank near the railway station at Newtownards; T.H.C.
Garden escapes, and nowhere permanently established.

30. Oxalis corniculata *Linn.*

Lane leading to Carr's Glen near Belfast; *T. Darragh.*
Escaped from cultivation.

31. Oxalis stricta *Linn.*
Near Belfast (*Geo. O'Brien*, 1842), and at Lisnegarvey near Lisburn (*S. Pim*); *Cyb. Hib.*
Wildings having a garden origin.

32. Linum usitatissimum *Linn.* FLAX.
Frequently met with in fields and waste ground, but only as a result of husbandry.

33. Medicago sativa *Linn.* LUCERNE.
Stranmillis, and Holywood; *Flor. Belf.* Abundant on railway bank close to Larne; S.A.S., 1869, and T.H.C., 1877.
A fodder plant nowhere native in this country.

34. Medicago falcata *Linn.*
Waste ground at the foot of the Milewater; *Millen, Phyt., O.S., vol. V.*
One of the numerous new plants which came up at that place spontaneously after 1847 when the railway was constructed.

35. Melilotus officinalis *Willd.* MELILOT.
On the Curran of Larne; *Cyb. Hib.* Sandhills at Portrush; R.Ll.P., 1887.

36. Melilotus arvensis *Willd.*
Seashore at Donaghadee (*Maffett*); *Cyb. Hib.* In great abundance on the railway bank at several points between Kilroot and Larne; *R. Tate*, and S.A.S., 1869. T.H.C., 1878, and R.Ll.P., 1886. Railway bank close to Cultra railway station; S.A.S., 1873.
Long established, and perhaps to be considered as naturalised in the district. These gravel-loving plants are, however, very uncertain.

37. Melilotus alba *Lamk.*
A number of plants in the gravel pits north of Giants' Ring; S.A.S., 1872 (still found there). Holywood; R.Ll.P.
An alien, not expected to be permanent.

38. Melilotus parviflora *Desf.*
Near Holywood; R.Ll.P., 1887.
An annual that has appeared at Holywood, with a number of other casuals, several of which are exotics.

39. Trifolium ornithopodioides *Linn.*

On a gravelly bank sloping to the sea at the north end of the Kinnegar at Holywood ; *Templeton.*

Not now found at the Kinnegar, or elsewhere in the district—a casual.

40. Lotus tenuis *W. & K.*

Sparingly on railway bank near Larne ; S.A.S., 1869.

Not found recently, the gravel bank on which it grew having been removed. From the suspicious *habitat* and surroundings, excluded as an alien.

41. Vicia tetrasperma *Moench.*

Ballyronan, and Ballycastle ; *Flor. Ulst., Supp.* Cornfield at Giants' Ring, and cornfield by the Lagan canal below Shawsbridge ; *C. Dickson.*

Mr. Dickson's plants were rightly named, but it is not certain that the others were correct. In any event, an introduced tillage plant.

42. Vicia bithynica *Linn.*

Railway bank near Whitehouse, and at the Milewater ; *Millen, Phytologist, O.S.*

An introduced alien.

43. Vicia sativa *Linn.* VETCH.

Borders of fields, and waste places in cultivated ground, frequent as a casual, but nowhere permanent.

44. Lathyrus aphaca *Linn.*

Near Belfast ; *Millen, Phyt., O.S.* Shady places in the grounds of Stranmillis ; *Jos. Murphy*, 1864, and known there until 1872.

Casual—not native in Ireland.

45. Onobrychus sativa *Lamk.* SAINFOIN.

Cherryvalley near Comber ; *Flor. Belf.* Fields near Whitehouse ; *W. Macmillan.*

A fodder plant spread from fields where cultivated.

46. Prunus insititia *Linn.* BULLACE.

In hedges, evidently planted ; *Templeton.* Rostrevor wood ; *Ir. Flor.* Ballynascreen, and Drumachose ; D.M. Colin Glen ; *Flor. Ulst.* Hedge near Purdysburn ; *G. C. Hyndman.* Frequent in many parts of Antrim and Derry ; *Cyb. Hib.* Knock Glen, and hedges at Drumbo, Ballymaleidy, and between Ballywalter and Greyabbey ; S.A.S.

In some instances growing spontaneously, but in such cases derived from an introduced stock.

47. Prunus cerasus *Linn.* DWARF CHERRY.
Among native trees on the shores of Lough Neagh "'fruit red, astringent, bitter, and not relished by birds"; *Templeton.* Ballynascreen, and Curlyburn; D.M.
An introduced bush, apparently much scarcer now.

48. Spiræa salicifolia *Linn.*
By Enagh Lough, and by the Roe at Pellipar near Dungiven; D.M. Roadside near Ballynahinch, and hedge near Drumbo; *Flor. Ulst.* Hillsborough Park; S.A.S.
An ornamental shrub escaped from cultivation.

49. Spiræa filipendula *Linn.* DROPWORT.
In some quantity on railway bank at the level crossing a half mile northwest of Knock station; *W. H. Patterson, M.R.I.A.*, 1868.
A rare plant in Ireland, and confined to the west. It held its ground here for some ten years, but has now disappeared, and must be regarded as one of those waifs which agricultural seed, or other means, accidentally spread abroad.

50. Fragaria elatior *Ehr.* HAUTBOY STRAWBERRY.
Near the old castle in Shane's Castle Park; *Cyb. Hib.*
Originally planted—not naturalised.

51. Rosa cinnamomea *Linn.* CINNAMON ROSE.
Ogilvie's demesne in Lower Cumber, but planted; D.M. Naturalised near Clady, Co. Derry; *Flor. Hib.*
The term "naturalised" has been used in a most elastic sense by some botanists. This rose may, perhaps, be sometimes nature-sown, but is not naturalised. The two phrases are not synonymous.

52. Rosa lutea *Link.* YELLOW ROSE.
Bellahill near Carrickfergus; *Wade Rar.*
A Continental rose not native in Britain—planted.

53. Pyrus communis *Linn.* PEAR.
Hedge at Thronemount near Belfast; *Flor. Belf.*
Planted.

54. Epilobium roseum *Linn.*
Casual, *vide* page 54.

55. Epilobium tetragonum *Linn.*

Frequent in moist places; *Flor. Ulst.* Side of field near Holywood; R.Ll.P.

The plant of Flora of Ulster, and of the earlier Floras was *E. obscurum*. The specimen from Holywood seems right, but must be considered as casually introduced.

56. Œnothera biennis *Linn.* EVENING PRIMROSE.

Gravelly shore of the lough at Massereene Park; *Smith, Gossip about Lough Neagh.*

Introduced as an ornamental plant, and has spread to the shore.

57. Sedum album *Linn.*

Roofs of houses in Antrim; *Moore, Nat. Hist. Rev., vol. VI.,* 1859. Walls in Massereene Park; *Cyb. Hib.*

Planted.

58. Sedum micranthum *Bast.*

On a wall at Greenmount gate south of Antrim; *Cyb. Hib.* (still there). Like the preceding, an introduced ornamental plant.

59. Sedum dasyphyllum *Linn.*

Near Ballyclare, Carrickfergus, and Ballycastle, and in the Island of Rathlin; *Flor. Ulst., Supp.* Rocks in Glenariffe; *Tate, Jour. Bot.,* 1870.

Some of the localities may be due to incorrect naming, others have the true plant, but in such cases introduced.

60. Sempervivum tectorum *Linn.* HOUSELEEK.

Tops of thatched houses in Co. Derry; D.M. On houses, walls, and cottages; *Flor. Belf.* and *Flor. Ulst.* Derriaghy; *Hind.* Ballygomartin; S.A.S. Conlig, and Comber; T.H.C.

Planted in every instance.

61. Ribes grossularia *Linn.* GOOSEBERRY.

Limestone quarry at Keady; D M. Naturalised in hedges; *Flor. Belf.* In different places; *Flor. Ulst., Supp.* Dunleidy Glen near Dundonald; T H.C. Rocky heath on Rathlin Island; S.A.S.

Additional localities could be enumerated, but always as one of the waifs and strays of cultivation.

62. Ribes nigrum *Linn.* BLACK CURRANT.

In different parts of the district; *Flor. Ulst., Supp.*

Escaped from cultivation.

63. Ribes rubrum *Linn.* RED CURRANT.

Shores of Lough Neagh near Antrim; *Templeton.* By the Roe near Limavady; D.M.

Escapes from the fruit garden.

64. Saxifraga umbrosa *Linn.* LONDON PRIDE.

Among heath on east side of Divis Mountain; S.A.S. Colin Glen, and by the Lagan canal near Belvoir; T.H.C.

Many more stations could be given as instances of garden escapes.

65. Saxifraga geum *Linn.*

Naturalised in Cultra demesne; *Flor. Ulst., Supp.*

Not naturalised, but planted.

66. Petroselinum sativum *Hoffm.* PARSLEY.

On the old walls of Londonderry; D.M.

Escaped from kitchen garden.

67. Carum carui *Linn.* CARAWAY.

Waste ground about the ruins of Greyabbey; *Ir. Flor.* Glen of the Roe above Dungiven, and meadow between Magherafelt and Castledawson; D.M. By the Quoile below Inch Abbey, roadside near Selsban, and on waste ground near Toome station; S.A.S. Whitepark near Ballintoy; R Ll.P.

Escapes from cultivation.

68. Meum athamanticum *Jacq.* BALDMONEY.

On the lawn at Maryville, 1808, and found plentifully on the lawn at Cranmore, 1815; *Templeton.*

A casual—did not remain. Mr. Templeton says "introduced."

69. Peucedanum ostruthium *Koch.* MASTERWORT.

Old hedges in the townland of Ballydolaghan, Co. Down; *Flor. Hib.* Ditch bank in the parish of Ballintoy; *Cyb. Hib.* Near Cushendall; *Rev. S. A. Brenan*, 1872. On the western boundary of the district near Cookstown; *W. Macmillan*, 1884. Abundant in a meadow by the path to MacArt's Fort, Cave Hill; S.A.S., 1886.

An introduced plant, brought with seed, and never permanent.

70. Pastinaca sativa *Linn.* PARSNEP.

Near the foot of Crumlin River, but perhaps not indigenous; *Templeton*, 1799. Abundant in sandy fields in the townland of Margymonaghan in Magilligan; D.M.

Certainly introduced, not now found at Crumlin, and perhaps gone from Magilligan.

71. Chærophyllum sativum *Lamk.* CHERVIL.
Waste ground on the commons along the seashore at Glenarm; *Ir. Flor.*
A casual which has, long since, disappeared, and is nowhere native in Ireland.

72. Coriandrum sativum *Linn.* CORIANDER.
Gravelly waste ground at Stockman's Lane, Malone; S.A.S.
A casual, not found now.

73. Viburnum lantana *Linn.* WAYFARING TREE.
Colin Glen (*Orr*, and *Thompson*); *Flor. Ulst.* Hedge at Oldpark near Belfast; *R. Tate.*
An accidental plant at Oldpark, in a hedge which has been cut down, and the plant thus lost. The Colin Glen record was an error, the plant found there being *V. opulus.*

74. Lonicera xylosteum *Linn.*
Walls at Bellahill; *Flor. Ulst., Supp.*
A garden plant.

75. Centranthus ruber *D. C.* RED VALERIAN.
Near Newcastle; *Flor. Ulst., Supp.*
An escape from garden culture.

76. Valeriana pyrenaica *Linn.*
Near Antrim; *Rev. W. S. Smith*, 1885.
An introduced Continental shrubbery plant.

77. Valerianella carinata *Loisel.* CORNSALAD.
Ditch bank on right side of the main road about a half mile east of Dundonald, and abundant on ditch banks on bye road leading thence to Holywood Hills; S.A.S., 1871.
Still maintains its position as a colonist at this place.

78. Petasites fragrans *Presl.* FRAGRANT BUTTERBUR.
In shade at Newtownbreda, Sydenham, Knock, and Stranmillis; S.A.S. Ormeau Park, and Malone; T.H.C.
Introduced to shrubberies, and now run wild, and spreading.

79. Petasites albus *Gaertner.*

Shady bank at Ballymenoch near Holywood; R.Ll.P.
Like the preceding an ornamental plant in copses—escaped.

80. Achillea tomentosa *Linn.*
Near Newcastle (*Miss Keown*); *Ball, Ann. Nat. Hist.*, 1839.
A casual.

81. Anthemis arvensis *Linn.* CORN CHAMOMILE.
Near Lambeg; *Templeton*, 1800. Shores of Strangford Lough; *Cyb. Hib.*
Introduced with seed.

82. Anthemis cotula *Linn.* STINKING CHAMOMILE.
In a field near Comber; *Templeton*, 1807. Mountcollier near Belfast, and fields near Kilroot; *Flor. Ulst.*
Introduced with agricultural seed.

83. Matricaria parthenium *Linn.* FEVERFEW.
Malone; *Templeton.* Knocknagoney, and Groomsport; *Flor. Belf.* Holywood, and Cavehill; *Flor. Ulst., Supp.*
Semi-wild localities for this species might be greatly multiplied, as it occurs very frequently in waste places, and on ditch banks near cottages. It is an ancient remedy for a number of ailments.*

84. Artemisia absinthum *Linn.* WORMWOOD.
At Lambeg; *Flor. Ulst., Supp.* Rocks, and ruins of an old house near the shore at St. John's Point; S.A.S., 1866.
Escapes from cultivation.

85. Artemisia campestris *Linn.*
At the foot of the Milewater near Belfast; *W. Millen.*
A casual which has long since disappeared.

86. Doronicum pardalianches *Linn.* LEOPARD'S BANE.
Ballymacash; *Phyt. O.S., vol. V.* Hedge bank near Ballylesson, waste ground by Drumbo Presbyterian church, and in the Braid near Broughshane; S.A.S. Beech Hill near Antrim; *Rev. W. S. Smith.*
All garden escapes.

87. Senecio viscosus *Linn.*

*"It is very good for them that are pursie, or troubled with the shortness of wind, and for melancholick people, and such as be sad and pensive and without speech." *Dodoens Herbal*, 1619.

In profusion on rubbish heaps at the foot of the milewater; *W. Millen*.

This was one of the most abundant and persistent of the remarkable colony of aliens which came up when the Ballymena railway was constructed in 1847. It was the last survivor of the group, and in defiance of all changes and disturbance held its place, and flourished for nearly forty years, and only disappeared when the site it occupied was recently built over. It is also singular that it failed to spread beyond its original bounds, although there are *habitats* quite as suitable in the immediate vicinity.

88. Senecio saracenicus *Linn.*

Ruins of Dunluce Castle; *Templeton*. Banks of a ditch at Kilkeel; *Ir. Flor.* Balloch; Co. Down; *Flor. Ulst., Supp.* By a stream near Cushendall coastguard station; *Dr. J. S. Holden*. In the last named locality; S.A.S., who was informed by old people on the spot that it was not planted, but "always grew there."

No doubt originally planted not far from the present stations.

89. Carduus nutans *Linn.* MUSK THISTLE.

Railway bank two miles from Carrickfergus; *Flor. Ulst.* A single plant by the Drumgavna River in the parish of Balteagh, Co. Derry; D.M.

The Antrim plant was a casual that appeared soon after construction of the railway. The Derry plant was, no doubt, a fleeting waif not seen again.

90. Silybum Marianum *Gaert.* MILKTHISTLE.

West of Lisburn, and at the Curran of Larne; *Templeton*. Magilligan; *Samp's. Surv. Derry*. Ram's Island; *F. Whitla*. Ballynascreen; D.M. Between Newtownards and Greyabbey, and between Larne and Ballyclare; *Flor. Ulst., Supp.* Portbradden, Toomebridge, and three miles south of Toome; S.A.S. Magheramorne, and Glendun; R.Ll.P.

Now growing spontaneously in these and other places, but always having a garden origin.

91. Tragopogon porrifolius *Linn.* SALSIFY.

Railway banks at Sydenham; *Flor. Belf.* Churchyard at Glenarm; *Flor. Ulst., Supp.* Abundant on railway banks from Whitehouse to Macedon; S.A.S., 1873. In plenty at Knock station; T.H.C., 1880. Railway bank at Marino; R.Ll.P., 1886.

An erratic plant with a decided preference for railway tracks, appearing occasionally, as shown above, on all the Belfast lines. It is also very capricious in the times of its appearance, there being intervals of years when it would seem to have left, or to be leaving its old haunts entirely. It reappears suddenly, however, either at the old spot, or in new ground.

It was unknown to Templeton, and is without doubt of modern introduction, but as to how it came nothing is known. The headquarters of the salsify have always been in the vicinity of Sydenham station, and the porter who was appointed to that station when the line was made stated that the plant was there when he came to the place.

92. Crepis biennis *Linn.*

In profusion in a meadow at the southwest side of the town dam of Belfast Waterworks, 1880, and in great abundance by the railway north of Whiteabbey, and in adjoining meadows and pastures, 1884 ; S.A.S. By the railway at Dunadry near Antrim ; *A. G. More.* By the railway at Marino, and in fields near Cairncastle ; R.Ll.P., 1886.

An alien of quite recent introduction, which continues to spread beyond precedent. It would seem to have been imported with grass seed.

93. Hieraceum auranticum *Linn.*

Old walls at the ruins of Portavo ; *Dr. H. Burden.* Stony margin of the clear-water basin of the Belfast waterworks ; S.A.S. (now lost).

Escapes from garden culture.

94. Campanula latifolia *Linn.* GIANT BELLFLOWER.

In great abundance in the woods of Gillhall, especially by the Lagan at the western extremity of the demesne ; *C. Dickson*, 1883.

Cultivated in gardens, and introduced to the wood. In the wild western portion of the grounds it has spread amazingly.

95. Campanula rapunculoides *Linn.*

In the sandhills about half a mile north of Newcastle station, and not very far from the railway line ; S.A.S., 1871 (about fifty plants seen).

A casual accidentally introduced, and apparently now lost.

96. Campanula rapunculus *Linn.* RAMPION.

Banks by the old Lisburn road near Ballydrain ; *Flor. Ulst., Supp.*

Not a native of Ireland, and not now found near Ballydrain.

97. Vinca major *Linn.* GREATER PERIWINKLE.

Roadside near Banbridge ; *G. C. Hyndman.* Ditch bank at north end of Islandmagee ; S.A.S.

Garden strays.

98. Vinca minor *Linn.* LESSER PERIWINKLE.

Rockport, and Derriaghy ; *Flor. Belf.* Abundant at the ruins of the old church at Kilroot : S.A.S.

Garden plants which have run wild.

99. Chlora perfoliata *Linn.* YELLOWORT.
Railway bank near Kings' Moss; *Millen (spec.)*.
A railway casual which came up soon after the line was constructed, and has, long since, died out.

100. Limnanthemum nymphœoides *Linn.*
In rivers and watercourses; *Samp's. Surv. Derry*, 1814. Naturalised in the Lagan above the Botanic Gardens; *Flor. Hib.* Lagan canal between first and second locks; *Flor. Belf.* In the Lagan, and in various parts of Lough Neagh; *Flor. Ulst., Supp.* Bog meadows; *Richd. Hanna.*
An aquatic plant often grown in ornamental waters.
Not native in Ireland, but was planted in the Lagan by Mr. J. Campbell, the first, or one of the earliest curators of the Belfast Botanic Gardens.

101. Polemonium cæruleum *Linn.* JACOB'S LADDER.
Near Bellaghy, etc.; *Flor. Ulst., Supp.*
A garden escape.

102. Cuscuta epilinum *Weihe.* FLAX DODDER.
Rathlin Island; *Miss Gage.* Flax fields at Moneymore, etc.; *Flor. Ulst , Supp.* Flax fields at Orlock Point; T.H.C.
An unwelcome alien imported with flaxseed.

103. Borago officinalis *Linn.* BORAGE.
On the Cave Hill among limestone; *Templeton.* On the banks of the Bann in Co. Derry; *Scientific Tourist.*
Garden escapes.

104. Symphytum tuberosum *Linn.*
Rubbish heaps at the Antrim road, streams near Whitewell, and on Carnmoney Hill above the Church; *Millen, Phyt. O.S.* Plentiful in a wood at Clandeboye; *Rev. Geo. Robinson*, 1887.
The *habitats* assigned to this plant by Millen are unlikely. It cannot, however, be tested, as he did not keep specimens, but the plant is not now found in any of his stations, and was probably misnamed. There is no uncertainty as to the Clandeboye station, and the *habitat* is so natural that if it were made clear that the plant was not designedly introduced it might rank as a native.

105. Atropa belladonna *Linn.* DEADLY NIGHTSHADE.
Benvarden near Ballycastle (*Miss Hincks*); *Hind, Phyt. O.S., vol. V.*
A garden plant.

PLANTS EXCLUDED.

106. Orobanche minor *Sutton.*
Abundant in a clover field near Ballylesson; *C. Dickson*, 1881.
Imported with seed.

107. Antirrhinum majus *Linn.* SNAPDRAGON.
Old walls at Lisburn; *Templeton.* Ruins of Greyabbey; *Ir. Flor.*
Old walls at Woodburn; *Flor. Belf.* Old walls at Carrickfergus; *Flor. Ulst.*
Not native, but well established—perhaps naturalised—on old walls.

108. Antirrhinum orontium *Linn.*
Sandy fields at Struell near Downpatrick; *Miss Mulgan*, 1868.
A cornfield casual.

109. Mimulus luteus *Linn.*
Shane's Castle Park, Carrickfergus commons, by the stream at Glynn, and in abundance by the Bann at Coleraine; S.A.S. At the western boundary of the district near Cookstown; *Wm. Macmillan.*
An immigrant plant from North America which is establishing itself quite freely in wet places.

110. Veronica peregrina *Linn.*
Plentiful in the fruit garden of The Lodge near Belfast, July, 1856, especially abundant amongst parsley; *Hind, Phyt. N.S., vol. II.* 1857-8. Belfast, and Londonderry; *Flor. Ulst., Supp.*
An alien from north Continental parts, first observed in Ireland, in County Tyrone, in 1836. It certainly has not spread here, and is perhaps extinct about Belfast.

111. Mentha viridis *Linn.* SPEARMINT.
Said to have been found near Belfast; *Flor. Hib.* Wet waste ground at Ballygowan; *Flor. Belf.* Near Belfast, etc.; *Flor. Ulst., Supp.*
Escapes from cottage garden cultivation.

112. Mentha rotundifolia *Linn.*
Waste ground at Bangor, Crawfordsburn, and Islandmagee; *Flor. Belf.* In County Derry; *Flor. Ulst.* In Glendarragh, and on the river bank below Crumlin; *Cyb. Hib.* Abundant by the shore one and a half miles southwest of St. John's Point, and sparingly on waste ground by the shore near Mountstewart; S.A.S.
Assumed to be escapes from cultivation.

113. Verbena officinalis *Linn.* VERVAIN.
On Saintfield road near Ballygowan; *Flor. Belf.*

One plant close to a cottage—casual. Templeton notes this species as about gardens.

114. Chenopodium vulvaria *Linn.* (C. OLIDUM *Curt.*).

On my own dunghill, in flower Sept. 21, 1808, *Templeton.*

A casual; formerly cultivated in gardens, and at one time in much repute in medicine on account of its abominable smell.*

115. Polygonum bistorta *Linn.* SNAKEWEED.

On the back lawn of Moira Castle; *Templeton.* Cherryvalley near Comber; *Grainger.* Plentiful on limestone at Springhill near Moneymore; D.M. Thornvale near Belfast, lower end of Colin Glen, and meadow at Castle Dobbs; *Flor. Ulst.* Meadows near the old Shane's Castle; *Cyb. Hib.* Greyabbey demesne; *B.N.F.C.*, 1871. Marino (but cultivated in a garden close by); *W. H. Patterson.* Abundant on the lawn at Kilwaughter; S.A.S.

Usually a garden escape, but may be naturalised in some of the above stations.

116. Polygonum fagopyrum *Linn.* BUCKWHEAT.

Fields, and broken ground about Ballyleidy, Co. Down; *Ir. Flor.* Escaped from cultivation, nowhere native in Britain.

117. Euphorbia cyparissias *Linn.* CYPRESS SPURGE.

Crawfordsburn; *Flor. Ulst., Supp.*

A garden plant.

118. Euphorbia lathyrus *Linn.* CAPER SPURGE.

By the side of an old road near Carnmoney; *Flor. Ulst., Supp.* Introduced at Ballytweedy; *Rev. W. S. Smith.*

A garden escape.

119. Humulus lupulus *Linn.* HOP.

About the ruins of Greyabbey; *Ir. Flor.* Roadside at Belfast where Mountcharles is now built, and again near Cranmore, also by a rivulet that runs from Ballymenoch to the sea; *F. Whitla.* Newforge, and roadside between Dunmurry and Colin Glen; *Flor. Belf.*

Escapes from cultivation—not established.

120. Cannabis sativa *Linn.* HEMP.

* All the herbe stinketh like rotten corrupt fish, or like stinking fish broth, or like a ranke stinking goate . . . . and is taken of some to be that stinking herbe, that of Plautus is named *Nautea* (nausea); *Dodoens Herbal*, 1619.

Near Antrim; *Rev. W. S. Smith*. Borders of a cornfield at Trooper's lane station; T.H.C.
A casual probably originated from birdseed.

121. Ulmus suberosa *Ehrhart*. ELM.
In woods, glens, and plantations, common, but so much planted that it seems impossible to distinguish between spontaneous and artificial growths.

122. Salix fragilis *Linn*. BEDFORD WILLOW.
Var. α S. DECIPIENS *Hoffman*.
Hedge near Limavady; D.M.
Var. β S. RUSSELLIANA *Smith*.
Near Kilkeel; *Mack, Catal*. By the Roe above Dungiven; D.M. Shane's Castle; *Cyb. Hib*.
Probably planted in these localities.

123. Salix undulata *Ehrhart*.
Neighbourhood of Coleraine, and by the Roe near Limavady; D.M. Between Coleraine and Magilligan; *Flor. Hib*. Abundant near Ballymoney; *Cyb. Hib*.
Planted for basketwork.

124. Salix rubra *Hudson*.
Hedges about Newtownards, and between Stewartstown and Moneymore; *Templeton*. Banks of the Faughan above Oaks Lodge; D.M. Glendermot in County Derry (var. *Forbyana*), and abundant near Ballymena, and Cloghmills (*Forbyana*); *Cyb. Hib*.
Most probably in all cases planted.

125. Populus alba *Linn*. WHITE POPLAR.
Sydenham, and Falls road; *Flor. Belf*.
Often seen in hedges and copses, but planted.

126. Fagus sylvatica *Linn*. BEECH.
Not native anywhere in Ireland.
This is such an abundant and valuable member of our sylvan flora that it seems unnatural to stigmatise it as an intruder, but the fact that it has not received an Irish name would, of itself, indicate that its introduction is of comparatively recent date.

127. Castanea vulgaris *Lamk*. CHESTNUT.
The sweet chestnut is often seen by roadsides, and in copses, but only where planted.

128. Pinus sylvestris *Linn.* SCOTCH FIR.
Vide page 134.

129. Stratiotes aloides *Linn.* WATER SOLDIER.
Belfast watercourse (*Templeton*); *Flor. Hib.* Whitehouse milldam; *W. Millen.* Springfield dam; *Flor. Belf.*

The origin of the plant in the Stranmillis watercourse was perhaps accidental, in the milldams, there can be little doubt, it was intentionally introduced. Now extinct in all three stations.

130. Elodea canadensis *Mich.* WATER THYME.
Vide page 135.

131. Orchis pyramidalis *Linn.*
A single specimen on gravelly pastures at Magilligan; D.M. Two plants on sandy shore at Ballyholme; S.A.S., 1873.
Stray plants accidentally brought, and not established.

132. Iris fœtidissima *Linn.* GLADDON.
Riverside at Newcastle, and woods at Greyabbey; *Ir. Flor.* Near Coleraine; D.M. Plentiful in a meadow by the stone cashel at Cushendun; *B.N.F.C.*, 1884.
Esteemed as a medicinal plant, and frequent, therefore, in gardens, from whence the wildings have originated.

133. Colchicum autumnale *Linn.* MEADOW SAFFRON.
Rich meadows about the ruins of Greyabbey; *Ir. Flor.*
A garden escape—not found now.

134. Acorus calamus *Linn.* SWEET FLAG.
Lakes at Hillsborough, and Ballynahinch, but most probably planted there; *Templeton.* In the Lagan canal at various points between Lisburn, and Moira station, being specially abundant from Blaris to Kesh bridge, 1866, also more sparingly in the lake at Ballynahinch, 1875; S.A.S.

Not an indigenous plant in Britain, and no doubt planted here, as suggested by Templeton. The lake in Hillsborough Park is less than two miles distant from the point in the canal where the *Acorus* is most plentiful, and the plant could easily have been derived thence. The great, creeping rootstock of the sweetflag renders its eradication difficult, and thus it has existed here for, at least, a century. In the canal it is so abundant as to be used for bedding cattle.

135. Lemna polyrrhiza *Linn.*
In a flaxhole in marshy pastures below the second lock of the Lagan canal; *Dr. S. M. Malcomson*, 1884.

A modern introduction, brought by accident, or perhaps placed out from an aquarium. It certainly was not in this place earlier than 1880, or 1881.

136. Phalaris canariensis *Linn.* CANARY GRASS.
In a potato plot near Londonderry, introduced with manure; D.M. Holywood, and Knock; *Flor. Belf.* Waste ground at Bangor, sandpit at Belmont, and dungheap at Belfast; S.A.S. Magheralin; C.H.W. Introduced with birdseed.

137. Bromus secalinus *Linn.*
Among wheat at Cranmore; *Templeton*, 1796. Ballyronan, and Lambeg; *Flor. Ulst.* A casual at Holywood; R.Ll.P., 1886.
Doubtless in every case introduced with seed.

138. Bromus arvensis *Linn.*
Springfield meadows; *Templeton*, 1803. Sandy cultivated land at Malone; *Cyb. Hib.*
An agricultural casual.

139. Hordeum murinum *Linn.*
Occurs as a casual at Holywood; R.Ll.P.
A casual only. The station placed in District 12 by Cybele Hibernica is really in District 5.

140. Lolium italicum *Braun.* ITALIAN RYE-GRASS.
Meadows, and roadsides; *Flor. Belf.*
This grass is a frequent escape from tillage, and propagates spontaneously in waste ground, and on roadsides.

141. Lolium linicola *Sond.*
Collected in the counties of Down, and Derry by D.M.; *Cyb. Hib.* Field by Belfast road a little west of Newtownards; *Rev. E. F. Linton.*
This, like the preceding, is a result of agricultural operations.

β *Plants erroneously recorded.*

1. Thalictrum majus *Smith.*
Mourne Mountains; *Flor. Hib.*
Not *T. majus*, but the mountain form of *T. minus.*

2. Myosurus minimus *Linn.*
North of Ireland; *Brit. Flor., eds.* 4 and 5.
A fraud—the party who is responsible for the record was imposed on by a person who obtained specimens from a Botanic Garden, and sent them to him as from the county of Antrim.

3. Ranunculus Drouettii *Schultz.*
Castle Espie, and Belfast Waterworks; *Flor. Belf.*
This was *R. trichophyllus.*

4. Ranunculus Baudottii *Godron.*
Brackish water at May's fields ; *Flor. Belf.*
Locality now dry ground, and the plant was *R. heterophyllus.*

5. Ranunculus tripartitus *D.C.*
At the Shawsbridge weir (*Templeton*) ; *Flor. Ulst.*
It is difficult to see why it was attempted to introduce this species to the Irish flora, certainly Mr. Templeton is not responsible! The Shawsbridge plant was named by him *R. fluviatilis.* This name was in his day applied to forms of *R. peltatus,* and this was no doubt the plant he intended.

6. Ranunculus circinatus *Sibthorp.*
Loughmourne ; *Flor. Belf.* Lough Neagh ; *Flor. Ulst., Supp.*
In neither instance correct, but in both cases, with little doubt, *R. trichophyllus.*

7. Ranunculus Lenormandi *Schultz.*
Moneymore ; *Flor. Ulst., Supp.*
Was, no doubt, *R. hederaceus.*

8. Ranunculus reptans *Linn.*
Shores of Lough Neagh ; *Templeton.*

Erroneous determination, *vide supra*, page 4.

9. Ranunculus hirsutus *Curtis.*

Warrenpoint (*Dickie*), Skeigoniel (*Millen*), and base of Magilligan rocks (*Moore*) ; *Flor. Ulst.*

The first and third of the above stations were admitted to be erroneous. The second was, without doubt, *R. bulbosus.*

10. Trollius europæus *Linn.* GLOBE FLOWER.

Reported as on the Cave Hill ; *J. Sim, Phyt., N.S., vol. III.* Near Giants' Causeway (*Whitla*) ; *Flor. Ulst.*

Never found on the Cave Hill, and the Giants' Causeway station seems to have been published on very slight grounds. In the interleaved copy of Flora Hibernica, annotated by Whitla, there is, at Tollius, a pencil note "Giants' Causeway, *F. W.—Torrilis.*" Mr. Whitla's note is dubious, but there is no doubt that *Trollius* does not grow at the Causeway.

11. Corydalis claviculata *D.C.* WHITE FUMITORY.

Near Warrenpoint; *Flor. Ulst., Supp.*

This has not been confirmed, and is assumed to have been *Fumaria pallidiflora.*

12. Nasturtium sylvestre *R. Brown.*

Side of the Lagan—" fruitstalk deflexed, siliqua bending upward " ; *Templeton*, 1807. Roadside between Lisburn and Hillsborough (*Orr*), Ballynahinch (*Oulton*), and abundant on the shores of Lough Neagh (*Moore*) ; *Flor. Ulst., Supp.*

The last of these notes was admitted to be erroneous, and it is certain that the two former were wrong also. The note on this plant in Irish Flora is without meaning.

13. Barbarea præcox *R. Brown.*

Several localities are mentioned in Flora Belfastiensis, but all belong to *B. intermedia.*

14. Cochlearia anglica *Linn.*

By the shore between Belfast and Holywood ; *W. Millen, Phyt., O.S., vol. V.* By the Foyle near Londonderry ; D.M.

Millen's plant was *C. officinalis*, and that found at Derry was most probably the same.

15. Viola hirta *Linn.*

At Fortwilliam, and near the Forth River, also in the old graveyard at Shane's Castle ; *Flor. Ulst., Supp.*

The Forth River plant was *V. odorata*, and no doubt the plants seen at the other stations were the same.

16. Viola lutea   *Hudson.*

Railway bank near Bloomfield; *Flor. Belf.* Carrickfergus commons (*Ferguson*), and sandhills at Newcastle (*Dickie*); *Flor. Ulst.* Shores of Lough Neagh at Shane's Castle (*Moore*); *Cyb. Hib.*

V. lutea is a rare plant in Ireland, and there seem to be no reliable northern stations for the true plant. The above quoted records should be divided between *V. tricolor* and *V. Curtisii.*

17. Polygala calcarea   *Schultz.*

By the roadside on ascent to Divis Mountain from Ballygomartin schoolhouse (*Orr*); *Flor. Ulst.* Cave Hill (*Hind*); *Flor. Belf.* States of *P. vulgaris.*

18. Polygala austriaca   *Crantz.*

Rocky *debris* on Black Mountain; *Flor. Ulst., Supp.*

Like the preceding a form of *P. vulgaris.*

19. Cerastium alpinum   *Linn.*

Sides of mountain rivulets about Glenarm, and Ballinleg; *Ir. Flor.* In Belfast Museum copy of Flora Hibernica there is a MS. note by Mr. Whitla stating that *C. alpinum* grows abundantly on the left side of the road from Larne to Glenarm, and that he pointed it out to Mackay in 1833.

In both cases erroneously determined.

20. Hypericum montanum   *Linn.*

On Mallagh Hill in the County of Down, about eight miles east of Belfast (*Orr*); *Cyb. Hib.*

This plant has been published as a member of the Irish flora on the sole authority of the very precise note above quoted, nevertheless the record must be rejected for very good reasons. There is no Mallagh Hill eight miles east of Belfast, but Meulough Hill (pronounced Malliagh) is about five miles south by east of Belfast. This small hill is no doubt the place intended, and it has been repeatedly, but in vain, searched for the plant in question. Petrological considerations also conflict decidedly with the record. It is not to be credited that a limestone loving plant would avoid all the great limestone areas of Ireland to settle down on the grits and slates of the region east and southeast of Belfast, and this, also, may be reckoned amongst the numerous errors published on the same authority.

PLANTS EXCLUDED. 293

21. Geranium rotundifolium *Linn.*
Fields near Antrim (*Orr*), and fields near Cookstown (*Whan*); *Flor. Ulst.*
Not confirmed, and authorities insufficient.

22. Geranium pusillum *Linn.*
Cave Hill (*Oulton*), Doagh (*Millen*), and shore at Bath Lodge (*Hincks*); *Flor. Ulst.* Between the Antrim road and the base of the Cave Hill (*Orr*); *Cyb. Hib.*
A very rare plant in Ireland, and too critical to be accepted on any save the best authority.

23. Geranium columbinum *Linn.*
Ballydown near Killinchy (*Hyndman*), Malone (*Orr*), and *debris* of quarries at Belfast mountains (*Thompson*); *Flor. Ulst.* Sandy ground at Lambeg (*Orr*); *Cyb. Hib.*
This, like the preceding, is a plant of the limestone, and *prima facie* improbable for this district. Possibly it may have been found as a casual in some of the above stations, but much more likely to have been misnamed.

24. Ulex nanus *Forster.* DWARF WHIN.
Not found in Ireland, and the numerous notes relating to it must be transferred to *U. Gallii*, that being the plant intended by Irish botanists.

25. Medicago denticulata *Willdenow.*
Sandy seashore near Donaghadee (*Maffett*); *Flor. Belf.*
A bad specimen of a melilot wrongly determined.

26. Trifolium scabrum *Linn.*
At Ballybristol, County Derry; *Samp's. Surv. Derry.*
Authority unreliable.

27. Trifolium resupinatum *Linn.*
Fields near Purdysburn, and fields near Giants' Causeway; *Flor. Ulst., Supp.* A specimen of this plant in *herb. Babington* was sent by Wm. Thompson, and though given to that gentleman as a Belfast plant its origin is doubtful.
Too questionable in all the stations quoted.

28. Trifolium filiforme *Linn.*
Ballyronan (*Whan*), Kinnegar, and along the shores of Belfast Bay; *Flor. Ulst.*
All errors—small forms of *T. dubium* (*minus*) incorrectly diagnosed.

PLANTS EXCLUDED.

29. Poterium sanguisorbia *Linn.* LESSER BURNET.
Rasharkin (*Moore*); *Flor. Ulst.*
Subsequently corrected by Moore to *Sanguisorbia officinalis*, the greater burnet.

30. Rubus Bloxamii *Lees.*
Black Mountain, Belfast; *Bab. Brit. Rubi.*
A specimen from Black Mountain, sent to Prof. Babington, was so named, but an apparently similar plant from the same spot was not confirmed. It must therefore be held as uncertain.

31. Rubus rudis *Weihe.*
Cotton moss, Bangor, and Forth River; *Flor. Belf.*
May occur in the district, but these stations must be expunged as incorrect.

32. Rubus glandulosus *Bellardi.*
In counties of Antrim and Derry (*Moore*); *Flor. Ulst.*
Not subsequently confirmed.

33. Rosa inodora *Fries.*
Near Belfast (*herb. Nat. Hist. Soc.*); *Flor. Ulst.*
Some error—there is no specimen in the Belfast Museum herbarium.

34. Chrysosplenium alternifolium *Linn.*
Near Belfast (*Templeton*); *Mack. Catal*, and *Flor. Hib.* Ballymaghan Glen; *Flor. Belf.* Colin Glen (*Templeton*); *Flor. Ulst.* By the river at Glendun, and naturalised on a ditch bank near Belfast Botanic Gardens; *Cyb. Hib.*
Not certainly found in Ireland, and all the above stations are erroneous. Mr. Templeton's name is attached to the first record, but this is probably an error as this species seems to be absent from his MS. list.

35. Petroselinum segetum *Koch.* CORN PARSLEY.
Shore near Portstewart; *Flor. Ulst., Supp.*
Very doubtful, and has not been confirmed.

36. Œnanthe pimpinelloides *Linn.*
By the Lagan canal below the first lock; *Templeton*, and *Flor. Hib.*
This was *Œ. Lachenalii*.

37. Daucus gummifer *Lamk.*
Near Holywood (*Millen*); *Flor. Ulst.*

Not found at Holywood now, and authority insufficient for a difficult plant.

38. Torilis infesta *Sprengel.*
Near Comber; *Flor. Belf.*
This proved to be luxuriant *Chærophyllum anthriscus.*

39. Galium sylvestre *Pollich.*
Rathlin Island; *Miss Gage.* Whiterock (*Orr*); *Flor. Ulst.* Lurigedan, and dry basaltic rocks at Fairhead (*Moore*); *Cyb. Hib.*
Only to be looked for on limestone. The plants found at Belfast, and Rathlin were *G. saxatile*, and there can be little doubt that those of Fairhead and Lurigedan were the same.

40. Galium uliginosum *Linn.*
Frequent; *Flor. Ulst.* In the County of Antrim (*Hind*); *Cyb. Hib., Supp.*
The note in *Flor. Ulst.* "frequent" proves that the plant was not understood. Mr. Hind's note is too vague, and leaves the impression that no definite station was known, and that the localization in Antrim was more or less haphazard, or from distant recollection.

41. Gnaphalium luteo-album *Linn.*
At Lambeg; *Flor. Ulst., Supp.* Belfast (*Thompson*), and in sandy cultivated ground at Malone (*Orr*); *Cyb. Hib.*
Authorities insufficient, and perhaps all the notes erroneous.

42. Artemisia maritima *Linn.*
Rathlin Island; *Miss Gage.* Near the beach at Holywood, and at Ballycastle; *Flor. Ulst., Supp.*
The plant really seen was *A. vulgaris.*

43. Arctium majus *Schkuhr.*
Knockagh, Trench, Islandmagee, Groomsport, and Tullygirvan; *Flor. Belf.*
All erroneous—forms of *A. intermedium* and *A. nemorosum.*

44. Centaurea jacea *Linn.*
In a field near Drumbridge; *Templeton.* Gravelly bank at Enagh Lough (*Moore*); *Flor. Hib.*
Templeton's plant which has been quoted in Flora Hibernica, Flora Belfastiensis, and Flora of Ulster was probably the form of *C. nigra* with large ray-flowers. Dr. Moore corrected the name of his Derry plant, and placed it, as a variety, under *C. nigra.*

45. Centauria scabiosa *Linn.* GREAT KNAPWEED.
Near Holywood ; *Flor. Ulst., Supp.*
Some error.

46. Carduus eriophorus *Linn.*
Waste, mountainous, sandy ground near Londonderry; *J. Nuttall,* quoted in Mackay's Catalogue, Irish Flora, and Flora Hibernica.

Not found elsewhere in Ireland, Nuttall's plant was some form of *C. lanceolatus.*

47. Crepis hieracioides *W. & K.*
Flinty rocks along the shore on each side of Glenarm ; *Ir. Flor. (sub Hieracium molle).*

Not Irish—plant really observed was *C. paludosa.*

48. Hieracium nigrescens *Willdenow.*
On rocks in Newtownards Glen (*sub H. pulmonarium*) ; *Templeton,* and copied into Mackay's Catalogue, Irish Flora, and Flora Hibernica.

Impossible to say with certainty which of the hawkweeds was intended, but was either *H. murorum,* or *H. vulgatum.* Both of these grow in the vicinity.

49. Erica stricta *Andr.*
North of Ireland (*Dr. Lloyd,* 1834), *spec. in herb. Hooker.* Cited also in De Candolle's Prodromus (*vide Britten, Jour. Bot.,* 1872, *p.* 25).

Not found in the North of Ireland, unless, perchance, a garden plant. Without a satisfactory history, specimens themselves are misleading in some hands.

50. Pyrola rotundifolia *Linn.*
At the Headwood (head of the Sixmilewater) ; *Templeton,* 1794. Scrabo Hill ; *Scientific Tourist.* Derry ; *Prof. Murphy, Mag. Nat. Hist.,* vol. 1. Near Garvagh ; *Flor. Hib.*

There is only one station for this plant in Ireland, and so far as yet known it is not found anywhere in the north. Templeton's *Pyrola* was afterwards corrected to *P. media,* and the Garvagh plant was admitted to be *P. minor.* The above quoted stations may be divided between these two species.

51. Monotropa hypopitys *Linn.* YELLOW BIRDSNEST.
St. Catherine's, Co. Down; *Ir. Flor.* Woods at Greyabbey ; *Flor. Ulst.*

St. Catherine's, Co. Dublin, was intended by the Irish Flora. The announcement of Greyabbey was probably only a conjectural localization of the station noted in Irish Flora.

52. Erythræa latifolia *Smith*.

County of Down (*Drummond*); *Flor. Hib.* Shore at Bangor (*Orr*), and Portstewart (*herb. Nat. Hist. Soc.*); *Flor. Ulst.*

Not confirmed, and all believed to be forms of *E. centaureum*. There is no specimen in the Belfast Museum herbarium.

53. Erythræa littoralis *Fries*.

Rocky ground below Bangor; *Templeton*, 1804. Shore west of Groomsport (*Millen*); *Flor. Ulst.*

Mr. Whitla states that Templeton admitted his plant to be a form of *E. centaureum*, and doubtless Millen collected the same form.

54. Myosotis sylvatica *Ehrhart*.

Cave Hill, and other places; *Templeton*. Wolfhill (*Thompson*), and south side of Cave Hill (*Orr*); *Flor. Ulst.*

This was a form, or perhaps variety, of *M. arvensis*.

55. Orobanche major *Linn.* (O. RAPUM *Thuill.*).

Rathlin Island; *Miss Gage*.

O. rubrum, only, is found on Rathlin.

56. Scrophularia umbrosa *Dumort.* (S. EHRHARTI *Stev.*).

Bog meadows (*Mateer*), and by the Lagan near Edenderry (*Thompson*); *Flor. Ulst.*

Errors—the authorities cited were not critical botanists.

57. Scrophularia scorodonia *Linn.*

Cave Hill, and Crawfordsburn; *Flor. Ulst., Supp.*

This was no doubt simply *S. nodosa*.

58. Mentha sylvestris *Linn.* HORSEMINT.

In ditches by the wayside near Rostrevor; *Flor. Hib.*

Doubtless a misnomer—not an Irish plant.

59. Calamintha acinos *Clairv.* BASIL THYME.

Northeast Ireland, very rare: *Hooker's Students' Flora, ed. III.*

A lapsus calami—should read southeast Ireland.

60. Lamium galeobdolon *Crantz.* WEASELSNOUT.

Near Comber and other places in the northern counties (*Templeton*); *Flor. Hib.*

PLANTS EXCLUDED.

Incorrect—Templeton's MS. specifies The Dargle, but no northern locality is mentioned. Mr. Whitla marks opposite this species—"Templeton, in his garden."

61. Galeopsis dubia *Leers.* (G. OCHROLEUCA *Lamk.*).
At Shankhill; *Flor. Ulst., Supp.*
Mistaken indentification.

62. Galeopsis ladanum *Linn.*
Near Ballycastle; *Flor. Ulst., Supp.*
May have occurred as a casual, but more probably a misnomer.

63. Teucrium scordium *Linn.* WATER GERMANDER.
"This grows in the County of Down, *Mollyneux*"; *Harris's Down*, 1744.
Some error—not found here. Not known so far north in Britain.

64. Ajuga alpina *Linn.*
Summit of Benevenagh (*Moore*); *Flor. Hib.*
No doubt the mountain form of *A. reptans* (*var. pseudo-alpina*). *A. alpina* is not a British plant.

65. Statice limonium *Linn.*
The segregate form is not found in Ireland, and the numerous localities mentioned in *Flor. Belf.* and *Flor. Ulst.* refer to *S. bahusiensis*.

66. Statice occidentalis *Lloyd.*
Near Belfast; *Millen, Phyt., O.S. (as S. binervosa).*
An error, the plant was same as the preceding.

67. Statice Dodartii *Gir.*
Near Belfast; *Millen, Nat. Hist. Rev., vol. 1. (as S. spathulata).*
This, and the two preceding were one plant under separate names.

68. Plantago media *Linn.*
Pastures at Holywood; *Flor. Belf.* At Newcastle, and on Carnmoney Hill; *Flor. Ulst., Supp.*
A form of *P. lanceolata* with broader leaves, and longer flower spikes.

69. Chenopodium polyspermum *Linn.*
Cultivated ground at Holywood, Sydenham, and Knock; *Flor. Belf.*
Erroneous determination.

70. Chenopodium urbicum *Linn.*

Beside the coal yard at the junction of the roads to Carrickfergus and Templepatrick, and on Belfast dunghills ; *Templeton*, 1807.

Some form of *C. album.* Templeton refers to *E. B.*, fig. 717, and makes the critical remark "but spikes too long and too bare." This seems to show, not that the figure is wrong, but that the specimen compared with it was not the true plant. Templeton's special station was close to Whitehouse, and not near Carrickfergus as stated in Flora of Ulster.

71. Chenopodium ficifolium *Smith.*

Near Belfast, sparingly (*Orr*) ; *Cyb. Hib.*

Possibly occurred as a casual, but more probably an error of the collector.

72. Chenopodium murale *Linn.* SOWBANE.

Ruins of a mud cottage by the Lisburn road (*Thompson*); *Flor. Ulst.* Same locality (*Orr*) ; *Cyb. Hib.*

Error, or casual—most probably the former, the authorities cited not being sufficiently careful.

73. Chenopodium hybridum *Linn.*

Same spot, and same conditions as preceding species, and excluded for similar reasons.

74. Salicornia radicans *Smith.*

Strand above Narrow-water ; *Ir. Flor.*, *Flor. Hib.*, and *Flor. Ulst.* Holywood, and Cherryvalley near Comber ; *Flor. Belf.*

The plant found at these places was the procumbent variety of *S. herbacea.* The Narrow-water error is attributed to Templeton, but *S. radicans* does not appear on his list, and he named the plant from Narrow-water as *S. procumbens.*

75. Atriplex marina *Linn.*
Dundrum Bay ; *Flor. Ulst.*
Not confirmed.

76. Rumex palustris *Smith.*

Bog meadows, Portmore, and banks of the Bann at Portglenone; *Templeton.* Plentiful on the Derry coast between Portrush and Portstewart (*Moore*) ; *Flor. Ulst.*

A plant misunderstood, and incorrectly determined. Dr. Moore subsequently stated that the Derry plant was an error.

77. Rumex acutus *Linn.* (R. PRATENSIS *M. & K.*).
A few plants at Newforge (*Thompson*, 1846); *Flor. Ulst.*
An unsatisfactory plant, and such a record should only be received on authority of a critical botanist.

78. Polygonum maculatum *Gray.* (P. NODOSUM *Reichb.*).
A number of localities mentioned in Flora Belfastiensis, but the plant was really *P. lapathifolium.*

79. Polygonum maritimum *Linn.*
Shore at Holywood; *Flor. Ulst., Supp.*
This was *P. Raii*, a near ally.

80. Euphorbia hiberna *Linn.* MAKINBOY.
Near Belfast (*Templeton*); *Eng. Bot.*, and *Eng. Flor.*
Specimens were sent to London, from the garden at Cranmore, to be figured in English Botany. By inadvertance they were noted by Sowerby as wild. If Templeton claimed this as a Belfast plant there would be some note to that effect in his MS. catalogue. No such note can be found.

81. Ceratophyllum submersum *Linn.*
Ballynahinch Lake; *Templeton.* Same locality (*Templeton* and *Mackay*); *Flor. Hib.*, and *Flor. Ulst.*
A misnomer—the plant in *herb.* Hyndman, named *C. submersum*, from the lake at Ballynahinch is *C. demersum.* The same was the case with the plant of Templeton, and of Mackay.

82. Salix acuminata *Smith.*
Vide page 130.

83. Salix ambigua *Ehrhart.*
Frequent on the Belfast hills (*Stewart*); *Cyb. Hib.*
Erroneous determination—an unsatisfactory plant.

84. Salix lapponum *Linn.* (var. STUARTIANA *Sm.*).
Shores of Lough Neagh near Antrim; *Flor. Ulst., Supp.*
A misnomer—var. of *S. repens.*

85. Betula verrucosa *Ehrhart.* WHITE BIRCH.
Various localities are stated in *Flor. Belf.* and *Flor. Ulst.*, the aggregate species is intended:—*vide* remarks under *B. glutinosa*, page 132.

86. Juniperus communis *Linn.*
The records under this name in our Floras seem, as far as concerns the

northeast of Ireland, to have reference only to *J. nana*. In Donegal the true plant grows luxuriantly, but we have no knowledge of its occurrence in our district.

87. Juniperus sabina *Linn.* SAVIN.
" *Sabina folio Cupressi*, on Slieve Donard " ; *Harris's Down*, 1744.

Harris should have exercised more caution, and less credulity when informed that a local physician had frequently gathered "the right female savine" in the Mourne Mountains. The plant was, doubtless, *Juniperus nana*.

88. Cephalanthera pallens *Reich.* (C. GRANDIFLORA *Gray*).
In County Antrim (*Whitla*) ; *Flor. Hib.* At Duneane near Antrim (*Whitla*) ; *Flor. Ulst.*

Specimens of the plant so named have proved to be luxuriant examples of *C. ensifolia* ; *fide Cyb. Hib.*

89. Alisma natans *Linn.*
County Down (*Templeton*) ; *Flor. Hib.*

Some error by Mackay, Templeton in his MS. list quotes Wade, but mentions no locality in the north.

90. Juncus diffusus *Hoppe.*
Ballyronan, Co. Derry (*Whan*) ; *Flor. Ulst.*

Accepted by Flora of Ulster, and inserted on insufficient authority.

91. Juncus nigritellus *D. Don.*
On the mountains of Antrim, but rare (*Moore*) ; *Cyb. Hib.*

A doubtful plant, and should be rigidly confirmed.

92. Juncus compressus *Jacquin.*
Very common in Derry ; D.M. Frequent in wet places by the sea ; *Flor. Ulst.*

Like the preceding, an unsatisfactory plant. All the specimens we have seen were *J. Gerardi*.

93. Eriocaulon septangulare *Withering.*
Rathlin Island ; *Miss Gage*, also *Flor. Ulst.*

This sufficiently probable station was at once accepted by Irish botanists, but it now appears that there was some unaccountable error, and the plant nowhere occurs on Rathlin.

94. Potamogeton plantagineus *Du Croz.*
At Magilligan, and not unfrequent in bog ditches in County Antrim (*Moore*) ; *Cyb. Hib.*

Possibly correct, but this seems not so much a plant of peaty bogs, as of shallow lake margins, and pools of stagnant water in limestone regions. The broad leaved form of *P. natans* (*var. ovaliformis*) might, with slight examination, be mistaken for the present species.

95. Potamogeton lanceolatus *Smith.*

In the Moyola River, near Shane's Castle (*Moore*); *Cyb. Hib.* Sixmilewater below Muckamore (*Orr*); *Bennett, Jour. Bot.*, 1882, *p.* 20. Co. Down; *Hook. Stud. Flor. ed. III.*

Mr. Bennett's identification is unquestioned, but the authority for the locality is not considered accurate. This identical portion of the Sixmilewater has been searched for pondweeds by botanists sufficiently critical. The note in Cybele Hibernica is erroneous in two respects. The plant was misnamed, but this was subsequently corrected (*vide Cyb. Hib., Supp.*). There still remains some confusion in the statement of the locality, the nearest point of the Moyola to Shane's Castle being about eight miles. Perhaps the river Main, which flows into Shane's Castle Park, was intended. The reference in Students' Flora is based on the Sixmilewater note, but the stream is in Co. Antrim, not Down.

96. Potamogeton zosterifolius *Schum.*

Bow Lake near Saintfield (*Stewart*); *Flor. Belf.* In a pool on the right hand side of the road from Newtownards to Bangor (*Orr*); *Flor. Ulst.* Co. Down; *Hook. Stud. Flor.*

The plant of Flora Belfastiensis was a state of *P. obtusifolius.* The note in Flora of Ulster is considered of no authority, this plant, like many others noted by the same collector, remaining unverified; the note in Students' Flora is based on the same authority.

97. Potamogeton mucronatus *Schrader.*

In the Bann about Portglenone; "leaves finely serrate"; *Templeton.* River Bann, Co. Down (*Orr*); *Bennett, Jour. Bot.*, 1881, *p.* 312. Co. Down; *Hook. Stud. Flor., ed. III*

At page 150 (under *P. crispus*) reason is given for the belief that Templeton's plant was Hudson's *P. serratus.* Mr. Bennett is in error in supposing that the station he indicates was referred to in Cybele Hibernica or Flora Hibernica. Both references were founded on Templeton's note, quoted above, which locates his plant on the dividing line between Antrim and Derry. There are two Banns the Upper Bann which rises in the Mourne range, and flows through counties Down and Armagh, emptying into Lough Neagh at its southern end, and the Lower Bann which flows out of the northern end of that lough, and in its passage to the sea divides the counties of Antrim and Derry. The distance, as the crow flies, between Templeton's station, and the nearest point of the Bann in Co. Down is over 30 miles. There is no good authority for the occurrence of *P. mucronatus* in Ireland, and before being admitted there must be better evidence than has been adduced. The note in Hooker's Flora is based on that in Journal of Botany.

98. Potamogeton trichoides *Cham.*
Pools, Conlig Hill, Co. Down, 1844 (*Orr*); Bennett, *Jour. Bot.*, 1881, p. 312.

The reasons shown for excluding the preceding species apply here also. Not known elsewhere in Ireland, and Conlig Hill is too near Belfast, and too well known for the occurrence to have escaped observation. From what has gone before it will be seen, that if the reports referred to were accurate, one collector alone would have added three pondweeds to the Irish flora. The counties of Antrim and Down, in which these plants were supposed to be overlooked, form a district as well worked as any equal area in Ireland.

99. Carex Davalliana *Smith.*
Spongy bogs in the County of Down (*Sherard*); *Raii Syn.*, ed. *II.*, p. 270, 1696 (*Gramen cyperoides minus Ranunculi capitulo longiore*). Same localities repeated by Harris. Kirkiston bog, and frequent in boggy situations; *Templeton.*

The plant observed by Sherard, and by Templeton was *C. dioica*, a pardonable error seeing that these two sedges are very closely allied. In 1814 Templeton relinquished the erroneous name, and placed his plant correctly under *C. dioica*, yet the error was repeated in English Flora, and in the Irish Flora. Mackay, too, though in communication with Templeton, failed to make the correction when preparing his Flora Hibernica.

100. Carex axillaris *Gooden.*
Shady places by the stream in Crawfordsburn; *Flor. Belf.*
This was luxuriant *C. remota.*

[CAREX ELONGATA *Linn.*
The notes on this species in Flora Belfastiensis are erroneous. The plant was *C. paniculata* depauperated.]

101. Carex punctata *Gaudin.*
On Ballygowan bog; Corry, *Jour. Bot.*, 1882.
A deceptive plant which proves to be *C. binervis*, over-ripe.

102. Carex filiformis *Linn.*
In County Antrim (*Moore*); *Flor. Ulst.* By the side of Lough Neagh at Selshan (*Moore*); *Cyb. Hib.*
May be correct, but should have recent confirmation.

103. Calamagrostis lanceolata *Linn.*
Banks of the Lagan ; *Templeton* (as *Arundo calamagrostis E.B.*, 2159).
Not easy to conjecture what this was—certainly no *Calamagrostis*.

104. Avena fatua *Linn.* WILD OAT.
Cornfields, and waysides; *Flor. Belf.*
Not distinguished from stray plants of the cultivated oat.

105. Avena pratensis *Linn.*
By the New Lodge Road, Belfast (*herb. Belf. Nat. Hist. Soc.*); *Flor. Ulst.*
The specimen so named was *A. avenaceum*.

106. Poa alpina *Linn.*
On Slievegallion in Co. Derry (*Moore*) ; *Flor. Hib.*
Subsequently reduced by Dr. Moore to a form of *P. pratensis*.

107. Festuca uniglumis *Solander.*
Sandy banks at seaside near Cushendall (*J. White*), and on the lighthouse island (Copelands); *Templeton.*
White who is said to have contributed the localities to the Irish Flora is silent as to the Cushendall station, and it may be assumed to have been an early error, afterwards abandoned. The Copeland Island record not having been confirmed may he set aside as erroneous. The records of this plant have been much mixed. Flora of Ulster is entirely wrong. Mr. White did not claim to have found it in Copeland Island, but attributes this station to Templeton, equally Templeton did not say he found it at Cushendall, but that White did so. Cybele Hibernica follows the Flora of Ulster, but substitutes Thompson for Templeton as authority for the Cushendall station. *F. uniglumis* does not come so far north in Europe.

108. Festuca myuros *Linn.*
In the orchard at Cranmore; *Templeton.* Abundant in Co. Derry; D.M. Knockagh (*Dickie*), and Ballycastle (*Whitla*) ; *Flor. Ulst.*
No doubt in all these cases the plants referred to this species were really *F. scuiroides*. When the notes of Templeton, and of Moore were made the distinctions between these two plants were not so well known, and Templeton even formed the opinion that they were one species. *F. myuros* is rare in Ireland, and seems, from its distribution, to be a limestone plant.

109. Triticum pungens *Persoon.*
Seacoast at Holywood, Greypoint, and Ballyholme ; *Flor. Belf.*

The above localities are to be transferred to *T. junceum*.

110. Elymus arenarius *Linn.*
Bays at Dundrum, Ballywalter, and Ballyholme (*Millen*) ; *Flor. Ulst.*
Not to be found—published on insufficient authority.

111. Asplenium fontanum (*Linn.*) *Presl.*
Cave Hill near Belfast ; *Newnham, herb. Bot. Soc. Lond.*
Not probable even as an escape—doubtless some dealers imposition.

112. Adiantum capillus-veneris *Linn.*
Reported to have been found in a cave near the Giants' Causeway (*Tyerman*) ; *Cyb. Hib.* Also reported that recently fragments shot down from the roof of the large cave at the Causeway were identified as fronds of the maidenhair fern. Reported also as found at Bloodybridge south of Newcastle.
These reports have never been confirmed by any of the numerous *botanists* who visit the Causeway, or Newcastle, and they are extremely improbable. We are not aware of a single instance where *A. capillus-veneris* is found in Ireland growing on any rock other than limestone. The Roundstone, and Donegal plants, are not exceptions although growing among non-calcareous rocks. The reports in question are doubtless due to erroneous identification.

113. Chara fragilis var. HEDWIGII *Agardh.*
In a fresh water pool at Portrush (*Moore*) ; *Cyb. Hib.*
May be correct, but should not be included in our flora unless identified by one who is a special expert in this difficult group.

114. Campylopus Schwarzii *Schimper.*
Mourne Mountains (*Lett*) ; *Stewart, Mosses N. E. Ir., Supp.*
Erroneous determination.

115. Dicranoweissia crispula (*Hedwig*).
Ballygally Head ; D.M. Rocks, and boulders between Rostrevor and Hilltown (*Lett*) ; *Stewart, op. cit.*
Not yet found in Ireland, the specimens hitherto so named being *Weissia cirrata*.

116. Mollia hibernica (*Mitten*).
Basaltic rocks of Rathlin Island ; *Stewart, Proc. R.I.A.*, 1884, and *Mosses N. E. Ir., Supp.*
Specimen poor, and on review considered an error, or at least very doubtful

117. Barbula Hornschuchii *Schultz.*
Vide page 218.

118. Grimmia ovata *Web. et Mohr.*
On stones at summit of Sallagh Braes; *Stewart, Mosses N. E. Ir.*
An error, the plant of Sallagh Braes was *G. Donii.*

119. Weissia (Orthotrichum) **Drummondi** *Hooker.*
Vide page 229.

120. Tayloria serrata *Br. & Schimp.*
It is now known that the British plant is not *T. serrata*, but *T. tenuis Dickson.*

121. Bryum erythrocarpum *Schw.* (B. SANGUINEUM *Ludw.*).
Wall beyond Donaghadee; *Stewart, op. cit.*
This proves to be a form of *B. cæspiticium.*

122. Hypnum irriguum *Wilson.*
On stones in a stream at Magheramorne; *Stewart, op. cit.*
A doubtful specimen, and has not be re-found.

123. Hypnum Sendtneri *Schimper.*
Boggy ground on Rathlin Island; *Stewart, Proc. R.I.A.*, 1884.
Locality transferred to *H. intermedium*, the plant which hitherto has usually been named *Sendtneri* by British bryologists.

124. Hypnum crista-castrensis *Linn.*
Colin Glen, 1847 (*Orr*); *Moore, Jour. Roy. Dub. Soc.*, 1856, and *Synopsis of Irish Mosses.*
The authority for this, the single reputed Irish station, is not accounted sufficient.

125. Hypnum hamulosum *Froel.*
On limestone rocks in Crow Glen, and Black Mountain; *Stewart, Mosses N.E.Ir.*
An error, the plant is now identified as *H. molluscum* var. *gracile.*

126. Hypnum elegans *Hooker.*
Slieve Donard; *Stewart, op. cit.*
Not British—all Irish records to be transferred to *H. Borreri.*

127. Antitrichia curtipendula *Linn.*

Mountains near Belfast ; *Davies, Phyt. N.S. vol. III.*

May be right, but has not been found since, and awaits confirmation.

128. Neckera pennata *Linn.*

Colin Glen (*Orr*); *Moore, Phyt. N.S. vol. II.,* and *Syn. Ir. Mosses.*

The Irish flora has been *only apparently* enriched by the publication of this species, and of no. 124, as occurring in Colin Glen. They depend on the same authority, and neither has been found by any of the numerous bryologists who have searched the shady recesses of Colin Glen during the past 40 years.

129. Sphærocarpus terrestris *Smith.*

On a wet clay bank at Colin Glen (*Orr*); *Moore, Rep. on Ir. Hep., Proc. R I.A.,* 1876.

This is an additional plant catalogued as Irish in Moore's list on the strength of its occurrence in Colin Glen, reported by the same collector as in preceding case. Not found elsewhere in Ireland, and not confirmed in this locality, and could not be admitted in this present enumeration.

TOPOGRAPHICAL INDEX.

The distances here indicated are stated in English miles measured "as the crow flies."

Aghalee—by the Lagan canal two miles south of Ballinderry.
Agivey—a river which flows past Garvagh to the Bann.
Agnew's Hill—three miles north by west of Larne.
Andersonstown—by the Falls road two miles southwest of Belfast.
Annadale—two miles south of Belfast, near Newtownbreda.
Annahilt—one mile southwest of Hillsborough.
Annalong—by the seashore six miles south of Newcastle.
Ardclinis—a parish ranging from Carnlough to Redbay.
Ardmillan—on Strangford Lough, five miles southeast of Comber.
Ards—a peninsula bounded by the sea, and Strangford Lough.

Ballinderry—four and a half miles north of Moira.
Ballinleg—a hilly district to the south of Ballycastle.
Ballintoy—five miles northwest of Ballycastle.
Balloch—three miles south of Newcastle.
Ballyalloley—two miles southwest of Comber.
Ballyarnot—three miles north of the city of Londonderry.
Ballycorr—on the Sixmilewater ten miles northeast of Antrim.
Ballydrain—by the old Lisburn road four and a half miles from Belfast.
Ballyeaston—one mile west of Ballycorr (*vide supra*).
Ballygalley—on the coast three and a half miles north of Larne.
Ballygomartin—between Black Mountain and Belfast.
Ballyhalbert—east shore of Ards ten miles south of Donaghadee.
Ballyharrigan—about four miles west of Dungiven.
Ballyholme—shore of Belfast Bay one mile east of Bangor.
Ballyhornan—west side of entrance to Strangford Lough.
Ballykinler—sandhills one mile east of Dundrum.
Ballylesson—near the Lagan five miles south of Belfast.
Ballymacormick—at Groomsport northeast of Bangor.
Ballymaghan—in the hills above Sydenham.

Ballymenoch—a half mile northeast of Holywood.
Ballynafeigh—beyond Ormeau, one mile south of Belfast.
Ballynascreen—Draperstown is the chief place of this parish.
Ballynure—seven miles northwest of Carrickfergus.
Ballyoran—beyond Dundonald on road to Newtownards.
Ballypallady—seven miles east of Antrim.
Ballyronan—shore of Lough Neagh four miles southwest of Toome.
Ballysillan—northwest of Belfast on road to Ligoniel.
Ballytweedy—five miles southeast of Antrim.
Ballywalter—east shore of Ards seven miles south by east of Donaghadee.
Banagher—by Owenrigh river two miles south by west of Dungiven.
Bannfoot—the ferry where the Bann empties into the sea.
Bellahill—three miles northeast of Carrickfergus.
Belmont—about a mile above Sydenham station.
Belvoir Park—three miles south of Belfast at Newtownbreda.
Benbradagh—a mountain overlooking Dungiven to the northeast.
Bencrom—a mountain three miles southwest of Slieve Donard.
Benderg—west side of entrance to Strangford Lough.
Benevenagh—the mountain that overlooks Magilligan flats.
Benone—sandy flat bordering Lough Foyle at Magilligan.
Birky Moss—between Castlereagh and Moneyreagh.
Blackhead—at the western entrance to Belfast Bay.
Black Mountain—two miles due west from Belfast.
Blaris—midway between Lisburn and Hillsborough.
Bloodybridge—two miles south from Newcastle.
Blue Lake—at the source of Annalong river in the Mourne range.
Bog Meadows—marshy flats by the Blackstaff close to Belfast.
Bond's Glen—about seven miles south southeast from Londonderry.
Boughal Hill—nearly midway between Belfast and Glenavy.
Bow Lake—four miles north of Ballynahinch.
Braniel—a low hill three and a half miles southeast of Belfast.
Brean Mountain—about six miles south of Ballycastle.
Broughshane—northeast of Ballymena three and a half miles.
Brown's Island—midway between Donaghadee and Newtownards.
Bruslee—three miles east of Doagh.
Bryansford—nearly three miles northwest of Newcastle.

Cairncastle—about five miles north of Larne.
Carmavey—about five to six miles southeast of Antrim.
Carndaisy—a glen two and a half miles northwest of Moneymore.
Carnearny—a hill about four miles northeast of Antrim.
Carngaver—about 3 miles southeast of Holywood.
Carntogher—east by south of Dungiven seven to eight miles.
Carrickarede—a rocky islet four miles west of Ballycastle.
Carrickmannan—about two miles east of Saintfield.
Carrowreagh—about 2 miles northeast of Dundonald.
Carr's Glen—between Cave Hill and Squire's Hill.

Castle Chichester—on the shore half a mile below Whitehead.
Castlerobin—three miles north by west from Lisburn.
Castleward—at Strangford seven miles northeast of Downpatrick.
Castlewellan—a village four and a half miles northwest of Newcastle.
Causeway Water—flows out to the sea four miles west of Kilkeel.
Cave Hill—three miles north of Belfast.
Clady (Claudy)—nine miles southeast of Londonderry.
Clady River—flows northwest from Divis to Sixmilewater.
Clondermot—east side of the Foyle near Londonderry.
Clontygeragh—mountain district some six miles southeast of Dungiven.
Clough (Antrim)—seven miles north of Ballymena.
Clough (Down)—southwest of Downpatrick five and a half miles.
Cloughey—east shore of Ards eight miles south of Ballywalter.
Colin Glen—four miles southwest of Belfast.
Conlig—one and a half miles north of Newtownards.
Connswater—flows to the Lagan between Belfast and Sydenham.
Cotton Moss—two to three miles west of Donaghadee.
Craigauntlet—in the hills two and a half miles southwest of Holywood.
Craignashoke—about four miles northwest of Draperstown.
Cranfield—shore of Lough Neagh six miles west of Antrim.
Cranmore—two miles from Belfast on old Lisburn Road.
Creagh Bog—marshes on the west side of Toomebridge.
Creevy Lough—three miles northwest of Ballynahinch.
Creevytenant—a townland north by east of Ballynahinch.
Cregagh—west of Castlereagh Hill two miles from Belfast.
Crow Glen—glen between Wolfhill and Divis.
Cumber—district eight to nine miles N.E. and N.N.E. of Londonderry.
Curleyburn—joins the Roe one and a half miles northeast of Limavady.

Dart Mountain—bordering Tyrone, ten miles southwest of Dungiven.
Deers' Meadow—in Mourne range near the source of the Bann.
Derriaghy—two miles north of Lisburn.
Derry Lake—two miles north of Ballynahinch.
Derrymore—by Lough Neagh two miles southwest of Portmore Lough.
Dervock—four miles north by east of Ballymoney.
Desertcreat—border of Tyrone and Derry near Cookstown.
Desertmartin—three miles northwest of Magherafelt.
Diamond Mountain—between Donard and Luke's Mountain.
Divis—a high mountain three miles west of Belfast.
Doagh River—an affluent of the Sixmilewater.
Donald's Hill—six miles southeast of Limavady.
Donegore—three and a half miles east by north of Antrim.
Drumachose—a parish between Limavady and Keady Hill.
Drumadarragh—seven miles northeast of Antrim.
Drumbo—about six miles south of Belfast.
Drumbridge—on the Lagan five miles above Belfast.
Drumcro—by the Lagan two miles southwest of Moira.

TOPOGRAPHICAL INDEX. 311

Drumnasole—a hill about four miles north of Glenarm.
Dunboe—three and a half miles north by west of Coleraine.
Dundela—near Sydenham, two miles northeast of Belfast.
Duneane—a parish on the border of Co. Derry at Toome.
Dunleidy Glen—in the hills north of Dundonald.
Dunseverick—on the coast about four miles east of Giants' Causeway.
Dunsilly—two miles northwest of Antrim.

Eagle Mountain—in the Mourne range five miles northeast of Rostrevor.
Echlinville—in the Ards five miles south by east of Greyabbey.
Edenderry—by the Lagan four miles above Belfast.
Enagh Lough—three miles northeast from Londonderry.
Errigal Banks—on Agivey river two mi'es above Garvagh.

Faughanvale—a parish between Londonderry and Limavady.
Finnebrogue—two miles north of Downpatrick.
Forth River—Ballysillan to Springfield near Belfast.

Galgorm—one and a half miles west by south of Ballymena.
Garvagh—ten miles south of Coleraine.
Giants' Ring—near the Lagan four miles above Belfast.
Gillhall—woods on the Lagan two and a half miles west of Dromore.
Glenariffe—glen opening to Redbay south of Cushendall.
Glenaveagh—in Mourne mountains three miles east of Hilltown.
Glenballyemon—northwest of Lurigedon near Cushendall.
Glendarragh—a demesne by the river at Crumlin.
Glendivis—on a tributary of the Forth river west of Belfast.
Glendun—a glen to the north of Cushendall.
Glenedra—in the mountains seven miles south from Dungiven.
Glenmakerron—southeast of Ballycastle.
Glenmore—near Lambeg—which see.
Gleno—three miles south of Larne.
Glenravel—a glen between Carnlough and Cloghmills.
Glenshesk—a glen leading south from Ballycastle.
Glenwherry—mountain district between Larne and Kells.
Gobbins—seacliffs on east side of Islandmagee.
Greencastle—on Carlingford Lough four miles southwest of Kilkeel.
Grogan's Glen—believed to be the same as Crow Glen
Gunn's Island—two miles south of entrance to Strangford Lough.

Hannahstown—four miles from Belfast on road to Glenavy.
Harbour Island—in Lough Neagh three and a half miles south of Toome.
Hillhall—one and a half miles east of Lisburn.
Hilltown—foot of Mourne mountains eight miles east of Newry.

Inch Abbey—by the Quoile northeast of Downpatrick.
Inver—river; and small parish at Larne.
Irish Hill—five miles west northwest of Carrickfergus.
Islandderry—midway between Dromore and Waringstown.

Islandmagee—the peninsula southeast of Larne.
Jennymount—at the northern extremity of Belfast.
Keady Hill—four miles east by north of Limavady.
Kearney's Point—outer shore of Ards, opposite Portaferry.
Kells—five miles south by east of Ballymena.
Kenbane Head—about three miles northwest of Ballycastle.
Kilbroney—two to four miles northeast of Rostrevor.
Kilclief—at western side of entrance to Strangford Lough.
Kilcorig—three and a half miles northwest of Lisburn.
Kilmore—by Lough Neagh three miles north of Lurgan.
Kilwarlin—a district near Hillsborough.
Kilwaughter—about two miles west of Larne.
Killard Point—west side of entrance to Strangford Lough.
Killead—a parish situate between Antrim and Crumlin.
Killeen Glen—above Belmont northeast of Belfast.
Killinchy—near Strangford Lough six miles northeast of Comber.
Killough—on seacoast six miles south by east of Downpatrick.
Killyglen—two and a half miles north of Larne.
Killymurry—two miles west of Glarryford station.
Killyleagh—north by east of Downpatrick five and a half miles.
Kings' Moss—about three miles west of Whiteabbey.
Kinnehalla—four and a half miles southeast of Rathfriland.
Kirkcubbin—south by east from Greyabbey three and a half miles.
Kirkiston—in the Ards three miles south of Ballyhalbert.
Knockagh—a low hill three miles west of Carrickfergus.
Knockbracken—about one and a half miles east of Purdysburn.
Knockdhu—a hill about four miles north of Larne.
Knockinelder—by the seashore north of Quintin Castle.
Knocklayd—a mountain which overlooks Ballycastle
Knocknagoney—midway between Holywood and Sydenham.

Lambeg—on the Lagan one and a half miles below Lisburn.
Langford Lodge—by Lough Neagh three and a half miles west of Crumlin.
Lignapeiste—a river glen southwest of Dungiven.
Ligoniel—two and a half miles northwest of Belfast.
Long Lake—two and a half miles north north east of Ballynahinch.
Lough Aghery—midway between Ballynahinch and Dromore.
Lough Beg—expansion of Bann two miles north of Toome.
Loughbrickland—a lake three miles south by west of Banbridge
Lough Henny—about three miles west of Saintfield.
Loughinisland—a lake four miles west of Downpatrick.
Loughinshollen—district west of north end of Lough Neagh.
Lough Leagh (Clay Lake)—four miles east northeast of Crossgar.
Loughmourne—a lake three miles north of Carrickfergus.
Lough Shanagh—in Mourne range eight miles north of Kilkeel.
Luke's Mountain—on southeast side of Tollymore Park.

TOPOGRAPHICAL INDEX. 313

Lurigethan—the mountain that overlooks Cushendall.
Lylehill—a hilly district south of Templepatrick.

Magheralagan—a lake two and a half miles west of Downpatrick.
Magheralin—two miles southwest of Moira.
Magherascouse—a small lake three miles south of Comber.
Malone—a townland touching Belfast on the south.
Marino—shore of Belfast bay immediately below Holywood.
Mealough Hill—a low hill five miles south by east of Belfast.
Megabbery (Maghaberry)—two miles northeast of Moira.
Milewater—stream north of Northern Counties Railway terminus, Belfast
Millisle—on the coast two and a half miles south of Donaghadee.
Milltown—near the Lagan, three and a half miles south of Belfast.
Moneyreagh—three and a half miles west by south of Comber.
Monkstown—one mile northwest of Whiteabbey.
Monlough—a small lake five miles southwest of Comber.
Montalto—a demesne on south side of Ballynahinch.
Mountcollier—at northern suburbs of Belfast (now built up).
Mountstewart—five miles southeast of Newtownards.
Movilla—an ancient burial place at Newtownards.
Moygannon—a glen about two miles northeast of Warrenpoint.
Muck Island—off the coast north of the Gobbins.
Muckamore—one mile southeast of Antrim.
Muff Glen—six miles east by north from Londonderry.
Mullaghmore—a mountain six miles south by east of Dungiven.
Murlough (Antrim)—a bay to the southeast of Fairhead.
Murlough (Down)—sands south of Dundrum.

Narrow-water—two miles northwest of Warrenpoint.
Ness Glen—about seven miles southeast of Londonderry.
Newforge—on the Lagan three miles above Belfast.
Newtownbreda—three miles south by east of Belfast.

Oldforge—between Dunmurry and Drumbridge.
Oldstone—about two miles south of Antrim.
Orlock Point—three miles northwest of Donaghadee.
Ormeau—a southern suburb of Belfast.

Parkmount—west shore of the bay two miles from Belfast.
Peoples' Park—original name of Victoria Park—which see.
Pigeonrock Mountain—northwest of Kilkeel seven miles.
Plains—by the Lagan near Belfast (now built over).
Portaferry—east side of Strangford Lough five miles above entrance.
Portavo—two miles northwest of Donaghadee.
Portballintrae—west of Giants' Causeway.
Portbradden—east of Giants' Causeway three to four miles.
Portglenone—on the Bann river about nine miles below Toome.
Portmore—east side of Lough Neagh near Ballinderry

Potterswalls—three miles north by east of Antrim.
Purdysburn—four miles south of Belfast.
Quentin Castle—east shore of Ards three and a half miles from southern end.
Rademon—a demesne one mile west of Crossgar.
Ram's Island—in Lough Neagh west of Glenavy.
Rasharkin—in County Antrim not far from the Bann at Kilrea.
Rathmore—one and a half miles north of Dunadry station.
Rathmullen—on the coast five miles east of Dundrum.
Ravernet River—flows into the Lagan one mile above Lisburn.
Redbay—a sandy bay south of Cushendall.
Redhall—near Ballycarry five miles northeast of Carrickfergus.
Rhanbuoy—shore of Belfast bay a half mile above Carrickfergus.
Ringsallin Point—four miles east of Dundrum.
Rockport—on the shore three miles below Holywood.
Rocky River—joins the Bann near Hilltown.
Roe—flows to Lough Foyle by way of Dungiven and Limavady.
Roepark—about one mile south of Limavady.
Roughfort—about midway between Antrim and Whiteabbey.
Sallagh Braes—inland cliffs three to four miles northwest of Larne.
Salterstown—west shore of Lough Neagh six miles from Toome.
Saul—two miles northeast of Downpatrick.
Sawel—a high mountain nine miles southwest of Dungiven.
Scrabo—an isolated hill one mile southwest of Newtownards.
Seaview—west shore of the bay one mile below Belfast.
Selshan—shore of Lough Neagh five miles west of Ballinderry.
Seymour Hill—about a half mile south of Dunmurry.
Shanlieve—a mountain five miles northeast of Rostrevor.
Shanslieve—a mountain above Donard Lodge.
Shawsbridge—on the Lagan over three miles above Belfast.
Sheepfield—some three miles north by west of Newry.
Shimna River—flows from the mountains through Tollymore.
Shrigley—six miles north by east of Downpatrick.
Sixmilewater—flows from near Larne to Lough Neagh at Antrim.
Skerry Rock—in the Braid two to three miles northeast of Broughshane.
Slemish—a mountain seven miles east by north of Ballymena.
Slievebernagh—about three miles west of Slieve Donard.
Slieve Bignian—in the Mourne range, second in height to Donard.
Slieve Commedagh—on the north of Slieve Donard.
Slieve Croob—five miles southwest of Ballynahinch.
Slievedermot—above Rostrevor to the northeast.
Slievegallion—a mountain four miles northwest of Moneymore.
Slievemartin—above Rostrevor, to the east.
Slievemeel More—five miles west by south of Newcastle.
Slievemeel Beg—one mile west of the preceding.
Slievenabrock—above Newcastle, two miles to the west.
Slievenaglogh—three miles west by south of Newcastle.

Slievenagriddle—a hill three miles east of Downpatrick.
Slievenamady—a mountain above Donard Lodge.
Slievenanee—about seven miles southwest from Cushendall.
Slievetrue—four miles northwest from Carrickfergus.
Slogan Bog—about midway between Toome and Bellaghy.
Sluggada—a mountain nine miles southwest from Dungiven.
Sperrin Mountains -a great range south and southwest of Dungiven.
Springfield Glen—glen of the Forth river west of Be'fast.
Stormount—about two miles north of Dundonald.
Straid—six miles northwest from Carrickfergus.
Straidkilly—about two miles north by west of Glenarm.
Strangford—seven miles northeast of Downpatrick.
Swatragh—four and a half miles north of Maghera.
Tamlaghtard—the parochial name for Magilligan.
Tannaghmore—midway between Antrim and Ballymena.
Tardree—a low hill five miles northeast of Antrim.
Thomas Mountain—above Newcastle, to the southwest.
Thornisland—a bog two and a half miles southwest of Donaghadee.
Three Islands—in Lough Neagh three and a half miles south of Toome.
Tievedocharragh—a mountain four miles northeast of Rostrevor.
Tildarg—about eight and a half miles northeast from Antrim.
Tillysburn—shore of Belfast Bay one mile above Holywood.
Tircrevan—at foot of Benevenagh mountains, on north side.
Tollymore—a park two and a half miles northwest of Newcastle.
Tor Head—some four or five miles southeast of Fairhead.
Trench (Antrim)—southwest of Belfast three miles by Falls road.
Trench (Down)—east of Comber.
Trostan—a high mountain four miles southwest of Cushendall.
Tullybranagan—foot of the Mourne mountains at Newcastle.
Tullycairn—by the Lagan some three miles below Dromore.
Tullymurry—four miles southwest of Downpatrick.
Tyrella—by the seashore three miles east of Dundrum.
Umbra Rocks—by the shore of Magilligan.
Victoria Park—on Belfast bay, east side of Connswater.
Whitehead—at the western entrance of Belfast Bay.
Whitehouse—on side of the bay over three miles below Belfast.
White Mountain—midway between Dungiven and Draperstown.
Whitepark Bay—a little west of Ballintoy.
Whiterock—above Springfield, at foot of Black Mountain.
Whitewater—flows through Mourne park west of Kilkeel.
Whitewell—on the Antrim road four miles from Belfast.
Windsor—a southern suburb of Belfast.
Windy Gap—a glen in Black mountain one mile northeast of Hannahstown.
Wolfhill—on Crumlin road three to four miles northwest of Belfast
Wolf Island—four and a half miles east of Newtownards.
Woodburn—two miles northwest of Carrickfergus.
Yellow-water River—five miles north by west of Kilkeel.

ALPHABETICAL INDEX

To the Orders and Genera of the flowering plants and the higher cryptograms enumerated.

Synonyms are printed in italic type.

Achillæa,	77	Apium,	61	Blysmus,	156
Adoxa,	69	Aquifoliaceæ,	93	Boraginaceæ,	96
Ægopodium,	62	Arabis,	10	Botrychium,	189
Æthusa,	64	Araceæ,	147	Brachypodium,	178
Agrimonia,	41	Arctium,	81	Brassica,	11
Agrostis,	169	Arctostaphylus,	90	Briza,	175
Aira,	170	Arenaria,	23	Bromus,	178
Ajuga,	113	Armeria,	118	Brunella,	109
Alchemilla,	42	*Armoracia*,	9	Bunium,	62
Alismaceæ,	141	Arrhenatherum,	172	Butomus,	141
Alisma,	141	Artemisia,	79		
Allium,	142	Arum,	147	Cakile,	15
Alnus,	132	Arundo,	168	Calamagrostis,	169
Alopecurus,	167	Asperula,	71	Calamintha,	108
Alsine,	23	Asplenium,	186	Callitrichaceæ,	126
Amentiferæ,	128	Aster,	76	Callitriche,	126
Anacharis,	135	Athyrium,	186	Calluna,	90
Anagallis,	116	Atriplex,	120	Caltha,	5
Anchusa,	96	Avena,	171	Campanulaceæ,	89
Andromeda,	90			Campanula,	89
Anemone,	2	Ballota,	117	Caprifoliaceæ,	69
Angelica,	65	Barbarea,	9	Capsella,	14
Antennaria,	77	Bartsia,	104	Cardamine,	10
Anthemis,	78	Bellis,	75	Carduus,	82
Anthoxanthum,	167	Beta,	120	Carex,	155
Anthriscus,	67	Betula,	132	Carlina—see addenda	
Anthyllus,	38	Bidens,	80	Carum,	62
Apargia,	84	Blechnum,	188	Caryophyllaceæ,	20

INDEX.

Catabrosa,	176	Droseraceæ,	18	Glyceria,	174
Celastraceæ,	33	Drosera,	18	Gnaphalium,	76
Centaurea,	81	Dryas,	47	Gramineæ,	167
Centunculus,	117			Gymnadenia,	136
Cephalanthera,	139	Echium,	97		
Cerastium,	24	Elatinaceæ,	19	Habenaria,	137
Ceratophyllaceæ,	126	Elatine,	19	Haloragaceæ,	55
Ceratophyllum,	126	Eleocharis,	153	Hederaceæ,	68
Ceterach,	188	Elodea,	135	Hedera,	68
Chærophyllum,	67	Empetraceæ,	124	*Helosciadum*,	61
Characeæ,	192	Empetrum,	124	Heracleum,	66
Chara,	192	Endymion,	143	Hieracium,	85
Chelidonium,	7	Epilobium,	53	Hippuris,	55
Chenopodiaceæ,	119	Epipactis,	139	Holcus,	170
Chenopodium,	119	Equisctaceæ,	181	*Honckeneja*,	23
Chrysanthemum,	78	Equisetum,	181	Hordeum,	179
Chrysosplenium,	59	Ericaceæ,	90	Hottonia,	114
Cichorium,	83	Erica,	90	Hydrocharidaceæ,	134
Cicuta,	61	Erigeron,	75	Hydrocharis,	134
Circæa,	54	Eriophorum,	155	Hydrocotyle,	60
Cladium,	152	Erodium,	31	Hymenophyllum,	188
Cochlearia,	13	Eryngium,	60	Hyosciamus,	100
Comarum,	43	Erythræa,	93	Hypericaceæ,	27
Compositæ,	74	Euonymus,	33	Hypericum,	27
Coniferæ,	133	Eupatorium,	74	Hypochæris,	83
Conium,	68	Euphorbiaceæ,	124		
Convolvulaceæ,	95	Euphorbia,	124	Ilex,	93
Convolvulus,	95	Euphrasia,	104	Inula,	76
Corylus,	133			Iridaceæ,	140
Cotyledon,	57	Festuca,	176	Iris,	140
Crambe,	15	Filago,	76	Isoetes,	191
Crassulaceæ,	56	Filices,	183		
Cratægus,	51	Fœniculum,	65	Jasione,	89
Crepis,	85	Fragaria,	43	Juncaceæ,	143
Cruciferæ,	9	Fraxinus,	93	Juncus,	143
Cryptogamme,	183	Fumariaceæ,	8	Juniperus,	133
Cynoglossum,	96	Fumaria,	8		
Cynosurus,	176			Knautia,	73
Cyperaceæ,	152	Galeopsis,	111	Kœleria,	172
Cystopteris,	186	Galium,	70		
		Gentianaceæ,	93	Labiatæ,	107
Dactylus,	176	Gentiana,	94	Lamium,	109
Daucus,	66	Geraniaceæ,	29	Lapsana,	83
Digitalis,	102	Geranium,	29	Lastrea,	184
Dipsacaceæ,	73	Geum,	48	Lathræa,	101
Dipsacus,	73	Glaucium,	7	Lathyrus,	39
Draba,	21	Glaux,	116	Lavatera,	27

INDEX.

Leguminosæ,	34	Montia—see addenda		Pinus,	134
Lemnaceæ,	147	Myosotis,	98	Plantaginaceæ,	118
Lemna,	147	Myrica,	132	Plantago,	118
Lentibulariaceæ,	113	Myriophyllum,	55	Plumbaginaceæ,	118
Leontodon,	83	Myrrhis,	67	Poa,	173
Lepidium,	13			Polygalaceæ,	18
Lepturus,	179	Naidaceæ,	152	Polygala,	18
Ligusticum,	65	Nardus,	168	Polygonaceæ,	122
Ligustrum,	93	Narthecium,	143	Polygonum,	123
Lilliaceæ,	142	Nasturtium,	9	Polypodium,	183
Linaceæ,	32	Neottia,	138	Polystichum,	185
Linaria,	102	Nepeta,	109	Populus,	131
see corrigenda		Nitella,	193	Potamogetonaceæ,	148
Linum,	32	Nuphar,	5	Potamogeton,	148
Listera,	138	Nymphaceæ,	5	Potentilla,	43
Lithospermum,	98	Nymphæa,	5	Primulaceæ,	114
Littorella,	119			Primula,	115
Lobelia,	89	*Obione*,	121	*Prunella*—see Brunella	
Lolium,	180	Œnanthe,	64	Prunus,	40
Lonicera,	70	Oleaceæ,	93	Psamma,	168
Lotus,	37	Onagraceæ,	53	Pteris,	188
Luzula,	145	Ononis,	35	Pulicaria,	76
Lychnis,	21	Ophioglossum,	190	Pyrola,	91
Lycopodiaceæ,	191	Orchidaceæ,	135	Pyrus,	51
Lycopodium,	191	Orchis,	135		
Lycopus,	107	Origanum,	108	Quercus,	132
Lycopsis,	96	Orobanchaceæ,	100		
Lysimachia,	115	Orobanche,	100	Radiola,	32
Lythraceæ,	52	*Orobus*,	40	Ranunculaceæ,	1
Lythrum,	52	Osmunda,	189	Ranunculus,	2
		Oxalidaceæ,	32	Raphanus,	15
Malaxis,	140	Oxalis,	32	Resedaceæ,	16
Malvaceæ,	26			Reseda,	16
Malva,	26	Papaveraceæ,	6	Rhamnaceæ,	33
Marsilaceæ,	190	Papaver,	6	Rhamnus,	33
Matricaria,	78	Parietaria,	127	Rhinanthus,	104
Meconopsis,	7	Parnassia,	60	*Rhodiola*,	56
Medicago,	35	Pedicularis,	104	Rhynchospora,	153
Melampyrum,	103	Peplis,	52	Rosaceæ,	40
Melanthaceæ,	143	Petasites,	74	Rosa,	48
Melica,	172	Phalaris,	167	Rubiaceæ,	70
Mentha,	107	Phleum,	167	Rubus,	44
Menyanthes,	94	Phragmites,	168	Rumex,	122
Mercurialis,	125	Picris,	84	Ruppia,	151
Mertensia,	97	Pilularia,	190		
Milium,	168	Pimpinella,	63	Sagina,	21
Molinia,	173	Pinguicula,	113	Sagittaria,	141

Salicornia,	120	Sisymbrium,	11	Triglochin,	142
Salix,	128	Sium,	63	Trioda,	172
Salsola,	119	Smyrnium,	68	Trisetum,	171
Sambucus,	69	Solanaceæ,	99	Triticum,	179
Samolus,	117	Solanum,	99	Tussilago,	74
Sanguisorbia,	41	Solidago,	75	Typhaceæ,	145
Sanicula,	60	Sonchus,	84	Typha,	145
Sarothamnus,	35	Sparganium,	146		
Saxifragaceæ,	57	Spergula,	26	Ulex,	34
Saxifraga,	57	Spergularia,	25	Ulmaceæ,	128
Scabiosa,	72	Spiræa,	41	Ulmus,	128
Scandix,	66	Stachys,	111	Umbelliferæ,	60
Schoenus,	152	Statice,	118	Urticaceæ,	127
Scilla,	142	Stellaria,	24	Urtica,	127
Scirpus,	154	Suæda,	119	Utricularia,	113
Scleranthus,	26	Subularia,	14		
Sclerochloa,	174	Symphytum,	96	Vaccinium,	91
Scolopendrium,	187			Valerianaceæ,	72
Scrophulariaceæ,	101	Tanacetum,	79	Valeriana,	72
Scrophularia,	103	Taxus,	133	Valerianella,	72
Scutellaria,	108	Teucrium,	112	Verbascum,	101
Sedum,	56	Thalictrum,	1	Veronica,	104
Selaginella,	192	Thlaspi,	13	Viburnum,	70
Senebiera,	15	*Thrincia,*	83	Vicia,	38
Senecio,	79	Thymus,	108	Violaceæ,	16
Sherardia,	70	Tolypella,	193	Viola,	16
Silaus,	65	Torilis,	66		
Silene,	20	*Tormentilla,*	43	Zannichellia,	151
Sinapis,	12	Trifolium,	36	Zostera,	152

INDEX TO THE COMMON ENGLISH NAMES.

Italic type denotes plants of the excluded list.

Adder's Tongue,	190	Bilberry,	91	Broom,	35
Agrimony,	41	Bindweed,	95	Broomrape,	100
,, Hemp,	74	,, Black,	124	Buckbean,	94
Alder,	132	Birch,	132	Buckthorn,	38
,, Black,	34	,, *White*,	300	*Buckwheat*,	286
Alexanders,	68	Birdsnest,	138	Bugle,	113
Alkanet,	96	,, *Yellow*,	296	Bugloss,	96
Allgood,	120	Bittersweet,	100	*Bullace*,	276
Allseed,	32	Blackberry,	44	Burdock,	81
Arrowgrass,	141	Blackheads,	145	Burnet-greater,	41
Arrowhead,	141	Black Horehound,	112	,, *Lesser*,	294
Ash,	93	Black Saltwort,	116	Butterbur,	74
,, Mountain,	51	Blackthorn,	40	,, *Fragrant*,	280
Avens,	48	Bladder Fern,	186	Buttercup,	2
,, Mountain,	47	Bladderwort,	113	Butterwort,	113
		Blaeberry,	91		
Baldmoney,	279	Bluebell,	163	*Cabbage*,	272
Barberry	271	Bog Asphodel,	143	Calamint,	108
Barley,	179	Bogbean,	95	Campion,	21
Beakrush,	153	Bog Myrtle,	132	,, Bladder,	20
Bearberry,	90	Bog Orchis,	140	,, Sea,	20
Bedstraw,	71	Bogrush,	152	*Canary Grass*,	289
Beech,	287	*Borage*,	284	*Caraway*,	279
Beech Fern,	183	Bracken,	188	Carrot,	66
Beenettle,	10	Brakes,	188	Catchfly,	20
Beet,	120	Bramble,	44	Catmint,	109
Bellflower,	90	Bree's Fern,	185	Catsear,	83
,, Giant,	283	Briar,	48	Catsfoot,	77
Bent,	168	Broad Fern,	184	Catstail Grass,	167
Bentgrass,	169	Brome Grass,	178	Celandine,	7
Benweed,	80	Brooklime,	105	Celery,	61
Betony,	111	Brookweed,	117	Centaury,	91

INDEX.

Chamomile,	78	Daisy, Michaelmas,	78	Fumitory, *White*,	291
,, Corn,	281	,, Oxeye,	78	,, *Yellow*	271
,, Stinking,	281	Dandelion,	84	Furze,	34
Charlock,	12	Danesblood,	69		
Cherry,	40-41	Darnel,	180	Garlic,	142
,, *Dwarf*,	277	Deadnettle,	109	Gentian,	94
Chervil-wild,	67	Devil's Bit,	73	Germander,	105
,, Garden,	280	Devil's Churnstaff,	124	,, *Water*,	298
Chestnut,	287	Dewberry,	47	Gilgowan,	78
Chickweed,	24	Docken,	122	Gipsywort,	108
Chicory,	83	Dogstail Grass,	176	*Gladdon*,	288
Cinquefoil,	43	*Dropwort*,	277	Glasswort,	120
Clover,	36	,, Water,	64	*Globeflower*,	291
Clubmoss,	191	Duckweed,	147	Goldenrod,	75
Clubrush,	154	Dyer's Weed,	16	Goldilocks,	4
Coltsfoot,	74			Gooseberry,	278
Columbine	271	Eglantine,	50	Goosefoot,	119
Comfrey,	96	Elder,	70	Gorse,	34
Coriander,	280	,, Dwarf,	69	Goutweed,	62
Corn Bluebottle,	82	Elecampane,	76	Grass of Parnassus,	60
Corn Cockle,	21	Elm,	128-287	Grasswrack,	152
Cornsalad,	72	Eryngo,	60	Gromwell,	98
Cotton Grass,	155	Espiebawn,	78	Ground Ivy,	109
Couch Grass,	179	Eyebright,	104	Groundsel,	79
Cowbane,	61				
Cowberry,	91	Farmer's-plague,	62	Hairbell,	90
Cowslip,	115	Fennel,	65	Hairgrass,	170
Cow-Wheat,	103	Fescue Grass,	176	Hard Fern,	188
Cranberry,	91	*Feverfew*,	281	Harts'-tongue,	187
Cranesbill,	29	Field Madder,	71	,, Scaly,	188
Cress,	9	Figwort,	103	Hawkbit,	83
,, Bitter,	10-272	Filmy Fern,	188	Hawksbeard,	85
,, Rock,	10	Flag,	140	Hawkweed,	85
,, Wall,	11	,, *Sweet*,	288	Hawthorn,	51
,, Wart,	15	Flax,	32-275	Hazel,	133
Crested Hairgrass,	172	,, Purging,	32	Heather,	90
Crosswort,	71	Fleabane,	76	Heath Grass,	172
Crowberry,	124	,, Blue,	75	*Hellebore*	271
Crowfoot,	2	Flixweed,	11	Helleborine,	139
,, Corn,	271	Flowering Fern,	189	Hemlock,	68
,, Water,	2	,, Rush,	141	,, Water,	61
Cuckoo Flower,	10	Flower of Dunluce,	30	*Hemp*,	286
Cudweed,	76	Forget-me-not,	98	Hempnettle,	111
Currant-black,	278	Foxglove,	102	Henbane,	100
,, *Red*,	279	Foxtail Grass,	167	Herb Robert,	31
		Frogbit,	134	Hightaper,	101
Daisy,	75	Fumitory,	8	Hogweed,	66

Holly,	93	Marsh Fern,	184	Parsley, Corn,		294
,, Sea,	60	Mashcorns,	43	,, Fool's,		64
Honeysuckle,	70	Masterwort,	279	,, Hedge,		66
Hop,	286	Mat Grass,	168	Parsnep,		279
Hornwort,	126	Mayflower,	5	,, Water,		63
Horsebane,	64	Meadow Grass,	173	Pear,		277
Horsemint,	297	Meadow Rue,	1	Pearlwort,		21
Horsetail,	181	Meadow Saffron,	288	Pellitory,		127
Houndstongue,	96	Meadowsweet,	41	Pennycress,		13
Houseleek,	278	Medic,	35	Pennyroyal,		107
Hyacinth,	143	Melic,	172	Pennywort,		60
		Melilot,	275	Peppermint,		107
Ivy,	68	Mercury,	125	Pepperwort,		13
		Mignonette,	273	Periwinkle,		283
Jack-by-the-hedge,	11	Milkwort,	18	Pignut,		62
Jacob's Ladder,	284	Millet,	168	Pilewort,		4
Juniper,	133	Mint,	107	Pillwort,		190
		Moneywort,	115	Pimpernel,		116
Knapweed,	81	Moonwort,	189	,, Bastard,		117
,, Great,	296	Moschatel,	69	,, Yellow,		116
Knawel,	26	Mountain Fern,	174	Pine,		134
Knotgrass,	123	Mouse-ear,	24	Pink,		274
		Mugwort,	79	,, Cushion,		20
Lady Fern,	186	Mullein,	101	,, Maiden,		273
Lady's Finger,	38	Mustard,	12	Plantain,		118
Lady's Mantle,	42	,, Hedge,	11	,, Water,		141
Lady's Smock,	10			Polypody,		183
Leopard's Bane,	281	Napperty,	46	Pondweed,		148
Ling,	90	Navelwort,	57	,, Horned,		151
London Pride,	279	Nettle,	127	,, Tassel,		151
Loosestrife-purple,	52	Nightshade-Black,	99	Poplar,		131
,, Yellow,	115	,, Deadly,	284	,, White,		287
Lousewort,	104	,, Enchanters,	54	Poppy,		6
Lovage,	65	Nipplewort,	83	,, Horned,		7
Lucerne,	275			,, Welsh,		7
		Oak,	132	Primrose,		115
Makinboy,	300	Oak Fern,	184	,, Evening,		278
Male Fern,	184	Oat,	171	Privet,		93
Mallow,	26	Oat Grass,	172	Prushus,		12
,, Tree,	27	Oat-wild,	304			
Maple,	274	Orache,	120	Quaking Grass,		175
Marestail,	55	Orpine Livelong,	56	Quillwort,		191
Marigold-Bur,	80	Osier,	129			
,, Corn,	78	Oxtongue,	84	Radish,		15
,, Marsh,	5			,, Sea,		16
Marjoram,	108	Pansy,	17	Ragged Robin,		21
Marram,	168	Parsley,	279	Ragweed,		80

Rampion,	283	Scurvy Grass,	13	Strawberry,	
Ramsons,	142	Sea Blite,	119	Hautboy,	277
Raspberry,	44	Sea Kale,	15	Sturdy,	180
Ratstail,	119	Sea Lavender,	118	Sundew,	18
Redshank,	122	Sea Purslane,	121	Sweetbriar,	50
Reed,	168	Sedge,	157	Sweet Cicely,	67
,, Bur,	146	Selfheal,	109	Sweet Gale,	132
Reedgrass,	167	Shamrock,	37	Sweet Woodruff,	71
Reedmace,	145	Shepherd's Needle,	66	*Sycamore*,	274
Restharrow,	35	Shepherd's Purse,	14		
Robin-run-the-hedge,	72	Shoreweed,	119	Tansy,	79
Rockbrake,	183	Silverweed,	43	Tare,	38-39
Rocket-Sea,	15	Sitfast,	4	Teasel,	73
,, Yellow,	9	Skullcap,	108	Thistle,	82
Rockrose,	273	Sloe,	40	,, Milk,	282
Rose,	48	Smallreed,	169	,, Musk,	282
,, Cinnamon,	277	*Snakeweed*,	286	Thrift,	118
,, Guelder,	70	*Snapdragon*,	285	Thyme,	108
,, *Yellow*,	277	Sneezewort,	77	,, Basil	297
Rosebay,	53	*Soapwort*,	274	,, Water	135
Rosemary,	90	Soft Grass,	170	Timothy,	167
Rosenoble,	103	Sorrel,	122	Toadflax,	102
Roseroot,	56	,, Sheep's,	123	Toadrush,	144
Rowan Tree,	51	,. Wood,	32	Toothwort,	101
Rush,	143	Sowthistle,	84	Tormenting Root,	43
,, Dutch,	182	*Spearmint*,	285	*Towerwort*,	272
Rye Grass,	180	Spearwort Greater,	4	Trefoil,	36-37
,, *Italian*,	289	,, Lesser,	3	Tutsan,	27
		Speedwell,	104	Twayblade,	138
Saggon,	140	Spikerush,	153	Twigrush,	152
Sainfoin,	276	Spindle Tree,	33		
St. John's Wort,	27	Spleenwort,	186	Valerian,	72
Sallow,	128	,, Sea,	187	,, Red,	280
,, Blooming,	37	Spurge,	124	Venus's Comb,	66
Salsify,	282	,, Caper,	286	*Vervain*,	285
Saltwort,	119	,, Cypress,	286	Vetch,	38-276
Sandwort,	23	Spurrey,	26	,, Kidney,	38
Sanicle,	60	,, Knotted,	22	Vetching,	39
Savin,	301	,, Sea,	25	Violet,	16
Saxifrage,	57	Squill,	142	,, Dame's,	272
,, Burnet,	62	Starwort,	75	,, Water,	114
,, Golden,	59	,, Water,	126	Viper's Bugloss,	97
,, Meadow,	65	Stitchwort,	24		
Scabious,	72	Stonecrop,	56	Wakerobin,	147
,, Sheep's,	89	Storksbill,	31	*Wallflower*,	272
Scorpion Grass,	98	Strawberry,	43	Wall Pepper,	57
Scotch Fir,	134	,, Barren,	43	Wall Rue,	187

Watercress,	9	Well Ink,	105	Wood Anemone,	2
Waterlily,	5	Wheat Grass,	179	Woodrush,	145
Water Milfoil,	55	Whin,	34	Woodsage,	112
Water Pepper,	123	,, Dwarf,	293	Wormwood,	79-281
Water Purslane,	52	White Beam Tree,	51	Woundwort,	111
Water Soldier,	288	Whitlow Grass,	12		
Waterwort,	19	Willow,	128	Yarrow,	77
Waybread,	119	,, Bedford,	287	Yellowrattle,	106
Wayfaring Tree,	280	Willowherb,	53	*Yellowort*,	284
Weaselsnout,	297	Windflower,	2	Yew,	153
Weld,	16	Wintergreen,	91		

INDEX TO THE MUSCI AND HEPATICÆ.

Names printed in *italics* are synonyms.

Amblyodon,	232	brevifolia,	217	alpinum,	236
dealbatus,	232	convoluta,	218	argenteum,	235
Andræa,	195	curvirostris,	216	*atropurpureum*	
alpina,	195	cylindrica,	218	bicolor, 236	[236
crassinervis,	196	fallax,	217	bimum,	235
petrophila,	195	Hornschuchii,	218	capillare,	236
Rothii,	196	revoluta,	218	cæspiticeum,	235
rupestris,	196	rigidula,	217	concinnatum,	234
Anœctangium,	225	rubella,	216	filiforme,	235
compactum,	225	spadicea,	217	inclinatum,	235
Mougeotti,	225	unguiculata,	219	intermedium,	235
Aneura,	268	Bartramia,	237	*julaceum*	235
pinguis,	268	arcuata,	238	murale,	236
Anisothecium,	203	calcarea,	238	pallens,	236
rubrum,	203	Œderi,	237	pendulum,	235
rufescens,	214	fontana,	238	*pseudo-triquetrum*	
squarrosum,	204	*Halleriana,*	238	roseum, 237	[236
Anomodon,	240	ithyphylla,	237	ventricosum,	236
viticulosus,	240	norvegica,	238		
Anthelia,	261	pomiformis,	237	Campylopus,	206
julacea,	261	Bazzania,	262	atrovirens,	207
Anthoceros,	270	trilobata,	262	brevipilus,	207
punctatus,	270	Blasia,	268	flexuosus,	207
Archidium,	201	pusilla,	268	fragilis,	206
alternifolium,	201	Blepharostoma,	261	pyriformis,	206
Asterella,	269	trichophylla,	261	setifolius,	207
hemisphærica,	269	Blindia,	205	Catharinea,	197
Atrichum,	197	acuta,	205	undulata,	197
Aulacomnion,	238	Breutelia,	238	Cephalozia,	262
		chrysocoma,	238	bicuspidata,	263
Barbula,	216	Bryum,	234	connivens,	263

divaricata,	262	flexicaule,	202	fascicularis,	223
Francisi,	262	homomallum,	202	funalis,	221
Ceratodon,	209	pusillum,	202	heterosticha,	223
purpureus,	209	tenuifolium,	201	hypnoides,	224
Cesia,	267			leucophæa,	222
crenulata,	267	*Encalypta*,	219	maritima,	221
obtusa,	267	*Entosthodon*,	232	microcarpa,	222
Chiloscyphus,	264	Ephemerum,	210	patens,	222
polyanthos,	264	serratum,	210	pruinosa,	22.)
Cinclidotus,	219			pulvinata,	221
fontinaloides,	219	*Fegatella*,	269	robusta,	221
Climacium,	255	Fissidens,	199	*Schultzii*,	222
dendroides,	255	adiantoides,	200	*spiralis*,	221
Cryphæa,	255	bryoides,	199	trichophylla,	221
arborea,	255	exilis,	199	*Gymnostomum*, 214,	216
heteromalla,	255	osmundoides,	199		
Cynodontium,	209	pusillus,	199	Hedwigia,	256
		taxifolius,	200	albicans,	256
Dichodontium,	208	Fontinalis,	255	*ciliata*,	256
flavescens,	209	antipyretica,	255	imberbis,	256
pellucidum,	208	squamosa,	255	Hepatica,	269
Dicranella, 203		Fossombronia,	268	conica,	269
cerviculata,	203	pusilla,	268	Herberta,	261
crispa,	203	Frullania,	259	adunca,	261
heteromalla,	203	dilatata,	259	Homalia,	254
secunda,	203	fragilifolia,	259	trichomanoides,	254
subulata,	203	germana,	259	*Hookeria*.	253
Dicranodontium,	206	tamarisci,	259	Hypnum,	240
Dicranoweissia,	207	Funaria,	232	aduncum,	242
cirrata,	207	calcarea,	232	albicans,	248
Dicranum,	207	fascicularis,	232	*arcuatum*,	252
Bonjeani,	208	hygrometrica,	232	*atrovirens*,	250
fuscescens,	208	obtusa,	232	Borreri,	252
majus,	207	Templetoni,	232	brevirostrum,	250
palustre,	208			*commutatum*,	241
scoparium,	208	Georgia,	196	confertum,	247
Scottii,	208	Brownii,	196	cordifolium,	244
Didymodon,	206	pellucida,	196	crassinervium,	246
cylindricus,	215	Grimmia,	220	cupressiforme,	251
denudatus,	206	acicularis,	223	cuspidatum,	253
nervosus,	212	apocarpa,	220	decipiens,	250
rigidulus,	217	aquatica,	223	denticulatum,	253
rubellus,	216	canescens,	224	depressum,	252
Diphyscium,	220	decipiens,	222	exannulatum,	242
Diplophyllum,	264	Donii,	222	falcatum,	242
albicans,	264	elatior,	222	filicinum,	241
Ditrichum,	201	elliptica,	222	fluitans,	242

giganteum,	244	Teesdalii,	246	pyriforme,	233
glareosum,	248	tenellum,	247	Leskea,	240
glaucum,	241	triquetrum,	250	polycarpa,	240
hamifolium,	243	*uncinatum*,	242	*sericea*,	249
illecebrum,	245	undulatum,	252	Leucobryum,	200
intermedium,	245	velutinum,	247	glaucum,	200
Kneiffi,	242	viride,	248	Leucodon,	255
loreum,	251	viviparum,	249	sciuroides,	255
lutescens,	249			Lophocolea,	264
lycopodioides,	243	*Isothecium*,	249	bidentata,	264
molluscum	251				
murale,	247	Jubula,	257	Marchantia,	269
ochraceum,	243	Hutchinsiæ,	259	polymorpha,	269
ornithopodioides,		Jungermania,	265	Metzgeria,	269
palustre, 244 [249		*affinis*,	265	conjugata,	269
parietinum,	250	bantriensis,	266	furcata,	269
Patientiæ,	252	barbata,	266	pubescens,	269
piliferum.	247	*cochleariformis*,	261	Mnium,	239
plumosum,	247	cordifolia,	265	*affine*,	239
populeum,	248	exsecta,	266	cuspidatum,	239
prælongum,	245	*Lyoni*,	266	hornum,	239
proliferum,	250	porphyroleuca,	266	pseudo-punctatum,	
proliferum,	240	quinquedentata,			239
pseudoplu-		riparia, 265	[266	punctatum,	240
mosum,	247	sphærocarpa,	265	rostratum,	239
pulchellum,	252	turbinata,	265	stellare,	239
pumilum,	246	ventricosa,	266	*subglobosum*,	239
purum,	245			undulatum,	239
resupinatum,	251	Kantia,	263	Mollia,	214
revolvens,	243	arguta,	263	brachydontia,	215
riparium,	241	trichomanis,	263	crispa,	214
rivulare,	248			crispula,	215
rusciforme,	247	Leersia,	219	æruginosa,	214
rutabulum,	248	contorta,	219	inclinata,	215
sarmentosum,	244	exstinctoria,	219	litoralis,	215
Schreberi,	250	laciniata,	219	microstoma,	214
scorpioides,	243	Lejeunea,	260	rutilans,	214
sericeum,	249	calcarea,	260	tenuirostris,	215
serpens,	241	hamatifolia,	260	tenuis,	214
speciosum,	246	*minutissima*,	260	tortuosa,	216
splendens,	250	ovata,	260	verticillata,	215
squarrosum,	251	serpyllifolia,	260	viridula,	214
stellatum,	241	ulicina,	260	Mylia,	265
stramineum,	244	Lepidozia,	262	Taylori,	265
striatum,	245	reptans,	262		
Swartzii,	245	setacea,	262	Nardia,	266
sylvaticum,	253	Leptobryum,	233	compressa,	267

crenulata,	266	Physcomitrium,	231	lucens,	253
emarginata,	267	pyriforme,	231	*Pterogonium*,	249-250
Funckii,	267	Plagiobryum,	234	*Ptychomitrium*,	224
hyalina,	267	Zierii,	234		
obovata,	267	Plagiochila,	264	*Racomitrium*,	221-224
scalaris,	267	asplenoides,	265	Radula,	260
Neckera,	254	punctata,	265	aquilegia,	260
complanata,	254	spinulosa,	265	complanata,	260
crispa,	254	Pleuridium,	201	*Rhabdoweissia*,	209
		alternifolium,	201	Riccia,	270
Oligotrichum,	197	axillare,	201	fluitans,	270
hercynicum,	197	subulatum,	201		
incurvum,	197	Pleurozia,	261	Saccogyna,	263
Oncophorus,	209	purpurea,	261	viticulosa,	263
Bruntoni,	209	Pleurozygodon,	225	Scapania,	263
crispatus,	209	æstivus,	225	curta,	264
striatus,	209	Pohlia,	233	nemorosa,	263
Orthopyxis,	239	albicans,	234	resupinata,	264
Orthotrichum,	226	annotina,	234	undulata,	263
affine,	226	carnea,	234	Seligeria,	204
cupulatum,	227	cruda,	234	calcarea,	205
diaphanum,	227	elongata,	233	Donii,	204
leiocarpum,	227	nutans,	233	pusilla,	205
Lyellii,	227	Polytrichum,	197	Sphagnum,	256
pulchellum,	228	aloides,	197	acutifolium,	258
rivulare,	228	alpinum,	198	cuspidatum,	258
rupestre,	226	commune,	199	cymbifolium,	257
saxatile,	228	gracile,	198	intermedium,	258
Sprucei,	226	juniperinum,	198	molle,	257
stramineum	228	nanum,	197	papillosum,	256
striatum,	227	piliferum,	198	rigidum,	257
tenellum,	228	strictum,	199	squarrosum,	258
		urnigerum,	198	subsecundum,	257
Pallavacinia,	268	Porella,	261	tenellum,	257
hibernica,	268	platyphylla,	261	Sphærocephalus,	238
Parotrichum,	254	rivularis,	261	palustris,	238
alopecurum,	254	Pottia,	210	Splachnum,	230
Pellia,	268	asperula,	211	ampullaceum,	230
calycina,	268	crinita,	211	pedunculatum,	230
epiphylla,	268	Heimii,	210		
Phascum,	210	intermedia,	211	Targionia,	270
acaulon,	210	Starkei,	211	hypophylla,	270
cuspidatum,	210	truncatula,	211	Tayloria,	231
patens,	231	viridifolia,	211	tenuis,	231
serratum,	210	Preissia,	270	Tetraplodon,	230
Physcomitrella,	231	commutata,	270	bryoides,	230
patens,	231	Pterygophyllum,	253	*mnioides*,	230

Thamnium,	254	mutica,	213	Weissia,		229
Thuidium,	240	papillosa,	213	Bruchii,		229
tamariscifolium,	240	*rigida*,	212	*controversa*,		214
Tortula,	211	ruralis,	213	Drummondi,		229
aloides,	212	stellata,	212	*mucronata*,		214
ambigua,	212	subulata,	212	phyllantha,		229
atrovirens,	212	*tortuosa*,	216	ulophylla,		229
ericæfolia,	212	Trichocolea,	261	*verticillata*,		215
inclinata,	215	tomentella,	261	vittata,		229
intermedia,	213					
lamellata,	211	*Ulota*,	229	Zygodon,		225
lævipila,	213			conoideus,		226
latifolia,	213	Webera,	220	*Mougeottii*,		225
montana,	213	sessilis,	220	Stirtoni,		225
muralis,	212	*Webera*,	223	viridissimus,		225

INDEX TO THE EXCLUDED PLANTS.

Acer campestre	274	Carduus eriophorus	296	Draba muralis	273
A. pseudo-platanus	274	C. nutans	282		
Achillæa tomentosa	281	Carex axillaris	303	Elodea canadensis	288
Acorus calamus	288	C. Davalliana	303	Elymus arenarius	305
Adiantum capillus-		C. elongata	303	Œnanthe pimpin-	
veneris	305	C. filiformis	303	elloides	294
Ajuga alpina	298	C. punctata	303	Œnothera biennis	278
Alisma natans	301	Carum carui	279	Epilobium roseum	277
Alyssum maritimum	272	Castanea vulgaris	287	E. tetragonum	277
Anthemis arvensis	281	Centaurea jacea	295	Erica stricta	296
A. cotula	281	C. scabiosa	296	Eriocaulon septan-	
Antirrhinum majus	285	Centranthus ruber	280	gulare	301
Antitrichia curtipen-		Cephalanthera pallens	301	Erythræa latifolia	297
dula	307	Cerastium alpinum	292	E. littoralis	297
Aquilegia vulgaris	271	Ceratophyllum sub-		Euphorbia cyparissias	286
Arabis perfoliata	272	mersum	300	E. hiberna	300
Arctium majus	295	Chara fragilis	305	E. lathyrus	286
Artemisia absinthum	281	Chærophyllum sativum	280		
A. campestris	281	Cheiranthus cheiri	272	Fagus sylvatica	287
A. maritima	295	Chenopodium ficifoli-		Festuca myuros	304
Asplenium fontanum	305	um	299	F. uniglumis	304
Atriplex marina	299	C. hybridum	299		
Atropa belladonna	284	*C. olidum*	286	Galeopsis dubia	298
Avena fatua	304	C. murale	299	G. ladanum	298
A. pratensis	304	C. polyspermum	298	*G. ochroleuca*	298
		C. urbicum	299	Galium sylvestre	295
Barbarea præcox	291	C. vulvaria	286	G. uliginosum	295
Barbula Hornschuchii	306	Chlora perfoliata	284	Geranium columbinum	293
Berberis vulgaris	271	Chrysosplenium alter-		G. phæum	274
Betula verrucosa	300	niflorum	294	G. pusillum	293
Borago officinalis	284	Cochlearia anglica	291	G. rotundifolium	293
Brassica oleracea	272	Colchicum autumnale	288	G. striatum	274
Bromus arvensis	289	Coriandrum sativum	280	Gnaphalium luteo-	
B. secalinus	289	Corydalis claviculata	291	album	295
Bryum erythrocarpum	306	C. lutea	271	Grimmia ovata	306
		Crepis biennis	283		
Calamagrostis lanceo-		C. hieracioides	296	Helleborus viridis	271
lata	304	Cuscuta epilinum	284	Helianthemum vulgare	273
Calamintha acinos	297			Hesperis matronalis	272
Camelina sativa	273	Daucus gummifer	294	Hieracium auranticum	283
Campanula latifolia	283	Dianthus deltoides	273	H. nigrescens	296
C. rapunculoides	283	Dicranoweissia cris-		Hordeum murinum	289
C. rapunculus	283	pula	305	Humulus lupulus	286
Campylopus Schwarzii	305	Diplotaxis tenuifolia	272	Hypericum elatum	274
Cannabis sativa	286	Doronicum pardali-		H. montanum	292
Cardamine impatiens	272	anches	281	Hypnum crista-	
				castrensis	306

INDEX.

H. elegans	306	Petroselinum sativum	279	S. rubra	287
H. hamulosum	306	P. segetum	294	S. undulata	287
H. irriguum	306	Peucedanum ostruth-		Saponaria officinalis	274
H. Sendtneri	306	ium	279	Saxifraga geum	279
		Phalaris canariensis	289	S. umbrosa	279
Iris fœtidissima	288	Pinus sylvestris	288	Scrophularia *Ehrharti*	297
		Plantago media	298	S. scorodonia	297
Juncus compressus	301	Poa alpina	304	S. umbrosa	297
J. diffusus	301	Polemonium cæruleum	284	Sedum album	278
J. nigritellus	301	Polygala austriaca	292	S. dasyphyllum	278
Juniperus communis	300	P. calcarea	292	S. micranthum	278
J. sabina	301	Polygonum bistorta	286	Sempervivum tectorum	278
		P. fagopyrum	286	Senebiera didyma	273
		P. maculatum	300	Senecio saracenicus	282
Lamium galeobdolon	297	P. maritimum	300	S. viscosus	281
Lathyrus aphaca	276	P. *nodosum*	300	Silene armeria	274
Lemna polyrrhiza	288	Populus alba	287	Silybum Marianum	282
Limnanthemum		Potamogeton lance-		Sphærocarpus terres-	
nymphœoides	284	olatus	302	tris	307
Linum usitatissimum	275	P. mucronatus	302	Spiræa filipendula	277
Lolium italicum	289	P. plantagineous	301	S. salicifolius	277
L. linicola	289	P. trichoides	303	Statice Dodartii	298
Lonicera xylosteum	280	P. zosterifolius	302	S. limonium	298
Lotus tenuis	276	Poterium sanguisorbia	294	S. occidentalis	298
		Prunus cerasus	277	Stratiotes aloides	288
Matricaria parthenium	281	P. insititia	276	Symphytum tuberosum	284
Medicago falcata	275	Pyrola rotundifolia	296		
M. sativa	275	Pyrus communis	277	Tayloria serrata	306
Melilotus alba	275			Teucrium scordium	298
M. arvensis	275	Ranunculus arvensis	271	Thalictrum majus	290
M. officinalis	275	R. Baudotti	290	Tilia europæa	273
M. parviflora	275	R. circinatus	290	Torilis infesta	295
Mentha rotundifolia	285	R. Drouettii	290	Tragopogon porrifolius	282
M. sylvestris	297	R. hirsutus	291	Trifolium filiforme	293
M viridis.	285	R. Lenormandi	290	T. ornithopodiodes	276
Meum athamanticum	270	R. parviflorus	271	T. resupinatum	293
Mimulus luteus	285	R. reptans	290	T. scabrum	293
Mollia hibernica	305	R. tripartitus	290	Triticum pungens	304
Monotropa hypopitys	296	Reseda lutea	273	Trollius europæus	291
Myosotis sylvatica	297	R. suffruticulosa	273		
Myosurus minimus	290	Ribes grossularia	278	Ulex nanus	293
		R. nigrum	278	Ulmus suberosa	287
Nasturtium sylvestre	291	R. rubrum	279		
Neckera pennata	307	Rosa cinnamomea	277	Valeriana pyrenaica	280
		R. inodora	294	Valerianella carinata	280
Onobrychus sativa	276	R. lutea	277	Verbena officinalis	285
Orchis pyramidalis	288	Rubus Bloxamii	294	Veronica peregrina	285
Orobanche major	297	R. glandulosus	294	Viburnum lantana	280
O. minor	285	R. rudis	294	Vicia bithynica	276
Orthothrichum Drum-		Rumex acutus	300	V. sativa	276
ondi	306	R. palustris	293	V. tetrasperma	276
Oxalis corniculata	274	R. *pratensis*	300	Viola hirta	291
O. stricta	275			V. lutea	292
		Salicornia radicans	299	Vinca major	283
Pastinaca sativa	279	Salix acuminata	300	V. minor	283
Petasites albus	280	S. ambigua	300		
P. fragrans	280	S. fragilis	287	Weissia Drummondi	306

Printed by A. Mayne & Boyd, Printers to Queen's College, Belfast.

www.ingramcontent.com/pod-product-compliance
Lightning Source LLC
Chambersburg PA
CBHW031423230426
43668CB00007B/414